Lecture Notes in Computer Science 868

Edited by G. Goos, J. Hartmanis and J. van Leeuwen

Advisory Board: W. Brauer   D. Gries   J. Stoer

Ralf Steinmetz (Ed.)

Multimedia:
Advanced Teleservices and High-Speed Communication Architectures

Second International Workshop, IWACA '94
Heidelberg, Germany, September 26-28, 1994
Proceedings

Springer-Verlag

Berlin, Heidelberg, New York
London, Paris, Tokyo
Hong Kong, Barcelona
Budapest

Ralf Steinmetz (Ed.)

Multimedia:
Advanced Teleservices and High-Speed Communication Architectures

Second International Workshop, IWACA '94
Heidelberg, Germany, September 26-28, 1994
Proceedings

Springer-Verlag
Berlin Heidelberg New York
London Paris Tokyo
Hong Kong Barcelona
Budapest

Series Editors

Gerhard Goos
Universität Karlsruhe
Postfach 69 80, Vincenz-Priessnitz-Straße 1, D-76131 Karlsruhe, Germany

Juris Hartmanis
Department of Computer Science, Cornell University
4130 Upson Hall, Ithaka, NY 14853, USA

Jan van Leeuwen
Department of Computer Science, Utrecht University
Padualaan 14, 3584 CH Utrecht, The Netherlands

Volume Editor

Ralf Steinmetz
IBM European Networking Center
Vangerowstrasse 18, D-69115 Heidelberg, Germany

CR Subject Classification (1991): C.2, H.4.3, H.5.1, D.4.4

ISBN 3-540-58494-3 Springer-Verlag Berlin Heidelberg New York

CIP data applied for

© Springer-Verlag Berlin Heidelberg 1994
Printed in Germany

Typesetting: Camera-ready by author
SPIN: 10479057   45/3140-543210 - Printed on acid-free paper

Preface

The first International Workshop on Advanced Teleservices and High-Speed Communication Architectures (IWACA) took place in München in March, 1992. It attracted more than 100 researchers and developers from all over the world. The discussions and interactions between the participants were very lively and productive. At that time we decided (1) to continue our effort in establishing such a forum for discussions and (2) to try to merge two closely related key topics, multimedia applications and communications. This volume contains the papers presented at the second IWACA workshop held in Heidelberg, Germany, September 26-28, 1994.

With respect to high-speed networks we focus on broadband ISDN in ATM-LANs as well as ATM being the wide area network of the future. Together with the appropriate protocols and services these form the **High-Speed Communication Architectures**. However, at the moment the development of broadband networks is still driven by technology, and it is most often used as a backbone network only. It is more challenging to conceive, implement and use these networks to support multimedia traffic. Therefore, we need to gain a deeper understanding of what the multimedia applications of the future will be. We "narrowed" the scope of distributed multimedia applications towards applications which always include the communication and presentation of audio and/or video data, namely **Advanced Teleservices**.

The scope and quality of most of the papers submitted to this IWACA workshop showed that the joint discussion of teleservices and the respective communications is a hot topic that requires further attention. In this sense each paper in this volume contributes to our overall knowledge and mutual exchange of experience between the multimedia application and the high-speed multimedia communication communities.

These proceedings cover
- resource management for distributed multimedia systems and ATM networks
- operating system issues related to the processing of multimedia data
- source modelling of ATM traffic
- protocol issues in ATM networks
- experiences in multimedia communications and personal communications
- audio and video information archives, retrieval and distribution applications and systems
- prominent teleservices projects
- tele-teaching
- conferencing
- synchronization, compression, and security as special issues in multimedia

Many people contributed to make this event and these proceedings a success. First, I want to mention the continuous inspiration, the huge numbers of hints and ideas provided by the driving forces of the first IWACA, namely Bob Ip, Siemens, and Alfred Weaver, University of Virginia, USA. The actual work in Heidelberg required a tremendous effort from my colleagues Helmut Coßmann and Werner Steinbeck. Both sacrificed hours and hours of their spare time, all being done concurrently with our daily work at the IBM European Networking Center in Heidelberg. Thanks to our management at IBM and Siemens we were able to make extensive use of the existing resources and infrastructure. Thomas Goebel, University of Mannheim, devoted all his available time in Heidelberg to make everything happen for IWACA. Thanks.

August, 1994   Ralf Steinmetz

Supporting/Sponsoring Societies

IEEE Industrial Electronics Society
IEEE Computer Society, TC on Multimedia Computing
Gesellschaft für Informatik (GI)
Informationstechnische Gesellschaft im VDE (ITG)

Program Commitee

Derek McAuley	University of Cambridge, UK
Simon Bernstein	Sprint International, Virginia, USA
Ernst Biersack	Institut Eurecom, Sophia-Antipolis, France
Guillermo Cisneros	GTI, Universidad Politécnica de Madrid, Spain
Andre Danthine	Institut D'Electricite Montefiore, University of Liege, Belgium
Jörg Eberspächer	Technical University of Munich, Germany
Wolfgang Effelsberg	University of Mannheim, Germany
Jose Encarnação	Technical University of Darmstadt, Germany
Edward A. Fox	Virginia Tech, USA
Bob Ip	Siemens, Munich, Germany (Co-Chair)
Charles N. Judice	Bell Atlantic, USA
Jürgen Kanzow	DeTeBerkom, Berlin, Germany
Paul J. Kühn	University of Stuttgart, Germany
Jörg Liebeherr	University of Virginia, USA
Philippe Loiret	France Telecom Expertel, Paris, France
Takeshi Nishida	NEC, Kanagawa, Japan
Stephen Pink	Swedish Institute of Computer Science, Stockholm, Sweden
Radu Popescu-Zeletin	GMD Fokus, Berlin, Germany
Venkat Rangan	University of California, San Diego, USA
Russell Sasnett	GTE, Waltham MA, USA
Otto Spaniol	Technical University of Aachen, Germany
Ralf Steinmetz	IBM European Networking Center, Heidelberg, Germany (Chair)
David Sutherland	British Telecom, UK
Alfred Weaver	University of Virginia, USA (Co-Chair)

Organizing Committee

Helmut Coßmann	IBM European Networking Center, Heidelberg, Germany
Thomas Goebel	IBM European Networking Center, Heidelberg, Germany
Werner Steinbeck	IBM European Networking Center, Heidelberg, Germany

Table of Contents

Self-Similar("Fractal") Traffic in ATM Networks

Nicolas D. Georganas

Multimedia Communications Research Laboratory
Department of Electrical Engineering
University of Ottawa
Ottawa, Ontario, Canada, K1N 6N5
E-mail: georgana@trix.genie.uottawa.ca

Abstract. Recent studies of high quality, high resolution traffic measurements in Bellcore Ethernets have revealed that this aggregate Ethernet traffic is *self-similar* ("fractal") in nature, quite different in "burstiness" features from traffic considered and studied up to now. Similar results were obtained in analysing VBR video traffic measurements. As ATM high-speed, cell-relay networks will most likely first make their impact as backbones interconnecting enterprise networks consisting of Ethernet and other LANs, their proper design and control is crucial. Thus far these networks have been studied under traffic assumptions that may not prove any more valid.This paper presents an analytical model for the study of an ATM buffer driven with (asymptotically) self-similar traffic.

1. Introduction

Recent studies of high-quality, high resolution traffic measurements have revealed a new phenomenon with potentially important ramifications to the modeling, design and control of broadband networks. These include an analysis of hundreds of millions of observed packets on several Ethernet LANs at the Bellcore Morristown Research and Engineering Centre [1,2], and an analysis of a few millions of observed frame data by Variable-Bit-Rate(VBR) video services [3]. In these studies, packet traffic appears to be statistically *self-similar* ("fractal") [4,5]. Self-similar traffic is characterized of "burstiness" across an extremely wide range of time scales: traffic "spikes" ride on longer-term "ripples", that in turn ride on still longer term "swells", etc.[1]. This "fractal" behaviour of aggregate Ethernet traffic is very different both from conventional telephone traffic and from currently considered models for packet traffic (e.g., Poisson, Batch-Poisson, Markov-Modulated Poisson Process, Fluid Flow models, etc. [6]). Contrary to common beliefs that multiplexing traffic streams tends to produce smoothed-out aggregate traffic with reduced burstiness, aggregating self-similar traffic streams can actually intensify burstiness rather than diminish it [1].

Self-similarity manifests itself in a variety of different ways: a spectral density that diverges in the origin, a non-summable autocorrelation function (indicating long range dependence), an Index of Dispersion of Counts (IDC) that increases monotonically with the sample time T, etc. [1]. A key parameter characterizing self-similar processes is the so called Hurst parameter, H, which is designed to capture the degree of self-similarity.

Asynchronous Transfer Mode (ATM), high-speed, cell-relay networks will most likely be first used as backbones for the interconnection of enterprise networks composed of several LANs, which may also carry VBR video traffic. A lot of studies have been made for the design, control and performance of such networks, using traditional traffic models. It is likely that many of those results need major revision, particularly those pertaining to cell-loss rates and congestion control, when self-similar traffic models are considered [8].

In this paper, we will present a model, introduced by Likhanov, Tsybakov and Georganas[10], for the performance analysis of an ATM buffer, driven by (asymptotically) self-similar input traffic. The resulting G/D/1 queueing model will be mapped into a M/ G/1 model where the service time is Pareto distributed with infinite variance . The latter model is then analyzed using classical queueing theory, to find the probability of buffer overflow, delay, etc. Our approach consists of first constructing a self-similar process from an infinity of on-off sources with Pareto service demands. We then consider the self-similar process arrivals as equivalent to the (Poisson) arrivals of activity-bursts of the said Pareto sources and easily analyze the resulting M/G/1 queueing system. To demonstrate the interesting behaviour of non-traditional queueing systems, we shall first start our exposé with the case of an infinite G/M/1 queue driven by Pareto traffic having zero average arrival rate ($\lambda=0$). We shall derive the delay distribution in this queue and its mean, without using Little's formula, which by the way is still valid in this case!

2. Self-Similar Stochastic Processes [1]

We consider a stochastic process $X = (X_1, X_2,...)$ with a constant mean $\mu = E\{X_i\}$, finite variance $\sigma^2 = E\{(X_i - \mu)^2\}$, and autocorrelation function

$$r(k) = \frac{E\{(X_i - \mu)(X_{i+k} - \mu)\}}{E\{(X_i - \mu)^2\}}, \quad k = 0, 1, 2,...$$

depending on k only.

Denote $X^{(m)} = (X_1^{(m)}, X_2^{(m)},...), \quad m = 1, 2, 3,...$

where

$$X_k^{(m)} = (X_{km-m+1} + ... + X_{km}) / m, \quad k > 0$$

The stochastic process X with $r(k) \sim k^{-\beta}$, $as \ k \to \infty$, $0 < \beta < 1$, is called (exactly) second-order self-similar with Hurst parameter $H = 1 - (\beta / 2)$, if

$$r^{(m)}(k) = r(k), \text{ for any m=1,2,..., (k=1,2,...)}$$

where $r^{(m)}(k)$ is the autocorrelation function for the process X.

The process X is called *asymptotically second-order self-similar* with parameter $H = 1 - (\beta / 2)$, if

$$r^{(m)}(k) \to r(k), \quad as \quad m \to \infty \quad (3.1)$$

In other words, X is *asymptotically second-order self-similar*, if the corresponding aggregated processes $X^{(m)}$ become indistinguishable from X, at least with respect to their autocorrelation functions.

The most striking feature of second-order self-similar processes is that their aggregated processes $X^{(m)}$ possess a nondegenerate correlation structure, as $m \to \infty$. This is in stark contrast to typical packet traffic models in the literature, all of which have the property that their aggregated processes $X^{(m)}$ tend to second-order pure noise, i.e., for all $k \geq 1$,

$$r^{(m)}(k) \to 0, \quad as \quad m \to \infty .$$

3. An Example of Non-Classical Analysis: G/M/1 Queue with Pareto Input Traffic (from [10])

Consider a G/ M/1 queueing system with infinite buffer. Let $x_1, x_2,...$ be the arrival times. Assume that the service time for each arrival is independent of other random variables and is denoted by s . We also assume that

$$\Pr\{s \geq t\} = e^{-\mu t}$$

The interarrival times $U = x_{i+1} - x_i$ are independent and Pareto distributed, i.e.,

$$\Pr\{U \geq u \} \approx u^{-\alpha}, \quad as \quad u \to \infty, \quad 0 < \alpha < 1$$

It is easy to see that $E\{U\} = \infty$. This means that the intensity of new arrivals in the time interval [0,t] is

$$\lambda(t) = \frac{E\{n(t)\}}{t} \to 0, \quad as \quad t \to \infty$$

where n(t) is the number of new arrivals in the interval [0,t].

Let N(t) be the number of customers waiting in the queue at time t. It can be shown that

$$E\{N(t)\} = 0, \quad as \quad t \to \infty$$

Let us introduce an additional arrival at time t. Let t_1 be the time when this arrival will start being served. By definition, the virtual delay is

$$d_v = \lim_{t \to \infty} E\{t_1 - t\}$$

In our case, it easy to see that $d_v = 0$.

We now consider the system at the time x_i of a new arrival. We denote $N_k = N(x_k)$. Consider the Markov chain $N_1, N_2, \dots$. Note that $E\{N_k\} > 0$.

Let p_i denote the conditional probability that i customers obtain service during the interval between two consecutive arrivals, given that at least i customers are waiting in the queue.

We have

$$p_i = \int\limits_0^\infty \rho(t) \frac{(t\mu)^i}{i!} \exp(-\mu t) dt,$$

where

$$\rho(t) = \frac{d \Pr\{U \le t\}}{dt} \cong t^{-\alpha-1}, \text{ as } t \to \infty.$$

The Markov chain $N_1, N_2, \dots$, is such that $N_i \ge 0$ and its average drift is negative for $N_i \ne 0$. Hence there exists a limit distribution

$$q_i = \lim_{k \to \infty} \Pr\{N_k = i\}, \qquad q_i > 0$$

We have

$$q_i = \sum_{j=0}^{\infty} p_j q_{i+j-1} \qquad\qquad (3.1)$$

As a solution of equation (3.1), q_i has a geometrical distribution

$$q_i = (1-x)x^i, \qquad 0 \le x \le 1$$

Substituting this distribution in (3.1), we obtain the equation for x:

$$x = L(x) = \int\limits_0^\infty \rho(t) e^{-\mu(1-x)t} dt \qquad (3.2)$$

The solution of (3.2) can be obtained graphically, as shown in Fig. 3.1

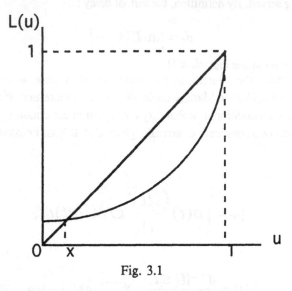

Fig. 3.1

Equation (3.2) has a unique solution in the interval [0,1) if $\lambda/\mu < 1$. Note that the equation (3.2) is used in classical G / M / 1 queueing theory [7]. In the considered case $\lambda/\mu = 0$. Hence, the derivative of L(u) at u=1 is infinite.

Let t_i be the time of beginning the service of customer i. Define the i-th customer delay by

$$d_i = t_i - x_i$$

Denote

$$\overline{d} = \lim_{i \longrightarrow \infty} d_i$$

Knowing q_i, it is easy to find $\overline{d}$:

$$\overline{d} = \frac{1}{\mu} \sum_{i=1}^{\infty} i q_i = \frac{x}{(1-x)\mu}$$

The probability distribution function of the delay is

$$\lim_{i \longrightarrow \infty} \Pr\{d_i \le t\} = 1 - x e^{-(1-x)\mu t}$$

We have thus found the delay distribution and average of a queue with zero arrival rate! Note that Little's formula is still valid, but it is trivial since in this case both the average number of customers in the buffer and the intensity of input traffic are zero.

Little's formula is $0 = 0\overline{d}$ and it cannot allow finding $\overline{d}$.

4. Self-Similar Traffic Model[10]

In [10] Likhanov, Tsybakov and Georganas construct a self-similar traffic model, as a superposition of an infinity of individual Pareto source models. In so doing, the model reflects the following features (formal proofs are given in [10]):

(1) An individual source behaves such that, at scarce random moments of time, it begins to generate bursts of a random-number of cells with rate R. The burst length distribution has a "heavy tail", decreasing as in a Pareto distribution[1, 8] with infinite variance. At the end of burst generation, the source becomes silent for a random time which is, as a rule, greater than the length of the cell generation interval;

(2) The number of individual sources, M, is large, so that it can be considered as infinity, but λ, the total intensity of the sources, is a finite and given value. In assuming this, the time between consecutive activity periods of a source tends also to infinity;

(3) The end points of individual source activity periods are proven to form a Poisson process with rate λ;

(4) The superposition of source traffic rates that gives the aggregate traffic rate is an (asymptotically) self-similar process with Hurst parameter H, where 0.5<H <1.

5. Analysis of an ATM Buffer

Assume that the above constructed (asymptotically) self-similar process, with Hurst parameter H (0.5<H<1), arrives at a buffered ATM channel. The corresponding queueing model is G/D/1. Since, however, the self-similar traffic is constructed from an infinity of on-off traffic sources with Pareto-distributed activity periods, with total arrival rate λ, the G/D/1 model is mapped into an M/G/1 model with the following statistics:

•Poisson arrivals with rate λ;
•Pareto service time distribution with infinite variance and parameter α=3−2H.

The model is now easily analyzed [10] to obtain the buffer length distribution. Its is shown that the probability of the number of "customers" in the buffer decreases algebraically, rather than exponentially as in classical traffic models. This simply indicates that the probability of buffer overflow will be higher than predicted from classical traffic theory applied to ATM networks and thus the onset of buffer congestion could be much earlier than thought before.

CONCLUSIONS

This paper gave some methodology for the construction of a self-similar traffic model and reported results detailed elsewhere [10]. Self-similar traffic arriving in an ATM buffer results in a buffer occupancy distribution with "heavy tail". This has a significant impact in ATM buffer congestion studies. Previous calculations and buffer dimensioning of ATM switches will have to be re-thought, in view of these new analytical results.

REFERENCES

1. W.E.Leland, M.S.Taqqu, W.Willinger and D.V.Wilson, "On the Self-Similar Nature of Ethernet Traffic (Extended Version)", IEEE/ACM Trans. on Networking, vol.2, No. 1, Febr. 1994, pp. 1-15 (also in Proc. ACM/SIGCOMM'93, San Francisco, 1993, pp.183-193).

2. W.E.Leland and D.V.Wilson, "High TimeiResolution Measurement and Analysis of LAN Traffic: Implications for LAN Interconnection", Proc. of IEEE INFOCOM'91, Bal Harbour, FL, 1991, pp.1360-1366

3. J.Beran, R.Sherman, M.S.Taqqu and W.Willinger, "Variable-Bit-Rate Video Traffic and Long-Range Dependence", IEEETrans. on Communications (to appear)

4. B.B.Mandelbrot, "Self-Similar Error Clusters in Communication Systems and the Concept of Conditional Stationarity", IEEE Trans. on Comm. Technology, COM-13, 1965, pp.71-90

5. B.B.Mandelbrot, "The Fractal Geometry of Nature", Freeman, New York, 1983

6. V.S.Frost and B.Melamed, "Traffic Modeling for Telecommunications Networks", IEEE Communications Magazine, Vol.32, No.3, March 1994, pp.70-81

7. L.Kleinrock, "Queueing Systems-Vol. I: Theory", Wiley, N.Y., 1975

8. D. Veitch, "Novel Models of Broadband Traffic", Proc. IEEE Globecom'93, Houston, Tx, Dec. 1993

9. H.J.Fowler, W.E.Leland, "Local Network Traffic Characteristics, with Implications for Broadband Network Congestion Management", IEEE J. Select. Areas in Commun., vol. 9, No. 7, Sept. 1991, pp.1139-1149.

10. N.Likhanov, B.Tsybakov and N.D.Georganas, "Analysis of an ATM Buffer with Self-Similar("Fractal") Input Traffic" , subm. to the IEEE JSAC special issue on the Fundamentals of Networking, 1994

A Service Kernel for Multimedia Endstations

Klara Nahrstedt and Jonathan Smith*
Distributed Systems Lab, University of Pennsylvania
e-mail: klara,jms@aurora.cis.upenn.edu

Abstract

Quality of Service (QoS) guarantees for *delay sensitive* networked multimedia applications, such as teleoperation, must be application-to-application. We describe a set of services, a *service kernel*, required at the end points, for multimedia call establishment with QoS guarantees. These services provide: (1) *Translation* among different domain specifications (layer-to-layer translation) and domain-resource specifications (layer-to-resource translation); (2) *Admission and Allocation* of resources; and (3) *Negotiation and Coordination* of QoS specifications among the distributed endpoints. For each service we propose architectural solutions.

We are testing the service kernel with an ATM-based telerobotics application.

1   Problem Description

Quality of Service (QoS) guarantees for delay sensitive networked multimedia applications must be application-to-application. Guarantees are achieved if: (1) the information is carried between end-points using delay-bounded communication protocols [6], [5], (2) the end-points use delay-bounded services of the operating system (OS) [4], [9], and (3) the application, OS and network are able to prepare and configure the environment for delay sensitive multimedia calls with QoS guarantees.

We identify a set of services required for QoS guarantees(in end-to-end multimedia establishment protocols) in a multimedia environment at the end-points. We call this set a *kernel*, because while additional services may be required (e.g., services for establishment of a video conference), these particular service are essential.

*This work was supported by the National Science Foundation and the Advanced Research Projects Agency under Cooperative Agreement NCR-8919038 with the Corporation for National Research Initiatives. Additional support was provided by Bell Communications Research under Project DAWN, by an IBM Faculty Development Award, and by Hewlett-Packard.

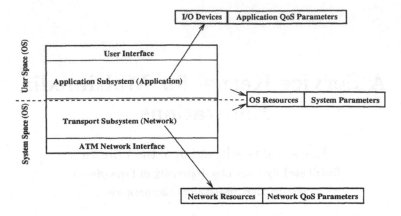

Figure 1: *End-Point Model with Application, Transport Subsystems and Operating System Specifications*

We believe that kernel services must provide the following functions: (1) *Admission and Allocation* of resources (e.g., task scheduling) for the local processor; (2) *Admission and Allocation* of network resources, such as bandwidth and pacing requirements; (3) *Negotiation and Coordination* among the other application end-points; and (4) *Translation* among different resource and domain (application, OS, network) specifications. For each, we propose architectural solutions.

2 · End-Point Architecture

The kernel services assume the end-point model of Figure 1. The communication stack consists of an *user interface*, an *application subsystem*, a *transport subsystem*, and a *network interface*. Both subsystems are embedded in a multi-user/multi-process OS environment.

An application identifies its specific requirements to the application subsystem using *application QoS parameters* (Figure 2). The application QoS parameter structure describes a multimedia stream in one direction. Hence, one has to keep in mind that for both directions (input and output) a multimedia stream description has to be given. The media quality component consists of an *interframe* specification and an *intraframe* specification. The interframe specification gives the characteristics of the homogeneous medium stream. If the individual samples in the stream differ in quality, intraframe specification has to occur. The parameters are stored in an application database.

The transport subsystem is configured with *network QoS parameters* (Figure 3), which describe the requirements on the quality of the network connection. The network QoS parameter structure (Figure 3) describes the QoS of data which is transmitted over one network connection. The network QoS parameters are stored in a network database at the end-point. Hence, the network database

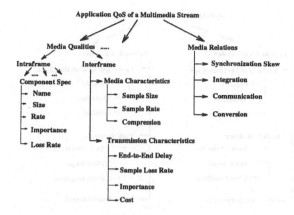

Figure 2: *Application QoS Parameters*

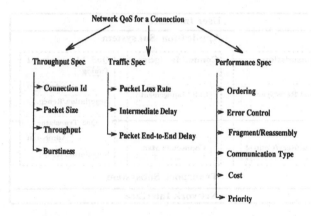

Figure 3: *Network QoS Parameters for a Connection*

includes as many network QoS descriptions as there are active connections for sending and receiving data.

The OS behavior is specified by the *system parameters* (Figure 4) which are stored in a system database. The system parameters mirror the requirements of the multimedia stream for the OS resources across both subsystems. They consist of *Task Specifications* for each medium and each connection, *Task Scheduler Specifications* and *Space Requirements*. The duration times of tasks are precomputed. The system database includes these parameters for both directions.

Based on this end-point model, the following three service groups are needed to support call/connection establishment with QoS guarantees:

- *Translation Services* provide mapping of parameters between communication layers (application QoS and network QoS) as well as between layer and resource (application/network QoS and system QoS). The functionalities of these services are given in Section 3.

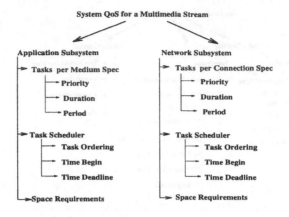

Figure 4: *System QoS for a Multimedia Stream*

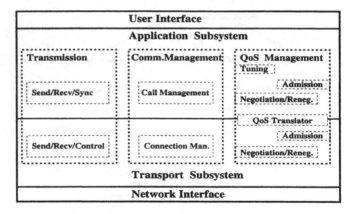

Figure 5: *End-Point Communication Architecture*

- *Resource Admission Services* provide a control mechanism for shared resource availability. Discussion of these services is presented in Section 4.

- *Information Distribution Services* implement layer-to-layer and peer-to-peer communication for call establishment with QoS guarantees. A brief discussion is presented in Section 5.

The translation, admission, and distribution services are added to call/connection QoS management. If an application requires guaranteed services, one or more of these services will be invoked. If no guarantees are required, traditional 'best-effort' call/connection establishment management is used. The resulting end-point communication architecture is shown in Figure 5. We are testing this end-point architecture and its QoS service kernel with an ATM-based telerobotics application. Its hardware setup and implementation are described in Section 6.

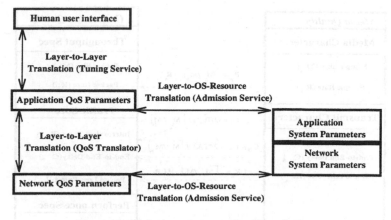

Figure 6: *Relations among the Translations*

3 Translation Services

Our description of the end-point implies that parameter sets have to be mapped onto one another. Mapping between communication layers leads to *layer-to-layer translation* functionality; mapping between layers and their corresponding system resources requires a *layer-to-resource translation* functionality. The relation among the translations is shown in Figure 6.

3.1 Layer-to-Layer Translation

The layering in Figure 1 indicates that we need layer-to-layer translation between the subsystems. Specific instances of such translation have been noted, for example, between ATM Adaptation Layer (AAL) and ATM Layer [8].

There are other layer interfaces where translation needs to occur. An important property of these translations, such as the one between application and transport subsystem, is *bidirectional translation*, which may cause problems of ambiguity. For example, in the case of video transmission, if bandwidth cannot be allocated for a video stream, the translation from the transport subsystem to the application subsystem may result in either a decrease of frame resolution quality, a decrease of frame rate, or both.

In our communication architecture (see Figure 5), the translation between the user and application is performed by the *tuning service*. This service maps the application QoS parameters onto audio/visual descriptions and vice versa. In our current implementation, a motion video file is used for visualization of the frame rate parameter. An important issue here is that the frame rate tuning must be based on the balance between the network capabilities and end-point OS capabilities[1] [2].

---

[1]Locally, the tuning service can display, for example, 15 frames per second, but when a network is involved between the video source and the destination, the frame rate may not be sustainable. The frame rate may decrease due to fragmentation/reassembly at the transport

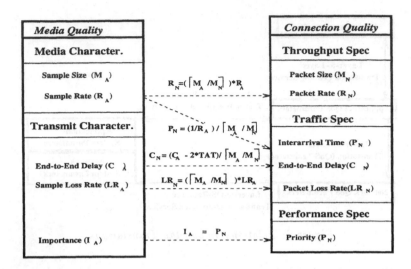

Media Quality		Connection Quality	
Media Character.		**Throughput Spec**	
Sample Size (M$_A$)		Packet Size (M$_N$)	
Sample Rate (R$_A$)	$R_N = (\lceil M_A / M_N \rceil)*R_A$	Packet Rate (R$_N$)	
Transmit Character.	$P_N = (1/R_A) / \lceil M_A / M_N \rceil$	**Traffic Spec**	
		Interarrival Time (P$_N$)	
End-to-End Delay (C$_A$)	$C_N = (C_A - 2*TAT)/ \lceil M_A / M_N \rceil$	End-to-End Delay(C$_N$)	
Sample Loss Rate (LR$_A$)	$LR_N = (\lceil M_A / M_N \rceil)*LR_A$	Packet Loss Rate(LR$_N$)	
		Performance Spec	
Importance (I$_A$)	$I_A = P_N$	Priority (P$_N$)	

Figure 7: *One-to-One QoS Translation*

The translation between the application and the transport subsystem is performed by the *QoS Translator* entity. The QoS Translator maps application QoS parameters onto network QoS parameters and vice versa. The translation includes at least three activities:

1. *One-to-one translation* involves a translation between the network connection quality and the medium quality. Figure 7 shows some cases of these transformations[2]. Other information transformations are:

   - *Media Relations* structure (Figure 2) in the application QoS provides the specification of *communication* (unicast/multicast/broadcast). This information gets copied to the *communication type* in the specification of the network QoS (Figure 3).

   - *Throughput* is computed from the *packet size* (M_N) and *packet rate* (R_N): $Throughput = R_N \times M_N$ This computation occurs after the translation from R_A to R_N (Figure 6).

     An ambiguity case, because of the bidirectional translation, may arise in cases when the network cannot guarantee the requested throughput and it suggests a lower throughput to the QoS Translator. The implications are as follows: The change in throughput influences first the packet rate, R_N. We assume that the packet size is fixed. The

---

subsystem. Therefore, one has to be mindful of what the tuning service promises and what the multimedia communication system can deliver.

[2] We specify single value (average) deterministic application QoS parameters, therefore the translated parameters (e.g., interarrival time P) are also single value deterministic parameters. We assume that the end-to-end delay C_N depends on TAT, which is processing time of application tasks to process a sample in the application subsystem. For loss rate LR_N, the transformation holds only for case where a sample and packet(s) are correlated.

changed packet rate, R_N', influences sample rate, R_A, and sample size (M_A) as follows:

$$M_A = \lfloor \tfrac{R_N'}{R_A} \rfloor \times M_N$$

$$R_A = \frac{R_N'}{\lceil \tfrac{M_A}{M_N} \rceil}$$

- *Fragmentation* is set TRUE when sample size is bigger than the packet size. If fragmentation occurs, it influences the computation of the end-to-end delay (C_N) for the packet as shown in Figure 6.

- *Ordering* is set TRUE, if continuous media with real-time behavior are sent. In non-real-time media behavior, the ordering requirement merely depends on the application's ability to handle out of order data.

- *Error Control* depends on the importance parameter and sample loss rate of the medium quality. If real-time behavior of the continuous media is required, its importance is high and sample loss rate is low, then a *forward error correction (FEC)* [7] mechanism is used in the communication protocol.[3] Otherwise, a different error correction mechanism (e.g., retransmission) can be specified, if supported by the communication protocol suite.

- *Cost* and *Burstiness* mappings are currently not supported.

2. *Integration* means interleaving (multiplexing) different media into one media stream which will be sent through one network connection. This implies that the different media qualities have to be merged into a new medium quality specification (many-to-one), as shown in Figure 8. After integration of the media qualities, one-to-one translation occurs between the resultant medium quality and the network QoS for the connection. It is important to point out that the resultant medium quality is the union with precedence of the media qualities being integrated. Therefore, integration should be done on media which have similar QoS requirements.

Because the translation is bidirectional, ambiguities can also occur in this case. Therefore, the QoS Translator passes to the application several possibilities and lets the user decide which medium will suffer in quality. In a more sophisticated system, a rule-based QoS Translator can be deployed which will make decisions based on rules given by the user *a priori*.

3. *Disintegration* means splitting of a medium stream into several streams which will be transmitted through several connections. This case occurs

---

[3] We use our own *RTNP – Real-Time Network Protocol*. It is a rate-based network protocol working above the AAL 4 layer. RTNP currently supports FEC if high reliability is required. Otherwise error correction is turned off. RTNP has error-detection and it reports to *RTAP (Real-Time Application Protocol)* about missing information. RTAP must resolve the conflict when missing information occurs.

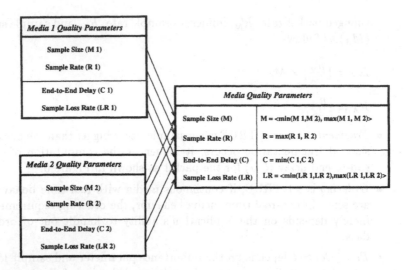

Figure 8: *QoS Integration*

when the medium stream carries different kinds of information (e.g., in a MPEG compressed video stream we have specification of I-frames, P-frames, and B-frames). Since the interframe medium quality specification includes the intraframe specification, the QoS Translator can perform one-to-one translation immediately between the intraframe component specification and the network QoS.

As an example, the architecture of the QoS translator process for a robotics stream, an audio stream, and an uncompressed video stream is shown in Figure 9. [4]

3.2 Layer-to-OS-Resources Translation

Each communication layer uses OS resources; hence, a mapping between the layer QoS parameters and system parameters is required. We consider translations between application QoS parameters and OS resources with respect to the application subsystem protocol (RTAP), as well as network QoS parameters and OS resources with respect to the transport subsystem protocol (RTNP). This mapping is done within the *admission services*. The application QoS parameters are mapped onto the system parameters (Figure 4): (1) task priorities, (2) task periods, and (3) buffer space requirements. Likewise, the network QoS parameters must be mapped onto the system parameters (Figure 4).

The *task priorities* are inherited from the importance of the medium quality and connection priority. The importance of priority inheritance for support of guarantees is experimentally shown in [2].

---

[4] In our telerobotics application, split of the sensor data occurs because the robotics stream carries 4 components (N, O, A, P), which have different meaning and importance for the movement of the robot arm.

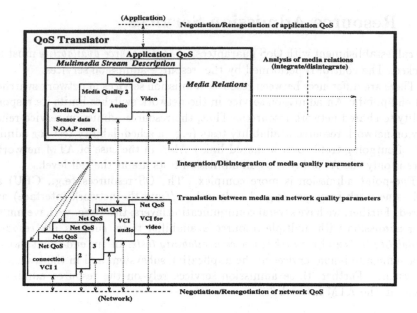

Figure 9: *The QoS Translator (Example)*

Task durations as well as specification of tasks in communication protocols are pre-computed and stored *a priori* in the system database. This parameter depends on the sample (packet size).

Task period is computed as the inverse of the sample rate (packet rate).

Task schedulers are computed from the the rates and priorities of the tasks in both subsystems (we use a mixed scheduling algorithm [11] which is composed of a rate-monotonic and deadline-driven algorithm). The important issue here is that the task scheduler in the network subsystem takes into account the task scheduler in the application subsystem. The reason is that the end-point has one processor Therefore, both task schedulers have to be in balance if guarantees are to be achieved.

The sample size (packet size), the fragmentation/reassembly, the integration, the error control, and disintegration determine the *space requirements* in the system QoS (both subsystems). In our communication protocols, there must be allocated for each direction at least $2 \times S_A$ space. The reason is that the application subsystem and the transport subsystem do not share the space where the sample is loaded. Therefore, at least one 'copy' operation occurs between the subsystems. Fragmentation/reassembly, FEC, and disintegration introduce an increase of space requirements at the transport subsystem. Integration can result in decrease of space requirements at the transport subsystem, but also may not.

4   Resource Admission Services

For call establishment with QoS guarantees, *shared resource availability* must be checked. The control is performed by the resource admission services.

There are differences between resource admission services in network switches and end-points: An admission service in the network switches limits its responsibility to shared network resources. Thus, the resource admission service relies only on network resource availability tests (e.g., a schedulability test for admission of outgoing traffic over a shared link.) Also, in the case of ATM networks there is only the ATM layer, so all admission tests are at the cell level.

End-point admission is more complex. The OS resources (e.g., CPU) as well as network resources (e.g., bandwidth to/from the network interface) are shared. Further, we have several communication layers; therefore, we have multi-layer admission with multiple resource availability tests, such as *OS resource availability tests* and *network resource availability tests*. In our end-point model, we assume admission services in the application subsystem and in the transport subsystem. Further, these admission services rely on the resource admission service in the ATM network.[5]

4.1   Assumptions for Admission Tests

For specification of schedulability tests (and a feasible task scheduler) in the admission services, it is important to specify:

1. *Types of Tasks*

   We examine applications with (i) *periodic tasks*, due to periodic production of media; and (ii) *deadline-driven tasks* for the movement of data from/to devices.[6] The deadline-driven tasks can be further classified into *hard-real-time deadline* tasks which process media streams such as tactile and kinesthetic data; *soft-real-time deadline* which process media streams such as audio and video streams; and *non-real-time deadline* tasks, such as QoS management tasks.

2. *Scheduling Algorithm*

   To schedule a set of periodic and deadline-driven tasks we choose a *mixed scheduling algorithm* [11] which combines rate-monotonic and deadline-driven scheduling algorithms. The rate monotony applies to the task processing media/connections according to the sample and packet production and the consumption rate at the devices. The deadline-driven algorithm applies to intermediate tasks such as moving data between devices, where the deadline is less than or equal to the period allocated to tasks responsible for producing/consuming the data.

---

[5]In the current implementation of the ATM network, there is no admission control in the ATM or AAL layer. Hence, we assume a lightly loaded ATM LAN, which provides an environment where the network resources are available.

[6]The periodic tasks are a subset of deadline-driven tasks, because the task period represents also a deadline.

4.2 OS Resource Availability Tests

The *resource admission service* at each subsystem level tests its own OS resources with a *schedulability test*. The final decision about the end-point OS resource CPU (i.e., if all tasks are schedulable) must be performed by the transport subsystem admission service which has more complete information about resource multiplexing at the end-point.

For the *schedulability test*, parameters of interests are: (1) task duration, e; (2) task period, p; and (3) context-switching time between two OS processes/threads cs.

From the schedulability condition for the mixed scheduling algorithm [11] $\sum_i \frac{e_i}{p_i} \leq 1$, where i is a number of tasks, we can derive the schedulability test in the application and transport subsystems:

- *Schedulability Tests in the Application Subsystem*

    Let us assume that $e_{o,i,r}^A$ specifies in application A the task r duration (processing time) of medium i sample (video/robotics data) in direction o (input/output). Let cs_j^A be the j-th context switching time between application A tasks. Let $min(P_{i,o})$ represent the minimal period among the media i sample periods P_i (inverse of sample rate) in direction o. The schedulability test in the application subsystem is:

$$T^A = \sum_o \sum_i \sum_r e_{o,i,r}^A + \sum_j cs_j \leq min(P_{o,i}) \qquad (1)$$

    Further, for each medium i in direction o, the following must hold:

$$\sum_r e_{o,i,r}^A \leq P_{o,i} \qquad (2)$$

- *Schedulability Tests in the Transport Subsystem*

    Let $e_{o,k,r}^{NET}$ denote the processing time of the task r performed over connection k packet in direction o in transport subsystem NET. Depending on the implementation of network tasks, cs_n^{NET} represents the n-th context switching between network tasks. The schedulability test in the transport subsystem is:

$$T^A + \sum_o \sum_k \sum_r e_{o,k,r}^{NET} + \sum_n cs_n^{NET} \leq min(P_{o,i}) \qquad (3)$$

$$\sum_r e_{o,k,r}^{NET} \leq P_{o,k} \qquad (4)$$

    The schedulability test - equation (3) - represents the global schedulability test at the end-point for CPU resource sharing.

Both tests assume that there is no interference of other applications and/or users. If an interference time D_I is present, it has to be added to the left side of the equations (1), (2), (3), and (4). The interference can be formally bounded as described in [10], but the current operating systems provide limited means to bound the interference [2] and provide a deterministic behavior. Because of an insufficient support of a determinism in OS, in our implementation we have limited the number of applications and users on the workstation. Further, the context switching contributes to unpredictable behavior [2], therefore the goal is to have a minimal number of context switching. The most predictable case is achieved when tasks are implemented in one process (no process context switching).

4.3 Network Resource Availability Tests

The network resource availability test is needed for end-to-end QoS guarantees. We discuss two tests: an *end-to-end delay test* and a *throughput test*. The final decision of the end-to-end QoS guarantees is performed at the remote (receiver) end-point in the application subsystem admission service, which has the complete information about the application-to-application behavior.

- *End-to-End Delay Test*

  For the *end-to-end delay test*, the parameters of interest are: (1) access and processing delay of a sample at the sender side D^S, which consists of the sum of (a) processing time of all application tasks for the sample and (b) processing time of all network tasks; (2) delivery and processing delay of a sample at the receiver side D^R which is computed similar to D^S, and (3) network propagation and queuing delay D^N. The final end-to-end delay test is performed in the application subsystem by the admission service. The sum of D^S, D^R, and D^N for medium i sample has to satisfy equation:

$$D_i^S + D_i^R + D_i^N \leq C_i^A \qquad (5)$$

- *Throughput Test*

  In the *throughput test* we test that: (1) the throughput over one connection must be less than the total bandwidth of the host interface in that particular direction; and (2) the sum of throughputs over all connection in each direction must be less than or equal to the total bandwidth of the host interface in each direction. For example, our ATM host interface has an effective bandwidth in each direction of 135 Mbits/second. Therefore, the sum of throughputs of all connections for sending data is checked against the 135 Mbits/second bound. The same test is done for connection over which we receive data. These tests are performed at the transport subsystem level. The throughput parameter is then translated into sample rate (sample size) and the schedulability test is done thereby determining if the network throughput can be propagated through the end-point to the application.

5 Information Distribution Services

Distributed networked applications have distributed resources, so QoS parameters as well as local decisions must be propagated between consecutive layers and between corresponding peers.

Layer-to-layer communication includes propagation of responses ('accept', 'reject', 'modify') about the acceptance of QoS between two consecutive layers. Communication between the layers is carried out by the tuning service at the user/application interface and by the QoS translator service at the application/transport interface. Further, if the QoS specification in every layer is different, translation is involved in the layer-to-layer communication as it was described above.

Peer-to-peer layer communication is performed by the *negotiation/renegotiation services*. In peer-to-peer negotiation two separate negotiations are supported:

- *Application QoS Negotiation*

  Application QoS negotiation happens between the application subsystems. It has some general properties, such as exchange of application QoS among the remote sites. It can also include some application-specific properties:

    - in our telerobotics experiment the application QoS negotiation is initiated at the operator side;
    - the operator specifies additionally to application QoS (Figure 2) also a *non-QoS information* (e.g., initial operation 'send me video frame' to evaluate the remote environment) which is sent to the remote robot.

- *Network QoS Negotiation*

  The network QoS negotiation is performed by the transport subsystem and includes: (1) exchange of the network QoS values, and (2) exchange of VCI mappings to specific connections supporting the media transmission.

A detailed description of the QoS negotiation service in a robotics environment is presented in [3].

6 Implementation Issues

The implementation of the service kernel is tested through our driving application – *telerobotics*. The hardware components of the experiment are shown in Figure 10. The end-point communication architecture (Figure 5) is implemented on the IBM RS/6000s. The connections between the robot control and the communication system is achieved through bit3 cards via bus-bus communication. This application puts new constraints on the system architecture of the end-points as well as on communication protocols and services in the network architecture.

As we pointed out earlier, this application has several specific properties which need to be considered in the implementation of the service kernel: (1)

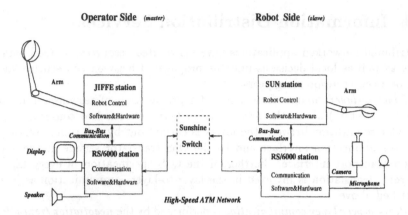

Figure 10: *Telerobotics Experimental Environment*

Telerobotics includes end-points (robots) without a human user, as well as end-points with a human user. Therefore initiation of negotiation, setup, and transmission is asymmetric. It is always started by the operator. (2) The quality requirements for sensory data are very high. (3) Video and audio are supporting information for the operator in order to have audio-visual support control over the workspace of the remote robot, and to make the proper decision in case of a failure or dangerous situation.

The services in the kernel are coordinated by the *QoS Broker* [1], an end-point orchestration entity which schedules the activities of the QoS management.

7 Summary and Conclusion

Translation, admission and negotiation services represent new services in multimedia communication systems and become a necessity for support of the call/connectio establishment management if QoS guarantees are required.

The advantages of these services are:

1. translation services allow domains to work in their preferred specification language;

2. admission services in the network are extended to the end-points which implies admission in upper layer protocols, and cooperation between upper and lower layer protocols to make global resource admission decisions; and

3. information distribution services provide communication between the layers and peers during call establishment.

It is important to emphasize that there are also other QoS services which will become necessary for QoS management during the transmission of continuous media. Examples of such services are *renegotiation* , *monitoring, notification* , etc. These services will soon become part of the service kernel.

References

[1] K. Nahrstedt, J. Smith, "The QoS Broker", *Technical Report*, MS-CIS-94-13, University of Pennsylvania, March 1994.

[2] K. Nahrstedt, J. Smith, "Experimental Study of End-to-End Issues", *Technical Report*, MS-CIS-94-08, University of Pennsylvania, February 1994.

[3] K. Nahrstedt, J. Smith, "QoS Negotiation in a Robotics Environment", *Proceedings of Workshop on Distributed Multimedia Applications and QoS Verification*, Montreal, Quebec, Canada, May 31-June 2, 1994.

[4] H. Tokuda, T. Nakajima, P. Rao, "Real-Time Mach: Towards a Predictable Real-Time System" *Technical Report*, Carnegie Mellon University, Pittsburgh, PA, 1993.

[5] C. J. Parris, D. Ferrari, "A Dynamic Connection Management Scheme for Guaranteed Performance Services in Packet-Switching Integrated Services Networks", *Technical Report*, UC Berkeley, October 1993.

[6] L.Zhang, S. Deering, D. Estrin, S. Shenker, D.Zappala, "RSVP: A new Resource ReSerVation Protocol", *IEEE Network*, September 1993.

[7] E.W. Biersack, "Performance Evaluation of Forward Error Correction in an ATM Environment", *IEEE JSAC*, May 1993, Vol.11, No.4, pp.631-640.

[8] J. Jung, D. Seret, "Translation of QoS Parameters into ATM Performance Parameters in B-ISDN", *IEEE INFOCOM'93 Proceedings*, Vol.II, San Francisco, CA, March 1993.

[9] A. Mauthe, W. Schulz,R. Steinmetz, "Inside the Heidelberg Multimedia Operating System Support: Real-Time Processing of Continuous Media in OS/2", *Technical Report*, IBM Research Center Heidelberg, Germany, 1992.

[10] K.W. Tindell, A. Burns, A.J. Wellings, "Guaranteeing Hard Real-Time End-to-End Communication Deadlines", *Technical Report Number RTRG/91/107*, Department of Computer Science. University of York, December 1991.

[11] C.L. Liu, James W. Layland, "Scheduling Algorithms for Multiprogramming in a Hard-Real-Time Environment", *Journal of the Association for Computing Machinery*, Vol.20, No.1, January 1973, pp.46-61.

Advance Reservation of Network Resources for Multimedia Applications

Wilko Reinhardt

Technical University of Aachen (RWTH Aachen), Dept. of Computer Science (Informatik IV)
Ahornstr. 55, 52056 Aachen, Germany
email: wilko@informatik.rwth-aachen.de

Abstract. To guarantee Quality of Services for continuous media like audio and video, intermediate nodes in networks must be able to reserve resources like buffer capacity, CPU time and bandwidth. Several methods have been introduced to verify that new connections can be accepted by the network without violating guarantees made for already established connections. One drawback of all proposed admission control algorithms is that they accept the reservation only at connection establishment time. It is not possible to reserve resources in advance. This paper introduces an algorithm that allows the user to reserve the desired QOS in advance. The paper further describes how the signalling of advance reservation can be realized using the Internet Stream Protocol ST-2.

1 Introduction

Communication support of distributed multimedia applications imposes several challenges on today's communication systems. The requirements of these applications reveal a lack of performance of traditional protocol stacks. Major reasons are contradictory demands made by the various media types. On one hand, isochronous data streams from live video and audio sources request high bandwidth, timely delivery of the data and low delay variation (jitter). On the other hand, asynchronous streams resulting e.g. from the transfer of still images or textual documents need a reliable transmission but can accept some delays and are insensitive to jitter. Broadband networks (e.g. B-ISDN, DQDB, FDDI, FDDI II, etc.) were developed in recent years, providing sufficient bandwidth to support the various requirements. Thus, the problem lies within the functionality and performance characteristics of the network and transport layers [5]. A new generation of protocols is necessary to bridge the functional and performance gap within these layers. XTP [10] and ST-2 [8] are two of the very promising approaches besides experimental approaches from several research laboratories [3,9]. The common part of all these protocols is the extended support of Quality of Service (QOS) parameters, allowing the user processes to define their exact requirements for the requested connection. In accordance with the QOS values the protocol functionality is configured and/or resources are reserved to provide a guaranteed service.

The 'real' support of QOS, i.e.. not just OSI's 'request and pray' approach is a key issue for sufficient communication support of multimedia applications. Currently, several extended QOS semantics are under discussion:

- *Threshold QOS:* The service user specifies 'worst case' values which are acceptable as the lowest quality. If values fall below these thresholds the user process will be informed so that it can decide to accept the lower quality or abort the connection. The QOS concept introduced by the OSI 95 project [1] proposes a more rigorous semantics, called *compulsory QOS*: If only one of the 'worst case' values is exceeded the connection will be aborted immediately.

- *Guaranteed QOS:* [2] introduced the guaranteed QOS semantics which has a rather strong legal flavour. When a service provider guarantees a performance value to the service user, it commits itself to provide that performance and to pay a penalty if the negotiated value can not be achieved. On the other hand the service user is obliged not to exceed the negotiated value.

Guaranteed QOS assumes that all intermediate nodes from the source to the destination allocate appropriate resources like bandwidth, buffer space and CPU time. These resources are exclusively used by the data streams the reservation was made for. A protocol for the negotiation of the QOS values and the corresponding reservations is needed. Examples of resource reservation protocols are ST-2 or RSVP [11]. The reservation protocol is a vehicle to transmit the user demands to the several network nodes. In general, it works independently from the routing and the admission control algorithms. The admission control algorithm is responsible to check if sufficient resources are available to serve the desired new connection without violating agreed performance parameters for already established connections.

A common drawback of all resource reservation protocols is that there is no possibility to reserve the resources in advance. This paper introduces an approach to solve this problem by presenting the necessary changes in the admission control procedure and the resource reservation protocol. Chapter 2 gives a brief introduction to ST-2 as an example for a reservation protocol. Chapter 3 describes an extension of intermediate nodes to manage the advance resource reservation. The necessary extensions of the resource reservation protocol are explained on the basis of ST-2 in chapter 4. The paper closes with a discussion of the approach in chapter 5.

2 ST-2: A Brief Overview

The experimental Internet stream protocol, version 2 (ST-2) as defined in RFC 1190 [8] is a revised version of the Internet stream protocol introduced in IEN-119 [4]. It provides especially realtime applications with guaranteed and predictable end-to-end level of performance across networks. Thus, multimedia applications are enabled to communicate on existing infrastructures and request QOS parameters according to their demands. Within the OSI reference model, ST-2 can be classified as a network layer protocol, and according to the ARPANET reference model as an Internet protocol. The main tasks performed are routing and signalling of data streams as well as the fast forwarding of data packets from the source host (called *origin* in the ST-2 terminology) to one or many remote hosts (called *targets*). Multicast communication is supported at the network layer by setting up a static routing tree from the origin to all targets, with the origin representing the root of the tree and the targets representing the leaves. Each intermediate node is an ST-2 router, forwarding the data downstream on the routing tree to the targets.

Data is transported over the tree as *streams*. RFC 1190 defines a stream as

- the set of paths that data generated by an application entity traverses on its way to its peer application entities,

- the resources allocated to support that transmission of data and

- the state information that is maintained describing the transmission of data.

The principle of ST-2 is to split connection management from data transfer. The stream set-up, maintenance and release are performed by the ST Control Message Protocol (SCMP). During connection establishment paths to all targets are installed. All intermediate nodes allocate, if possible, resources for the requested stream characteristics and establish connections to the next hops on the way to the targets. During the data transfer phase the data follow the established paths. In the header of each data packet the next intermediate node is identified by a hop identification number negotiated at connection set-up time, but it includes no control information. This leads to an efficient forwarding of all data packets since no control packets disturb the information flow along the path.

2.1 The Stream Protocol (ST)

The task of the stream protocol is to forward the data down the routing tree. ST provides a connection oriented but unreliable data transfer. Each ST header is 8 Bytes long, carrying 8 fields (Figure 1).

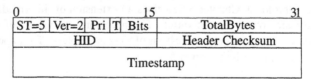

Figure 1: ST Header

ST operates as an extension of the IP protocol. Therefore, the first field in the header contains a special IP version number to indicate an ST packet. The current version number of IP is 4, therefore the RFC 1190 proposes the version number 5 for the ST protocol. The second field contains the ST version number. The priority field (*PRI*) is used to indicate packets that can be dropped by an intermediate node if a stream is exceeding its resource allocations. During connection set-up the ST agents negotiate whether or not the header contains a timestamp and its semantics. If the timestamp is included in the header, the *T* bit is set. The *TotalBytes* field contains the complete length of the ST packet including header, optional timestamp and user data. The *HID* field is the hop identifier. Each hop within the stream is identified by a unique identifier. The HID represents a stream and how the packet will be forwarded to the next hop. When the packet leaves the station, the HID will be updated with the HID that indicates the stream at the next node. All HIDs related to a stream are negotiated during connection set-up. For the calculation of the *HeaderChecksum* field which covers only the ST header the standard Internet checksum algorithm is used.

2.2 The ST Control Message Protocol

SCMP is responsible for the stream control. Within the telecommunication world's terminology SCMP would be called a signalling protocol. Its task is to create, maintain and release the routing trees used by ST. SCMP messages are encapsulated in ST packets and are distinguished from data packets by a HID that is zero. SCMP messages are sent from one ST agent to a single other in a reliable mode. Each message must be acknowledged immediately, otherwise the sender of the message re-transmits several times until a timer expires.

Connection set-up is handled as follows: The origin ST agent receives a list of targets from the application that are destinations of the stream. The desired QOS parameters of this stream are described in the flow specification (called *Flow Spec*). The ST Flow Spec structure contains, among other parameters, values for the average and minimum acceptable bandwidth, acceptable and maximum values for delay, delay jitter and error rate. After receiving a connect request the agent calls the routing function that returns a set of next-hop ST agents and the parameters of the intervening networks. Using these results the ST agent is able to decide whether the selected networks will support the requested Flow Spec parameters. After the routing decision the ST agent sends connect messages to all determined next-hop agents, containing a proposal for the HID, link references, the updated Flow Spec in accordance to the reservation at the node, and an updated target list (all target lists together will deliver the original list). This procedure is repeated until all targets are reached. The target agent requests an acknowledgement for the modified Flow Spec parameters from the application, and sends back an acknowledgement message to the origin.

3 Advance Resource Reservation

Most of the work in the area of resource reservation at the network layer has been done on reservation at connection establishment time. If the resources are available at that time they are reserved in accordance to the user demands. The word 'if' describes an element of uncertainty, which means that the resources are only reserved when they are available and not used by other connections. In the worst case, the connection request must be rejected or the connection will be established only with extremely reduced QOS parameters. Today, this is a common situation if we consider e.g. wide area data links. Users are not always guaranteed a suitable connection upon their request. In this situation users have to try again later or they need a lot of patience since the offered performance of the network is not sufficient. However, if we consider new applications providing e.g. video conferencing with several participants this situation is not satisfying. A date for the meeting has to be arranged so that all participant are available at the scheduled time. What happens if they are not able to set up a conference due to the fact that not enough resources are available from the network? If the users wait for better conditions on the network, a new date will have to be arranged with all problems of setting up a time schedule between several partners and the risk that it is not sure that the resources are available at the new date.

This paper proposes to integrate an advance resource booking mechanism into the reservation procedure. [7] states that it would be a helpful feature for the users if they were able to reserve the requested resources a certain time in advance. The initiator of a conference could set-up the data streams for the time the conference is scheduled. At that time the necessary resources are guaranteed from the network.

To realize this concept an adequate signalling protocol must be responsible for the negotiation of the requested resources between the source, intermediate nodes and all targets. Within the network nodes some more work has to be performed. Not only the resources used by currently active connections have to be managed but also requests for future reservations have to be examined. It has to be decided if the resources are available at the desired time and if so, the agreed performance values and the dates must be stored and managed. This work can be performed by a *resource manager*.

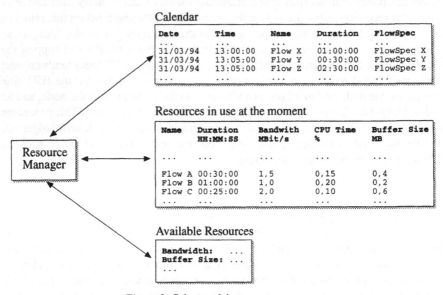

Figure 2: Scheme of the resource manager

Before a router reserves the resources, it has to check if sufficient resources are available to guarantee the desired QOS parameters. The router has to make sure that the guaranteed service parameters for existing connections are not violated. Therefore, an admission control algorithm is executed to determine whether already agreed parameters from existing connections and the requested performance values from the new connection can be guaranteed. Only in this case the new connection is accepted with the desired values.

If a new connection will violate the guarantees of other flows the desired QOS has to be reduced, or the connection request must be rejected.

The same procedure has to be performed if resources are allocated in advance. In this case the admission control algorithm must check if the new request can be accepted for the desired time without violating already confirmed reservations. Therefore, the functionality of a router has to be extended by a resource manager responsible for monitoring the available resources, the resources in use and the already allocated resources for the future. To remember the future allocation states, a *calendar* is needed that stores already confirmed reservations and their related parameters. These parameters include the date when the connection will be in place, their duration, a pointer to the name descriptor of the stream and a pointer to its agreed Flow Spec (Figure 2). All confirmed reservations are sorted in relation to their starting time. This makes it easier for the admission control algorithm to verify the situation at the particular time the new connection wishes to reserve resources in advance. The resource manager has to consider the situation at the starting time of the planed connection. It has to make sure that none of the performance guarantees for already confirmed future active connections are violated. Furthermore, it has to verify the connection request against all reservations that are made in the time period the new connection will be active. To make this verification process more efficient, connections should only be allowed to pre-reserve at certain times.

For the determination of this certain times the following considerations can be made: The starting time of a conference and its duration is determined by the user. In general, dates are made in steps of five or up to fifteen minutes. A date at e.g. 11:00:00 h is much more likely than a date at e.g. 11:02:34 h. Since the resource manager has to control all starting times of reserved connections the overhead can be reduced significantly when only a small number of possible starting times are allowed. Consider for an example a time period of two days within which advance reservations are possible. The necessary number of comparisons can be calculated as a function of the interval size. Time intervals in steps of 1 second cause 172.800 time periods within 48 hours, whereas an interval size of 5 minutes lead to 576 time periods. A further increase of the interval size up to 15 minutes reduces the number of time periods down to 192 periods.

If we further consider the already mentioned user behaviour, an interval size of five or fifteen minutes should be acceptable for advance reservations. This leads to the consequence that the resource manager should only accept starting times for reservation which are in accordance to these time steps. With this smaller amount of possible starting points it is possible to store the complete traffic situation for all time intervals. Therefore, the admission control algorithm is able to consider the traffic situation over the time period in question, without recalculating the reservations at the beginning of all time intervals. The number of comparisons is further reduced because the tests are only necessary within the duration of the connection. The duration of a video conference will for example be two hours. If we assume possible starting points in steps of five minutes the admission control must be called twenty-four times. Since this effort will not disturb the data transfer through the intermediate agents we can assume that the additional processing overhead is acceptable.

For the signalling of the advance resource reservation a protocol is needed that transmits the user request to all involved network nodes. Existing reservation protocols can be used if they are extended by adequate features. To realise the advance booking the connection setup of the reservation protocol can be used, since all negotiations are identical to the negotiations during the immediate connection establishment. The resource allocation within the intermediate nodes is generally independent from the reservation protocol.

The existing signalling protocols can easily be enhanced to provide the signalling for advance resource allocation. Two additional parameters have to be added to the respective FlowSpec of the reservation protocol:

- The starting time of the data flow across the connection and

- the duration of the connection from the starting event.

The starting time is only used for advance resource reservation, to indicate the date and time the new connection will start. The duration parameter is needed for all requested connections to evaluate if the required resources conflict with already reserved resources in the near future. The starting time can further be used to indicate the immediate connection setup. If the value is set to zero the path is established for immediate use.

4   Signalling of Advance Resource Reservation with ST-2

Only a few changes have to be made to use the ST-2 protocol for resource reservation. First the duration and starting time parameters as described above need to be integrated in the Flow Spec. The routing path is established immediately through a connect message, and all parameters (e.g. the HIDs, Flow Spec values, etc.) and resource requests are negotiated as in the conventional ST-2 version. (Figure 3) The only difference is, that all reservations are stored within the resource manager of each ST agent if the starting time parameter is not zero. The reservations are not performed immediately, but the resources are reserved for the time the connection will be used by the applications.

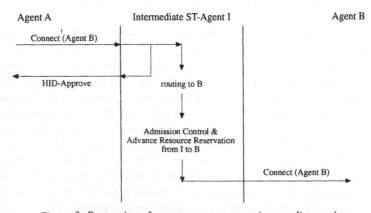

Figure 3: Processing of a connect messages at intermediate nodes

When the connection is to be activated by the users, a *FlowStart* PDU is sent from the origin to all targets. This PDU-type has to be integrated into SCMP. The FlowStart message indicates all intermediate ST agents at which resources allocated in advance for this connection that they will be used within the agreed duration (Figure 4). If the FlowStart message is not be received from the involved agents at the agreed time, the pre-allocated resources are released within a specified time interval and are available for other connections.

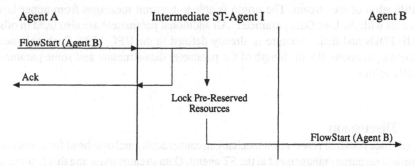

Figure 4: Processing of a FlowStart message at intermediate nodes

The FlowStart message should be integrated into the SCMP protocol since all SCMP messages are transferred in a reliable mode. An ST-2 agent that receives a FlowStart message immediately generates an ACK message to the previous hop. If the previous hop does not receive an ACK within a fixed time period it has to re-send the FlowStart message. Just to set a flag in the first packet of the data flow, indicating its start is not suitable since the transmission of ST packets is unreliable. If the first packet is lost or corrupted the intermediate agents will not be informed that the reserved resources will be used and release the reservation.

OpCode	Options	Total Bytes
RVLId		SVLId
Reference		LnkReference
SenderIPAddress		
Checksum		0
DetectorIPAddress		
Name Parameter		
UserData Parameter		

Figure 5: FlowStart PDU

Figure 5 depicts the structure of the FlowStart message. In addition to the mandatory SCMP parameters (OpCode, Options, TotalBytes, RVLId, SVLId, Reference, LinkReference, SenderIPAddress and Checksum) the DetectorIPAddress, Name and UserData Parameter are added. The DetectorIPAddress Parameter identifies the agent originating the message (in general the origin). It is copied from hop to hop and is primarily used for control purposes. The *name* data structure identifies the stream. It contains the origin's IP address, a unique stream identifier issued by the origin and a timestamp. The combination of the three parameters guarantees an globally unique identification of the stream. The origin is able to transmit messages from upper layer protocols with the UserData parameter. All additional parameters are also used in other SCMP PDUs and their structure is already defined in the RFC 1190. Each parameter contains a parameter ID, the length of the parameter data structure and some parameter specific values.

5   Discussion

The advance resource reservation mechanism causes additional overhead for connection set-up and resource management at the ST agents. Data streams requiring this feature, are typically long term connections transmitting video and audio streams for multimedia conferences. These streams require high bandwidth and real time delivery of data packets. New network technologies provide this high bandwidth, but we have to take into account that a lot of data streams with identical characteristics are transmitted, over several links, thus reducing the available capacity for a single stream. If we further consider that generally more than two partners are participating in a video conference an agreement for the date of this conference has to be arranged. If the connection can not be set-up due to limited available network resources this situation leads to a new time schedule for the meeting. A mechanism that provides the possibility for the users to define their requirement in advance seems to be a useful service provided by the network.

The overhead of the newly introduced advance resource reservation is limited to the connection establishment procedure. Most of the management PDUs for connection maintenance have to be exchanged anyway. Only one additional PDU that indicates the start of the data flow has to be added. Within the intermediate agent a calendar function has to monitor all reservations. The advantage of this proposal is that the data transfer is not disturbed by these negotiation, since all negotiations will be finished at the time the applications start transmitting their data.

The amount of memory that is needed to store information about all confirmed resources is another issue. It should be clear that there will be more 'intelligence' within future routers. In normal network nodes only the determination of the routes to the next hop are performed, while in future systems the nodes will be aware of the characteristics of data flows. An overhead for the processing of advance reservation should not influence the decision, whether or not it will be supported. In the end users should pay for the additional service.

The advance reservation mechanism causes several problems that have to be solved. Within this paper these problems can only be mentioned. They are described briefly with an idea how a solution may look like.

- *Time synchronisation*: All clocks within the involved nodes have to be synchronized with a central time so that the reserved resources are available at all nodes at the same time. A possible solution is the use of the Network Time Protocol [6].

- *Cancellation of the conference*: How long must the resources be reserved without use after a scheduled time? A delay has to be agreed within which the reservation is still valid. Within this time period the resource may be used by other connections, but with the proviso that the resources may be re-allocated when the delayed connection is activated. From my point of view users have to be charged for the effort this new feature causes. Therefore, the users have to pay a fee, if they don't use already reserved resources. The signalling of the cancellation could be performed with the disconnect message like a normal disconnect request in the original ST-2 version.

- *Failure within an intermediate agent*: In the time between advance resource allocation and the actual start of the data flow one or more intermediate agents may crash, thus corrupting advance booking within this node. This will interrupt the complete routing path. The intermediate agents should inform each other about their status as usual. ST-2 provides several protocol functions that allow the interchange of status information between intermediate nodes. The notify message e.g. is used if the routing function causes problems or if the resource allocation changes. The intermediate node that receives a notify messages decides whether to solve the problem locally, or to forward the message to the previous node or the origin.

- *Exceed of reservation duration:* The proposed mechanism requires the users to define the duration of the reservation. This is also a problem for traditional meetings that are not support by communication systems. Meeting room have to be reserved and a time schedule is agreed for the agenda. For the reservation of network resources it has to be investigated whether the connection is aborted immediately or the connection is further support with best effort QOS.

6   Conclusion

Within this paper an algorithm for advance reservation of network resources was introduced. It has been considered which extension are necessary for the common reservation procedures, and how they can be integrated into existing environments. An enhancement of the reservation protocol ST-2 was presented that allows to support advance reservation. As a result of the final discussion it can be stated that such a mechanism is a desirable feature for multimedia conference applications, but there are still problems to be solved.

References

1. A. Danthine: *A new Transport Protocol for the Broadband Environment*, in A. Casaca, ed., Broadband Communications, Estoril, Jan. 1992, C-4, Elsevier Science Publisher (North-Holland), Amsterdam, 1992.

2. D. Ferrari: *Client Requirements for Real-Time Communication Services*, RFC 1193, Nov. 1990.

3. D. Ferrari, D. Verma: *A Scheme for Real-Time Channel Establishment in Wide-Area Networks*, in IEEE Journal on Selected Areas in Communications, Vol. 8, No. 3, April 1990.

4. J. Forgie: *ST - A proposed internet stream protocol*, IEN 119, MIT Lincoln Laboratory, 1979.

5. B. Heinrichs, W. Reinhardt: *The DYCE Concept - Architecture and Implementation Strategies,* in V.B. Iversen, ed., Integrated Broadband Communication Networks and Services, Copenhagen Apr.1993, C-18, Elsevier Science Publisher (North-Holland), Amsterdam 1994.

6. D. Mills: *Network Time Protocol (Version 3), specification, implementation and analysis*, RFC 1305, 1992.

7. C. Partridge, S. Pink: *An implementation of the revised Internet Stream Protocol (ST-2)* in Interworking: Research and Experience, Vol. 3, 1992.

8. C. Topolcic (ed.): *Experimental internet stream protocol*, RFC 1190, 1990.

9. C. Vogt, R. Herrtwich, R. Nagarajan: HeiRAT: *The Heidelberg resource administration technique: Design philosophy and goals*, in N. Gerner, H.-G. Hegering, J. Swobodowa, ed. Communication in Distributed Systems, Springer Publisher, 1993.

10. *XTP Protocol Definition Revision 3.6*, 11 January 1992.

11. L. Zhang, S. Deering, D. Estin, S. Shenker, D. Zappalla: *RSVP: A new resource reservation protocol*, IEEE Networks, No. 5, 1993.

Media Scaling in Distributed Multimedia Object Services

Thomas Käppner and Lars C. Wolf

IBM European Networking Center, Vangerowstr. 18, D-69115 Heidelberg
Mail: {kaeppner, lwolf}@vnet.ibm.com

Abstract: Real-time support for multimedia streams in currently installed workstation environments has been based on resource management systems that provide mechanisms for streams with guaranteed or statistical quality of service (QoS) by admission control and resource reservation. In contrast, media scaling is a technique that dynamically adapts the load of media streams to the current availability of resources. Scaling can keep media streams meaningful to the user which would break during overload situations. Instead of interrupting the service for a stream when an overload situation is encountered, the quality of the stream is gracefully degraded when the resource load situation reaches a critical state. Since media scaling is a technique that dynamically takes actual resource load into account it can easily adapt to changing situations and has the potential to keep the system in a range of optimal load. In this article we show how media scaling can be integrated in a general system support for multimedia in order to simplify the implementation of scalable applications and support their concurrent utilization of scarce resources.

1 Introduction

Due to recent advances in computer technology, high performance workstations with digital audio and video capabilities are becoming available which leads to the integration of multimedia data with computing. This integration allows for scenarios in which computer systems support services such as video conferencing, news distribution, advertisement, and entertainment.

Due to its special nature the processing of multimedia data demands real-time support from the underlying computing platform in order to continuously transmit, synchronize, and present audio and video data streams within a distributed system.

Real-time support for multimedia streams in currently installed workstation environments has been based on resource management systems such as [7] that provide mechanisms for streams with guaranteed or statistical quality of service (QoS). After a strict admission control for the establishment of new streams, provision of guaranteed QoS is based on worst-case assumptions for resource usage, which results in resource underutilization in case of streams with variable bit rates. Statistical streams may experience breaks due to lack of resources.

In contrast, media scaling is a technique that continuously adapts the load of streams to the current availability of resources. In the case of resource overload the graceful degradation of stream quality leads to the situation where resources are shared so to allow the continuous flow of all streams. Media scaling is not depending on the upper bound of the bit rate for a given stream, thus it has the potential to support more streams than traditional resource management that offers only hard guarantees for a stream.

1.1 Related Work

Previous work on media scaling concentrated mostly on communication aspects of multimedia data. Its utilization has mainly been reported to avoid congestion in networks that can not properly be supported by resource management [4]. Jeffay et al. have developed a special queuing mechanism to adapt the bandwidth taken by video sent across packet-switched networks [4]. Fluent Technology has based a product for networked multimedia on a proprietary scaling scheme [6]. Tokuda et al. have implemented a dynamic QOS management for local area networks [1, 5], whereas our work focuses on the end-system. In principle, the scaling operations to adapt to changed system load can be implemented within the application, which forces programmers to construct their own mechanisms and leads to interworking problems between applications. In this article we show how media scaling can be integrated in a general system support for multimedia in order to simplify the implementation of scalable applications. Delgrossi et al. [2] have shown how to integrate media scaling and resource management into a multimedia transport system. In turn, the system we describe can utilize scaling-capable transport systems in order to accelerate scaling operations but does not depend on their availability.

1.2 Our Contributions

We enhance the meaning of media scaling from the network to the end-system level and show its significant value for the user. Media scaling provides two kinds of benefit. First, scaling has the potential to increase the number of streams a system can support simultaneously in comparison to systems using hard guarantees. This is due to its ability to handle and resolve a system's overload situation. Second, scaling keeps media streams meaningful to the user which would break during overload situations. Instead of interrupting the service for a stream when an overload situation is encountered, the quality of the stream is degraded when the resource load situation reaches a critical state.

Since media scaling is a technique that dynamically takes actual resource load into account it can easily adapt to changing situations and can keep the system in a range of optimal load.

Whereas end-system-transparent scaling within transport systems is constrained to scaling on a per-packet basis, scaling that involves changing the encoding of transmitted media at the sender can be based on a much finer granularity. In general changing the quality of transmitted video can relieve the congestion on resources without significantly being perceived by users. We show that distinguishing types of resources for scaling of media can be used in order to optimize the quality/load ratio.

Previous work on scaling of media streams has not considered the behavior of the system in case of several streams being scaled simultaneously, which raises the question of fairness and balancing between streams. We show how to coordinate down- and upscale operations of all scalable streams and guarantee the balanced sharing of resources.

We show how scaling can be integrated in a layer that provides system support for multimedia applications during development and run-time. The Distributed Multimedia Object Services offer an abstract interface to real-time multimedia objects. Multimedia applications create and combine multimedia objects such as audio and video streams, which can form an acyclic graph spanning several machines. Realization of multimedia objects happens in Distributed Multimedia Object Servers, which are responsible for the handling of continuous media data at a single site. Cross-site streams can be realized transparently for the client, automatically establishing real-time connections between sites via a multimedia transport system [3, 8].

This paper describes the principle methods of media scaling on end-systems and its integration into the Distributed Multimedia Object Services.

2 Scaling Mechanisms

Scaling, the ability to adapt to situations in which resource demand is larger than resource availability, requires mechanisms which observe resource usage and detect resource shortages as well as later 'recovered' resource availability – the resources have to be monitored.

2.1 Resource Monitoring

A resource monitor (RM) observes resource usage and resource load. If the RM detects that the resource load state (RLS) changes in such a way that a reaction is necessary, it provides an indication about the RLS and how critical the current situation is. We distinguish the following types of RLS:

1. a resource is in a state of stable and acceptable load,
2. the load of a resource is increasing and reaches a critical state,
3. a resource is overloaded, and
4. the load of a resource is decreasing and left an overloaded state.

This classification not only leads to indications of 'stable' states, but also allows, by the dynamicity of state 2, for a proactive generation of indications which yields earlier and faster reaction to resource shortages and prevents the system from reaching state 3. Similarly, indications for recovered resource availability should not be generated too early to inhibit permanent oscillation between different RLS.

Detecting the dynamicity of states 2 and 4 can be approximated by using a finer granularity for load regions. In addition to the definitely overloaded state 3, several load regions of increasingly high resource load are distinguished by dividing the base load states 1 and 3 into smaller regions $S_0...S_n$, S_0 being the stable and S_n the overloaded state and state transitions reflecting the dynamicity (see Figure 1). The indications generated by the RMs inform the resource users to reduce the load faster (and stronger) as the load region enters higher load regions.

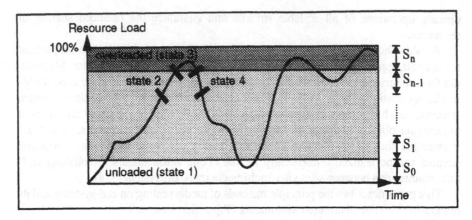

Fig. 1. Resource Load and Resource Load States

In principle resource monitoring is necessary for all resources that are concerned with the processing and transmission of multimedia data. These include resources like CPU, buffer space, and network, but also disks and the system bus. This paper concentrates on the former because they are important for all types of multimedia applications. Other resources, not being considered in this work are I/O related resources, especially file system resources, i.e., disks and their controllers, and the system bus.

2.2 Load Dimensions

In general, we distinguish two dimensions for resource usage:

- throughput and
- processing time requirements.

While they are not completely orthogonal (the time needed to copy data of some amount from one buffer to another depends on the size) they are also not completely correlated (if more time is spent for compression, the resulting space requirements may become smaller). Treating these two dimensions independently produces faster and more exact decisions than using only one measure reflecting the total acceptable load. Additionally, data stream properties can be adapted to available resources along these two dimensions. Two-dimensional scaling matches quality of streams and current load more accurately, yielding higher utilization and better overall quality of streams. Overload with regard to network or buffer resources is resolved by reducing throughput of streams, while CPU problems are tackled by decreasing processing time requirements.

As shown in Figure 2, the resource monitors for CPU, network, and buffer space provide the mechanisms for resource overload detection (in general, for resource state indication) and deliver the respective indications. Based on these indications a policy agent decides which stream, or set of streams, is affected by the resource overload. The policy agent informs the stream about the necessity to adapt to the changed resource availability.

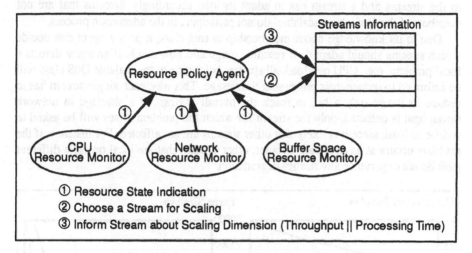

Fig. 2. Resource Monitors and Resource Policy Agents

The information provided by the policy agent indicates whether throughput or processing time usage must be adapted. The reaction of the stream depends on the particular stream handling thread and is determined by its capabilities as well as the data types. For instance, for a motion JPEG stream subsampling and quantization table parameters can be varied in order to adapt the load to either or both dimensions.

This is illustrated in Figure 3 where time and space requirements for the compression of a single JPEG picture are depicted over a range of quantization tables and 4 different settings of subsampling. If the throughput requirements of a stream have to be reduced, the quantization may be changed to a coarser degree or subsampling may be switched to a larger level. As can be seen from the measurements, if a stream operates at high quality with respect to quantization, reducing the quantization decreases frame size considerably. In lower quantization areas (below ≈ 80), switching to a different subsampling level yields better buffer space requirement reductions.

2.3 QoS Class for Scaling

The resource policy agent needs the information which streams to take into account when making scaling decisions. We have introduced the service class 'scalable QoS' as an extension to our resource management system [7] that already offered guaranteed and statistical QoS.

All streams belonging to this class may be affected by the decisions of the policy agent to ensure fairness to streams and properly balancing of resources. Membership to this class is voluntary, the stream creator decides to request the QoS class 'scalable' for a stream, mostly because the charged costs are lower than the costs for using a guaranteed service.

For streams in the 'scalable' QoS class, the policy agent and the stream agree on a certain behavior. The resource policy agent is responsible to deliver scaling indications to the streams and a stream has to adapt its load accordingly. Streams that are not members of the QoS class 'scalable' do not participate in the adaptation process.

Due to its knowledge about membership to this class, a policy agent can decide which streams should adapt their resource usage and how much. If an agent detects a local problem, e.g., CPU overload, all streams belonging to the scalable QoS class will be informed to reduce their processing time usage. This way, each single stream has to reduce its usage only a bit, to reach the overall savings. If a shortage in network throughput is detected, only the stream for which the problem arises will be asked to reduce its load, since it is likely that other streams are not affected. For instance, if the problem occurs at an overloaded router, other streams that are local or use a different path do not experience any quality degradation.

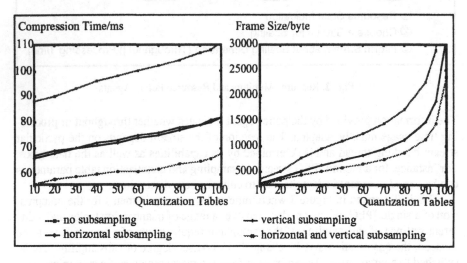

Fig. 3. JPEG Compression Time and Frame Size for different Quantization Tables

3 Scaling in Distributed Multimedia Object Services

Multimedia applications handle several kinds of media including continuous media such as audio and video. Due to its special nature the processing of these data demands real-time support from the underlying computing platform in order to continuously transmit, synchronize, and present audio and video data streams within a distributed system.

Multimedia streams are flowing through the system based on software and special hardware. Offering access to proper abstractions for hardware and data can make exclusive resources sharable between applications and allows easy application development and portability across supporting platforms.

The Distributed Multimedia Object Services (DMOS) support multimedia applications effectively during development and run-time by offering an abstract interface to real-time supported multimedia objects. Applications, as clients of DMOS, can create and combine multimedia objects such as audio and video streams to acyclic graphs potentially spanning several machines. Realization of multimedia objects happens in Distributed Multimedia Object Servers, which are responsible for the handling of continuous-media data at a single site. Cross-site streams can be realized transparently for the client, automatically establishing real-time connections between sites with the support of a multimedia transport system [3, 8].

3.1 Important Classes of Distributed Multimedia Object Services

The *LogicalDevice* class abstracts from hardware and software realizing a specific functionality of input, output, or filtering of multimedia data. Subclasses include Speaker, Camera, and Microphone each of which encapsulates all details of the underlying implementation.

The *Stream* class allows to combine a set of LogicalDevice objects in order to control the flow of data through the devices via a single interface. Control operations include start and stop operations and the acquisition and release of all resources necessary to process the data.

The *QualityOfService* class represents the kind of service the client is requesting. An instance of this class can be associated with a Stream in order to express the service class as guaranteed, statistical, or scalable, and the flow specification in terms of delay, throughput, and loss. The service class scalable is chosen when the type of application allows a temporary service degradation. However, the application itself is not involved in the process of scaling at all but relies on DMO servers. See Figure 4 for the client view of multimedia objects in a remote surveillance application.

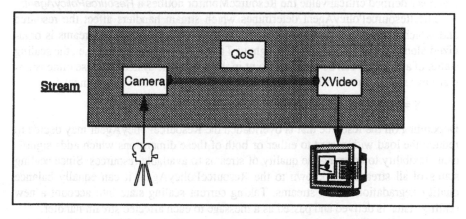

Fig. 4. View of Multimedia Objects for a Remote Surveillance Application

3.2 Scaling in Distributed Multimedia Object Servers

Every object that is created by a multimedia application is realized in a DMO server. However, other objects exist within a server that can not be accessed by clients. In order to provide real-time services the client-visible objects utilize a layer of stream handlers and management threads (see also Figure 5).

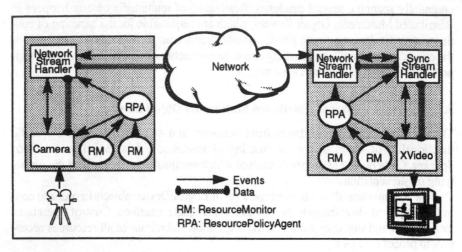

Fig. 5. Implementing Scaling in DMO Servers

Logical devices are mapped to stream handlers, i.e., a stream handler is a real-time thread that performs a specific task with regard to the processing of multimedia data. Stream handlers pass data packets along the stream and can exchange events or pass them to client-visible objects.

A *ResourceMonitor* is responsible for the status of a single resource. If the load reaches a defined critical value the ResourceMonitor notifies a *ResourcePolicyAgent*.

The ResourcePolicyAgent determines which stream handlers affect the resource and which of them belong to the service class 'scalable'. Scaling of streams is organized along the two dimensions throughput T and CPU requirements C, i.e., the scaling status of a stream S can be seen as a point in the coordinate system of these dimensions that have scales from 0 to 100 percent of the stream's respective requirements:

$$S = (T, C), \{0 \leq T, C \leq 100\}$$

Depending on the resource that is overloaded the ResourcePolicyAgent may decide to reduce the load with regard to either or both of these dimensions which adds significant flexibility to adapting the quality of streams to available resources. Since scaling states of all streams are known to the ResourcePolicyAgent it can equally balance quality degradation for all streams. Taking current scaling state into account a new scaling status is derived and passed as a message to each affected stream handler.

If its scaling status is changed by the scaling message a stream handler receiving such a message sends an event to the next stream handler up-stream in order to propagate the scaling status to the source of the stream. Note that there are two principle sources for the change of a scaling status: An event received from a stream handler residing down-stream containing the new sub status $S_s = (T_s, C_s)$ or a message from the ResourcePolicyAgent with $S_p = (T_p, C_p)$. A minimum status has to be derived in both cases from the most recently updated sub states as

$$S = (\min(T_s, T_p), \min(C_s, C_p)).$$

This filtering of scaling messages ensures that different perceptions of resource load for stream handlers, potentially being on different machines are properly resolved on their propagation path to the source of the stream.

When the source of a stream changes its scaling status the quality of the stream is adapted according to the new state S such that the stream's new specifications T and C with regard to throughput and CPU are met. Every stream handler that can serve as a source within a scalable stream has a built-in strategy to adapt the stream it generates. For instance the stream handler generating JPEG images from a camera has a considerable flexibility using a combination of subsampling and quantization table parameters (see Figure 3). This gracefully degraded quality is much more acceptable than a sudden drop of the frame rate or even break of the stream in case of packet losses.

4 Conclusion

Media scaling is a technique to adapt the amount of audio and video flowing through a system to available resources. Its usage has been reported to avoid congestion in networks that can not properly be supported by resource management. We have enhanced the meaning of media scaling from the network to the end-system level and have shown its value for the user. Media scaling not only has the potential to increase the number of streams a system can support simultaneously but it also keeps media streams meaningful to the user that would break during overload situations. Scaling of media streams in dependence of actual resource utilization can keep the system in a range of optimal load. The distinction of different load dimensions caused by a stream helps to adapt to available resource capacity more precisely.

Whereas end-system-transparent scaling within transport systems is constrained to scaling on a per packet basis, scaling that involves changing the encoding of transmitted media at the sender can be based on a much finer granularity.

If scaling is not supported by a system layer, applications have to duplicate the effort of implementing scaling mechanisms. Additionally, due to the lack of coordination, scaling mechanisms working concurrently in different applications would rather compete than cooperate. By coordinating down- and upscale operations of all streams on a given system, the balanced sharing of resources can be guaranteed.

We have shown how to integrate scaling in a layer that provides system support for multimedia applications during development and run-time. The Distributed Multimedia Object Services offer an abstract interface to real-time multimedia objects. Scalability of streams can be set by multimedia applications using a QualityOfService object that is associated with a stream. Realizations of streams and actual scaling operations across machines are handled by DMO servers.

References

[1]   S.T.-C. Chou, H. Tokuda: *System Support for Dynamic QOS Control of Continuous Media Communication.* Third International Workshop on Network and Operating System Support for Digital Audio and Video, San Diego, 1992.

[2]   L. Delgrossi, C. Halstrick, D. Hehmann, R.G. Herrtwich, O. Krone, J. Sandvoss, C. Vogt: *Media Scaling in a Multimedia Communication System.* First ACM Multimedia Conference, Anaheim, 1993.

[3]   L. Delgrossi, R.G. Herrtwich, F.O. Hoffmann: *An Implementation of ST-II for the Heidelberg Transport System.* Internetworking – Research and Experience, Vol. 5, Wiley, 1994

[4]   D. Hoffman, M. Speer, G. Fernando: *Network Support for Dynamically Scaled Multimedia Data Streams.* Fourth International Workshop on Network and Operating System Support for Digital Audio and Video, Lancaster, 1993.

[4]   K. Jeffay, D.L. Stone, T. Talley, F.D. Smith: *Adaptive Best-Effort Delivery of Digital Audio and Video Across Packet-Switched Networks.* Third International Workshop on Network and Operating System Support for Digital Audio and Video, San Diego, 1992.

[5]   H. Tokuda, Y. Tobe, S.T.-C. Chou, J.M.F. Moura: *Continuous Media Communication with Dynamic QOS Control Using ARTS with an FDDI Network.* ACM SIGCOMM 92, Baltimore, 1992.

[6])   P. Uppaluru: *Networking Digital Video.* 37th IEEE COMPCON, 1992.

[7]   C. Vogt, R.G. Herrtwich, R. Nagarajan: *HeiRAT: The Heidelberg Resource Administration Technique, Design Philosophy and Goals.* Kommunikation in verteilten Systemen, Munich, 1993.

[8]   L.C. Wolf, R.G. Herrtwich: *The System Architecture of the Heidelberg Transport System.* ACM Operating Systems Review, Vol. 28, No. 2, April 1994.

An Interactive Cable Television Network for Multimedia Applications

Nicos Achilleoudis, Jacqueline Lamour, Flavio Daffara

Laboratoires d'Electronique Philips
22 avenue Descartes - BP 15 - 94453 Limeil-Brévannes Cedex, France

Abstract. Nowadays there is a lot of interest in how to utilize the huge data capacity of existing cable television networks for providing a range of new services. In this report we will briefly look at what kinds of new services need to be provided and then go on to describe a network architecture that can facilitate their quick deployment with low initial infrastructure costs, while providing enough flexibility to adapt to future changes to both capacity requirements and network infrastructure.

1. Introduction

In many countries there are two wired telecommunications infrastructures that reach people's homes, namely the telephone networks and the Cable Television (CATV) networks. The telephone and cable companies have traditionally stuck to different markets. The former concentrated on telephony, data and other two-way services, while the latter concentrated on broadcasting services, mainly radio and television related, with perhaps some limited data capabilities in some countries (for example, Teletext).

Now, however, because of the merging of computers and television, as well as new political developments regarding deregulation, new opportunities are opening up for the telecommunications companies as well as the cable TV operators to expand their services into each other's traditional territories.

As far as the link to the home, (local loop) is concerned, both have invested a lot of time and money in wiring up entire communities, with twisted pair and coaxial cable respectively. Neither of them would want to make extensive changes to the local loop installations, at least not in the short term. Both would wish to keep the present installations in place, but adapt the ends of the cable connections to the requirements of new services. So the telephone companies will try to increase the capacity of their star-configured network, which is already bidirectional. On the other hand, the cable TV companies have enough bandwidth; their problem is how to move from a broadcast-oriented system to a 2-way network that can easily interface to the rest of the world.

In this report we will consider the case of the cable television companies. We will briefly look at what kinds of new services can be provided and after a brief description of a typical CATV network, we will go on to propose a network architecture that would enable these services to be deployed within a short time frame and with low initial infrastructure costs, transforming their unidirectional broadcast system into a bidirectional network.

2. New Services

This increase in capabilities will make it economically feasible for new kinds of services to be offered. These can be broadly divided into two categories.

The first category is that of interactive services. These are either greatly expanded versions of existing services, or completely new services, offered for the first time to users on a grand scale. Such services are:

- Pay-per-view (Near Video on Demand),

- Video on Demand,

- Teleshopping,

- Other Online information services,

- Tele-quiz and instant opinion polls,

- Multiplayer video games, or just interactive games.

The second class of services consists of services that have traditionally been provided by telephone companies and for which the cable operators could now act as alternative providers. Such services are:

- Normal Plain Old Telephone Service (POTS),

- Wireless telephony, either with home, building or neighborhood base stations, eg DECT (Digital European Cordless Telephony),

- Videophony and videoconferencing,

- Data communication services, like e-mail, Local Area Network (LAN) gateways to Wide Area Networks (WAN's), etc.

These services could be used in a traditional business structure as well as in telecommuting applications. These business uses would also be important in absorbing the initial expenses, as the technology will initially be expensive and businesses will be willing to pay much more than consumers for these advanced services.

For all this potential to be tapped, however, the cable networks need to acquire 2-way capabilities.

3. Description of a Typical CATV Network

Figure 1 shows a typical modern CATV network, with a central head-end that can serve up to 800,000 subscribers. In the normal direction of transmission, from the head-end to the subscriber premises, called the "forward" direction, the head-end output starts as an optical signal that is fed through several kilometers of star-connected optical fiber until it reaches an opto-electronic interface node, which can service up to 5,000 subscribers. At that point it is converted into an electrical signal and fed into coaxial cable arranged in a branch and tree architecture, with amplifiers along the way to preserve signal strength. The coaxial cable path is usually a few hundred meters in length, with a maximum of 2 km in some parts. The forward direction gets the lion's share of the bandwidth, with around 500 MHz available. The Signal-to-Noise Ratio (SNR) available is around 23 dB.

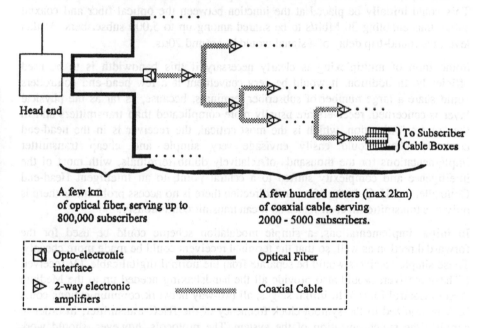

Figure 1 A typical Cable Television installation in Europe

As far as the other direction, or "return" path, goes, most networks either have or can be retrofitted with reverse amplifiers for the coaxial cable part. The optical path is also not much of a problem, because with typical installations a lot of redundant fibers are included, which can be utilized for a high-capacity return channel. The coaxial cable, however, has a return-channel bandwidth of only 20-40 MHz, depending on the country and the particular cable installation. To make matters worse, the SNR is only around 12 - 16 dB.

4. Possible System Architecture

There have been some proposals in the past by [2,3,4] on how to add two-way capabilities to a CATV network, with an overview offered by [1]. These, however, usually concentrated on lower bit-rate applications or on data-only applications. In addition, they had more hardware constraints because of the then available technology.

As mentioned in the previous section, the physical channels in the two directions of a cable system differ widely in performance. This leads to a very asymmetric channel capacity of around 3 Gbit/s in the forward direction (using 64-QAM) and around 30 Mbit/s in the reverse direction (using QPSK). Moreover, the present topology does not allow subscriber terminals to hear each other directly, so a central, possibly intelligent, network controller will have to provide this feedback to the terminals. This could initially be placed at the junction between the optical fiber and coaxial cable, thus enabling 30 Mbit/s to be shared among up to 5,000 subscribers. At this level, the round-trip delay of a signal would be around 20µs.

Some form of multiplexing is clearly necessary if this bandwidth is to be used efficiently. In addition, it would be very convenient if a few head-end controllers could share a large number of subscriber terminals, because, as far as the physical layer is concerned, receivers are usually more complicated than transmitters and in the "reverse" direction, which is the most critical, the receiver is in the head-end controller. One could easily envisage very simple and cheap transmitter implementations for the thousands of relatively dumb terminals, with most of the intelligence and complexity shifted to a central point, to an Intelligent Head-end Controller, or IHeC. In the "forward" direction there is no access problem, as there is only one transmitter in each IHeC, so it can transmit continuously.

In initial implementations, a simple modulation scheme could be used for the forward direction as well, so that the terminal receivers could be made more cheaply. These simple receivers would be separate from the normal digital cable TV receiver. (These receivers would also provide all the handshaking needed up to the Medium Access Control Layer). In initial stages, all two-way network communications could be synchronized to the symbol clock frequency of the IHeC transmitter, in order to simplify the synchronization of the system. The protocols, however, should work even with independent clocks in the subscriber transmitters. This way, in the future, different data rates could be used for the two directions and different terminals could operate their receivers in different channels, without the need to have these channels synchronized at the IHeC. This would allow the system, some time in the future, to get rid of the extra receiver on the terminal side and use instead only one receiver, the one used for the normal MPEG (Moving Pictures Expert Group) television decoding.

5. Description of the Medium Access Control Protocol

In our system, it was decided to split up the "return" bandwidth into channels of 8 MHz each. This gives us a raw data rate of around 12 Mbit/s, to be shared among up to 2,000 subscriber terminals. At lower bit rates perhaps a Carrier-Sense Multiple Access with Collision Detection (CSMA-CD) contention access system could have been used. But at these rates with short packet sizes, round-trip delay can reach up to 3 slot lengths, just because of the channel. So a hybrid access system is used, which can dynamically vary from a Reservation system to a Contention system, depending on traffic requirements, as well as hardware capabilities. In this system, each 8 MHz channel is divided into time slots which have one of two fixed lengths, as shown in Figure 2. The two lengths correspond to two different kinds of packets, the reserved-mode and contention-mode packets, as shown in Table 1.

Type of packet	Capacity per packet	Possible packet delays	Types of messages supported
Reserved-mode packet	Up to 48 bytes	Very low, deterministic at constant data rate.	All relatively high-rate streams, both constant and variable in rate.
Contention-mode packet	A few bytes	Random, depends on instantaneous network load and on contention algorithms used.	Bursty messages, usually of low rate.

Table 1 Slot types and their characteristics

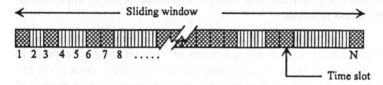

1 2 3 4 5 6 7 8 .....   N

Time slot

▨ "Reserved" Slot, very efficient, reservation required

▯ "Contention" Slot, not so efficient, no reservation required

Figure 2 Mixing of slot types

Detection and synchro symbols (length T.B.D.)	Packet header (6 bytes)	Useful Data (48 bytes)

Figure 3 Reserved slot packet structure

Reserved slots are allocated by the IHeC to specific terminal modems, either after a prior request by the terminal, or through an invitation by the IHeC. As only one terminal modem is allowed to transmit per slot, there is no question of packet collisions. As can be seen in Figure 3, the packets sent in a reserved slot have a useful data capacity of around 48 bytes in the lowest layer. This is compatible with the ATM (Asynchronous Transfer Mode) packet data size.

The slots that are not reserved are further divided into an integer number of sub-slots, called "contention" sub-slots. These can be used to carry fixed-length contention-mode packets, which have a useful data capacity of only a few bytes. They can carry certain signalling messages as well as very short data messages.

The reserved slots do not need to be grouped together in a frame structure. They can be arbitrarily placed within a sliding window. The reservations themselves can be either continuous or of a limited duration. The system tries to use the reserved time slots for as many data and signalling exchanges as possible. If a terminal that has reserved slots wants to send a new short message, or reserve additional slots, it can embed these requests into a part of a previously reserved slot. It will only resort to contention slots if it has not been transmitting for a while and hence has no reserved slots for embedding the information, or if it cannot wait for the next reserved slot. Another case in which contention slots can be used is for very short bursty messages.

The access, scheduling and flow control could then be controlled by the IHeC with algorithms similar to those used in point-to-point networks.

6. Applications and Performance Requirements

Different applications can have widely varying performance requirements. In Table 2 we list certain common applications and indicate how these could be implemented using our multiple access protocol.

Short bursty messages can be produced in an interactive TV system by an infra-red remote control or joystick, or other low-rate interactive terminal communications. In such cases, one needs to send bursts of a few bytes in length, perhaps once every few seconds. These signals, however, need to have a system response ranging from 0.25 seconds for something like an arcade game or Video-on-Demand, to a few seconds for something like an answer to a TV quiz program or Near-Video-on-Demand. Even less demanding, are applications like remote meter reading, for electricity, gas and water companies (the utilities companies).

Note, however, that in some cases strong peaks might occur. In Near-Video-on-Demand, for example, although the user requests will only consist of a few bytes per user per hour at maximum, the requests might occur over very close time intervals, say a lot of users might choose a movie within the last few minutes before it starts.

Speech and other constant rate connections can be implemented as regular reservations, occurring at fixed intervals. As the packets are presently of constant size, the buffering delay of the data mainly depends on the data rate. For example, for normal 64kbit/s speech, the delay is around 5 ms. In the future, as new speech-compression algorithms are introduced, we will need to have variable-length packets, if we are to keep speech delay to less than 5 to 10 ms. Moreover, reserved speech slots could be temporarily freed during periods of silence.

Application	Rate type	Rate	Time constraint	Slot used
Speech (Normal rate)	Constant	64 kbit/s	5 ms	Reserved 48-byte
Speech (DECT)	Constant	32 kbit/s	10ms	Reserved 48-byte
Speech	Variable	<32kbit/s	10 to 20 ms	Reserved 48-byte
Videophony	Constant or Variable	20 - 100's of kbit/s	100 ms for rate change	Reserved 48-byte
Computer network	Variable	As high as possible	As tight as possible	Contention[1]
File transfers	Variable	As high as possible	Very loose time constraints	Reserved 48-byte
Arcade style TV games	A few bytes per second	Speed not very high	Response within 0.25 seconds	Contention[1]
Video on Demand	Very bursty	Low	< 0.5 seconds	Contention[1]
Near Video on Demand	A few bytes per day	Low	A few seconds to several minutes	Contention[1]
Remote Metering	A few bytes per month	Low	Very loose time constraints	Reserved 48-byte

Table 2 Potential applications and their preformance requirements

Even better, one could implement truly variable-rate speech and video compression schemes, which could have a low average transmission rate, but would be capable of very high-speed short bursts when needed. These would match very well the trend towards variable-rate ATM systems and would make low bit rate videophony practicable.

---

[1] Tries to embed the message in an already assigned reserved slot, if one exists within its available time constraints.

7. Possible Plan for Commercial Deployment

Any introduction of such a network will have to take place gradually and with a high degree of compatibility and interoperability. Any subscriber terminal should be usable with any IHeC. Moreover, the subscribers should be able to buy modular components with only the features they need. Almost all models would have at least a simple modulator and demodulator. The differences would mainly be in the controlling hardware included with these boxes.

In a full telephone and interactive TV installation, for example, the home units would have an additional digital TV demodulator, for watching both interactive and normal TV. The subscriber terminal would have logic for both reservation and contention operation. If only limited interactive TV services are required, the user would get a network transceiver with only contention-based transmission capabilities and thus a simplification in the digital electronics required.

Also, businesses and other people that only need telephony services, for example, would only need to have the network interface portion, without the high-speed TV demodulator. Moreover, a single network interface could interface directly to a Private Branch eXchange (PBX), or even replace it. Such a high-traffic system would, itself, have a standard hardware interface for connecting to other digital data systems, such as DECT, TCP/IP (Transport Control Protocol / Internet Protocol), etc. The interface would actually conform to an existing computer bus standard, like PCMCIA (Personal Computer Memory Card International Association) or PCI (Peripheral Components Interconnect), for maximum economy of scale.

As cable traffic increases and technology evolves, the cable operators could encourage users to upgrade to new more spectrally efficient terminals, by providing incentives, like reduced usage rates for terminals that use higher constellations and thus reduce traffic on the cable network.

At the other end of the network, the head-end would also provide different levels of performance. For example, cheap IHeC models could provide very basic collision feedback, whereas more complicated head-end models could also provide information on the number of packets that collided in a given slot.

8. Conclusions

In this document, we have presented the work in progress to design a new interactive data communication system that can be cheaply and gradually deployed in existing cable television networks and yet be flexible enough to accommodate future advances in CATV architectures, as the fiber comes ever closer to the home, maintaining all along a high degree of performance and efficiency.

52

References

1   A.I. Karshmer, J.N. Thomas, "Computer Networking on Cable TV Plants," *IEEE Network Magazine*, Nov. 1992, pp. 32-40.

2   I. Kong, L. Lindsey, "CableNet: A Local Area Network Reservation Scheme", *Proc. of COMPCON 1982*, 6C.2.1-5, March. 1982.

3   M. Hatamian, E.G. Bowen, "Homenet: A Broadband Voice/Data/Video Network on CATV Systems," *AT&T Technical Journal*, vol. 64, no. 2, pp. 347-367, Feb. 1985.

4   A.I. Karshmer, J. Phelan, J. Thomas, "TVNet: An image and Data Delivery System using Cable T.V. facilities," *Computer Networks and ISDN Systems*, Elsevier Publishers, vol. 15, no.2, pp.135-151, Feb. 1988.

Eurescom IMS1 Projects
(Integrated Multimedia Services at about 1 Mbit/s)

Christian Bertin
IMS1 Project Leader
CCETT
4, rue du Clos Courtel B.P. 59
F-35512 Cesson-Sévigné Cédex

Abstract. This paper introduces the two Eurescom projects under the name IMS1 (Integrated Multimedia Services at about 1 Mbit/s) dealing with the specification and development of a general multimedia retrieval service (including VOD) for residential and business users over public networks. Laboratory demonstrator has been specified and developed. A full configuration using terminal and server with public network has been set up. A number of applications have been developed. Technical solutions for the Commercial Service have been investigated to guide the decision on the technical solutions to be implemented in the Laboratory Demonstrator. The technical choices made in the project are introduced in this paper.

1 Introduction to Eurescom IMS1 projects

In 1991, Eurescom launched the first IMS1 project (IMS1 stands for Integrated Multimedia Services at about 1 Mbit/s) whose objectives were to demonstrate the potential of multimedia retrieval services (including VOD services) over public networks and to assess the suitability of the public networks to support such services.

It is important to mention that in IMS1 we addressed only "interactive retrieval services" (Service between a human-operated terminal and a remote host). Messaging and conversational services are out of the scope of IMS1. It is assumed that more complex services can be built on top of the IMS1 service making use of some of the above mentioned services.

It is also important to mention that we were not addressing only VOD Services but a general retrieval service which should be adequate for both VOD services and any information service which make use of audio-visual sequences such as product catalogue, distance learning application, telegame, surveillance application, etc. in different business sectors.

The IMS1 service should allow the retrieval of different media types : rich text, graphics, audio, still and moving pictures and audio-visual sequences which can be retrieved through a terminal access link not exceeding 2 Mbit/s.

This project produced general specifications of the service and related equipment (server, terminals, production tools). 9 European public operators were involved in it. A focus was put on the future commercial network architecture. It was clear that several possibilities for the terminal access network were to be investigated (Copper pair with ADSL, ISDN Primary Rate, ATM in competition with CATV Networks), depending of the network development strategy of each operators. On the server side it was quite easy to agree on the use of ATM.

Based on the results of the first IMS1 Project (called IMS1 Phase I), a second Eurescom IMS1 project (numbered EU-P360 and called IMS1 Phase II) started beginning of 1993. Its objective was to develop laboratory demonstrators of a retrieval service over public networks. It should be finished in September 1994. 6 European Network Operators have decided to share the project developments (CSELT Italy, France Telecom France, DBP-T Germany, Telefónica I+D Spain, Telia Sweden and TLP Portugal).

2 Eurescom project EU-P106 (IMS1 phase I, General Specifications)

This project was divided in a number of tasks dealing with all different elements of the multimedia retrieval service. The results of these tasks are decsribed in the folowing sections.

2.1 Market Survey and Selection of Applications

The experts of this task identified a great number of application sectors which could make use of video retrieval service.

A set of criteria was set up to classify the application sectors and to select the most attractive applications for the laboratory demonstrators to be developed in the next phase of IMS1.

The most attractive or promising applications which were selected are the following ones :
- Visual Encyclopaedia
- Real Estate Information Service
- Tour Operator Travel Information

2.2 User Requirements

Many players are involved in the provision of a retrieval service, it was important to identify them. They are listed here :
- End-User
- Network Operator
- Service Operator
- Service Provider
- Information Provider
- Application and Content Producers

Knowing the retrieval service players, it was necessary to collect all their requirements and this was done. The following table gives us the list of the user requirements for each service player.

2.3 Service Architecture

A big question was raised about the distribution of intelligence between the server and the terminal.

A general model describing the operation of a retrieval application has been developed in an ETSI Technical Report called "Architectures for M&HIRS (Multimedia and Hypermedia Information Retrieval Service)". It is shown on the following figure :

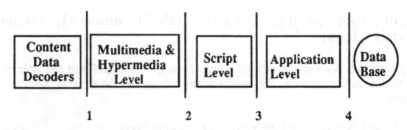

Fig. 1. Functional levels in applications

It introduces 4 interfaces where a network can be inserted and so four types of terminals :

1) Media Decoder Terminal which can decode various types of information (textual, graphics, audio, still and moving pictures). It knows only about data syntaxes and co-ordinates to display them.
2) M&H Terminal which can decode the various types of information included in objects with spatial and temporal synchronisation when presenting the embedded elements. The M&H terminal can react to end-user interaction by presenting, removing or modifying objects.
3) Script Terminal which can interpret a scripting language and relate with the application for interface with data base.
4) Application Terminal which includes everything escept the data base. It generates request to the Data Base and handle the answers from the Data Base.

We decided to carry on with the M&H terminal in the project for the following reasons :
- more and more CPU Power is available in the terminal, this allows to do more complex processing in the terminal,
- the higher the interface, the lower is the traffic between the server and the terminal after a certain time,
- the more processing is done in the terminal, the better is the response time
- a draft standard was emerging for the M&H Level (ISO MHEG) and it was the only one on which it was possible to agree among the IMS1 partners

It was also important to make use of other standards as much as possible when they are available and applicable : this is why we decided to use JPEG for still pictures, MPEG1 for audio-visual sequences, MPEG CC for the control of audio-visual sources.

2.4 Network Architecture

Based on the service architecture, an investigation was made on available and future networks suitable for such a service. The experts were asked to recommend a network solution for the Commercial Service and this can be seen on the following figure :

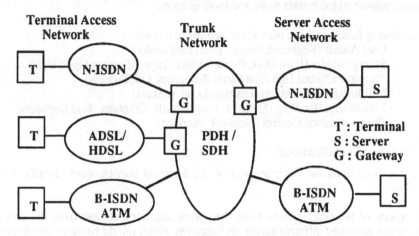

Fig. 2. Network Architecture for the Commercial Service

2.5 Protocol Stack

An interesting question was raised about the different QOS (Quality Of Service) required for the transport of the different media types.

In addition to this, there was another question about the multiplex to be used for the transport of the different flows. Two solutions were proposed and the discussion did not allow to reach an agreement on the one to be recommended for the commercial service.

Solution A : use of only one MPEG channel from the server to the terminal, multiplexing in the MPEG private data flows the Non Real Time Flows, with a return flow using another type of multiplexing.

Solution B : use of two channels between the server and the terminal : one MPEG Flow for Real Time Flows and another one for dialogue and Non Real Time Flows.

Another point of investigation was the definition of a NII (Network Independent Interface) which was considered as very useful mainly in the terminal where a number of network accesses may be used depending on the network operator strategy to introduce such a service.

2.6 Terminal Specifications

After the agreement on distribution of intelligence, it was quite easy to agree on the terminal functionalities. The solution to be implemented is clearly depending on the implementation time frame because new products, new add-on boards, new software packages appear on the market more and more quickly.

The following functions have been identified in the terminal :
- User Access (Keyboard, mouse, smart card reader)
- Storage Media (Hard Disk, Printer, Video Tape, Magneto-Optic Disk)
- Terminal Control Unit (Hardware Resources, OS, Process)
- Multimedia Presentation (Display, Loudspeakers)
- Multimedia Handling (MPEG1, Audio, Stills, Graphics, Text Decoders)
- Communication Control (Network Interface)

2.7 Server Specifications

All server functions for the provision of the retrieval service were identified and described.

The experts of the task, starting from the player requirements and from the service architecture proposed different server architectures which should be taken into account for the service.

The simplest configuration was the integrated server configuration where all server functions are located in the same machine.

Another configuration derived from the simplest one was the LAN server where the multimachine server is seen as an integrated server.

Another configuration is also proposed making use of a central application server associated with a number of remote content servers located close to the End-Users in order to reduce the network cost if it is too expensive. Additionally a content server can be associated to the central application server for less requested contents.

All these configurations are represented on the following figure :

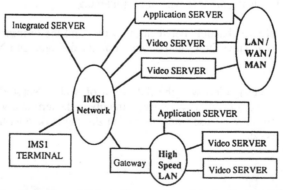

Fig. 3. Configuration of Server Connections for the Commercial Service

2.8 Production Tool Specifications

By production tool we mean media production and application production.

A production tool is something different from a terminal and a server.

A production tool includes all the terminal functions, because the developers want to see what they produced as soon as possible on the end-user terminal, without having to use the full networked configuration.

A production tool also includes server functions because the application developer wants to run the produced applications in the production environment as they will run in the server on the full networked configuration (Data Base Management System if necessary, File Management System, storage capacity, application interfaces as in the server).

In addition to this the production tool has specific functions : media capture, editing and encoding and application production. These specific functions usually require additional hardware and softwareIf there is no real time encoders for some media, this will have an impact on the Production Tool storage capacity and on the production process organisation.

So when the server and terminal platforms are different, when the operating systems are different, which one to choose for the production tool platform which will allows to run the production tool specific functions ?

To select a production tool environment, it is necessary to make a State-of-the-Art Survey of existing products, taking into account the terminal and server environment, and to delay as much as possible the deadline for the purchase of the production environment because new solutions appear very often on the market.

An IBM Compatible PC was considered as a suitable platform for a production environment because of the widespread distribution of such a platform offering at the same time a large number of competing products for production needs.

The PCU on this PC should offer at least the Intel 486 processing capabilities.

3 Eurescom project EU-P360 (IMS1 phase II, Laboratory Demonstrator)

After the first Eurescom IMS1 Project (EU-P106) dealing with general specifications, it was decided to carry on with a Laboratory Demonstrator Project within Eurescom.

Such a service project was too expensive to be included in the Eurescom budget for 1993, so it was proposed as an SIP (Special Interest Project). A Eurescom SIP is different from a Eurescom GIP (General Interest Project) in the following way : it is funded only by the Eurescom shareholders which support it and as a consequence the results belong only to the participating shareholders. This is maybe the reason why three partners from the first IMS1 project decided to withdraw from the second phase (BT from UK., Royal PTT research from the Netherlands and Telecom Eireann from Ireland).

After having decided to put the M&H Level of intelligence in the terminal, we can enter into more details with the terminal and server architectures corresponding to the use of MHEG in the terminal.

3.1 Server and Terminal Architectures for retrieval services.

The following figure, which represents server and terminal architectures will be useful to understand what are the relations between server and terminals.

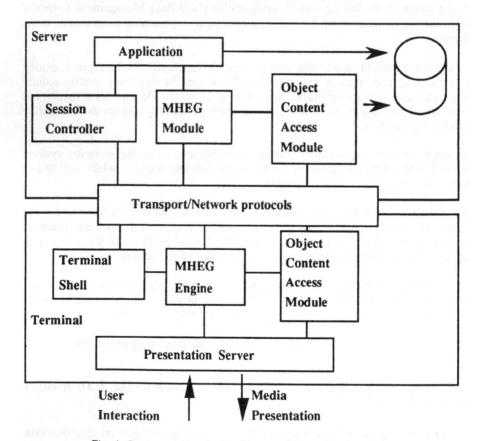

Fig. 4. Server and Terminal architectures for retrieval services

Let us introduce briefly the different elements which can be seen on the figure, starting with the terminal architecture:

Terminal Shell : This module is the local application, responsible for the dialogue with the end-user in the disconnected and connected mode. It allows the end-user to access a local directory of remote multimedia retrieval servers, to set up a connection with any of them and to clear the connection.

Presentation Server : This module is responsible for the media presentation and interaction device control. It may hide also the particular graphical user interface used on the terminal platform.

MHEG Engine : This module receives commands from the remote application, requests MHEG Objects from the OCAM module, interprets them and delegates the presentation to the Presentation Server. In the other direction, this module receives the user interaction from the presentation server and reacts as indicated in the MHEG Objects (presentation of other objects or return of results to the application).

OCAM (Object and Content Access Module) : This module hides the location of objects and contents to other terminal modules. It dialogues with other OCAM located outside the terminal to retrieve the objects or contents requested by the MHEG Interpreter or the Presentation Server.

Transport/Network Protocols : This module is responsible for data and message transmission between the terminal and the server in accordance with the QOS required by each type of data to be transmitted.

On the server side :

Transport/Network protocols : as in the terminal.

OCAM (Object and Content Access Module) : as in the terminal.

MHEG Module : This module which was not foreseen at the beginning of the project is necessary for the dynamic creation of contents such as a list of items resulting from a selection with a number of criteria filled in by the end-user (e.g. a product or a movie catalogue)

Application : This is one of the specific applications developed for the project.
An application consists of a program, maybe a data base, MHEG objects and data contents which can be encapsulated in MHEG Objects or not. They are not encapsulated in MHEG if they have real time constraints. An application may have access also to a data base (e.g. to store a product or a movie catalogue).

Session Controller : This module is responsible for sessions set-up between remote terminals and the server. It receives the call request from the terminal and activates the application which is called by the End-user. After that, it may handle billing information for the application. It may also monitor the call to clear it if there is no traffic for a certain period.

Now that all modules have been briefly introduced, let us describe a scenario with an end-user wishing to access an application hosted by a remote server to retrieve audio-visual information.

First, a connection has to be set up between the terminal and the application server. The end-user starts a dialogue with the Terminal Shell which gives access to a local directory of video retrieval applications or to a public directory. When the application is identified and the end-user has confirmed his wish to access this application, a call is set-up with the remote server hosting the application by the terminal.

A dialogue takes now place between the Terminal Shell and the Session Controller in the remote server, to launch the selected application in the server if there are several.

When the application in the server is activated, it communicates directly with the end-user through the MHEG Module in the server and the MHEG Engine in the terminal. Usually, the first MHEG command transmitted by the application will be a Run_Object command, maybe preceded by a set of Prepare_Object commands to make objects available to the terminal in advance, improving the response time later on.

The MHEG Engine will interpret these MHEG commands. A Run_Object Command has as a parameter an MHEG_Object_Id. The MHEG Engine has first to get this object and after to parse it because it is encoded in ASN.1 Format for interchange and finally to apply the required action on it which is likely to give it to the presentation server for presentation.

When an MHEG Object or a data content is needed for presentation or interaction with the user, a request is sent either by the MHEG Engine (when an MHEG Object is required) or by the Presentation Server (when a data content is required) to the OCAM (Object and Content Access Module). OCAMs communicate one with the other to identify the location of the requested element and after, the transfer can take place under the control of the receiving side.

This short scenario demonstrates that objects and contents can be distributed over the network in different server locations. They can be located either in the same server as the application or in different servers ordered in a hierarchical way (national server, regional or local servers). It is the responsibility of OCAM modules, using object and content directories to allow the transfer of the requested objects and contents from any location they could have been stored. It is up to the information provider to decide which strategy for content storage is the best for him.

User interaction is handled first by the presentation server and completely processed by it for basic interaction such as cursor tracking or character echoing.
For more complex user interaction such as button click, menu selection or end of character entry the user interaction is passed over to the MHEG Engine. Depending on the response to be given to the user interaction, the MHEG Engine can handle the user interaction or give it further to the application if it is not able to handle it. The MHEG Engine can process a user interaction independently of the application when the following conditions are met :
- the response to the user interaction can be coded with MHEG Commands,
- the set of MHEG Commands corresponding to the user interaction response has been linked to the user interaction event.

Let us consider in more details the presentation of an audio-visual sequence. When an AV sequence is requested for presentation, its transfer between OCAM is controlled by the "DSM CC" protocol (Digital Storage Media Control Commands) under standardisation in ISO MPEG.

After having introduced the terminal and server operation as we have designed it for the commercial service, we have now to focus on the solution we have implemented in the project taking into account the time frame of the project, the amount of available manpower and the availability of hardware and software pieces.

3.2 The Laboratory Demonstrator choices and configuration

The server platform used for the Laboratory Demonstrator was a SUN workstation.

The terminal platform used for the Laboratory Demonstrator was a IBM Compatible PC using the Microsoft Windows Operating System with a C-Cube board for MPEG1 decoding.

As far as the network is concerned, we started with the idea of putting terminal and server directly on public networks for the laboratory demonstrator. We looked for any board with sufficient efficiency to carry an MPEG1 sequence over a 2 Mbit/s link. But rapidly it became obvious that no such a board with public network interface was available for all software and hardware platforms to be used in the project (PC Windows for the terminal, SUN Workstation and PC for the server).

This is the reason why we decided to connect the server and the terminal on a very common network which is Ethernet, for which boards are available on every OS and hardware platforms. However for public network operators it was important to make also use of public networks, this is why we interconnected the two Ethernet networks through a 2 Mbit/s Leased Line.

On the following figure you can see the configuration of the IMS1 Laboratory Demonstrator which has been developed in the project.

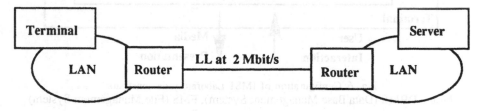

Fig. 5. IMS1 Laboratory Demonstrator Configuration

For practical reasons, we decided to use TCP/IP protocols over Ethernet and PPP (Point-to-Point Protocol) between routers. However it was clear that such a solution could not be recommended for the commercial service because it did not guaranty a sufficient QOS (Quality Of Service) for the transmission of synchronous data such as audio-visual data.

The main objective being to demonstrate the service, it was considered that the network solution used for the laboratory demonstrator was not so crucial.

Unfortunately we had not enough time to implement the full configuration described above in our Laboratory Demonstrator. It becomes rapidly obvious that it was impossible to have the development and integration of MHEG and OCAM done together in the project time frame with the available resources. We had to reduce our ambition and to focus on development around MHEG and its use in a networked environment as it is shown on the following figure :

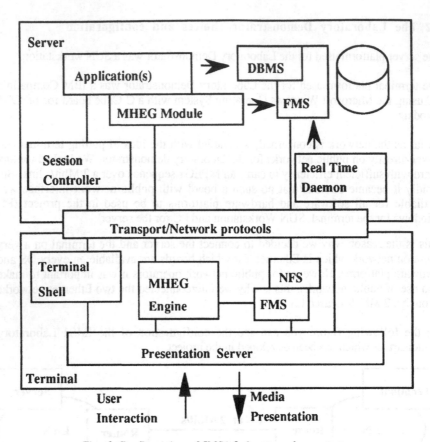

Fig. 6. Configuration of IMS1 Laboratory demonstrator
Note : DBMS (Data Base Management System), FMS (File Management System)

The applications are hosted by the servers. The MHEG objects and the data contents (text, graphics, still picture, audio and moving picture) are located in servers. To run an application, the user has just to know how to reach the hosting server (address in a newspaper, a personal directory or a public directory). All data to be presented to the user will be sent to the terminal from the server on request from the MHEG engine or associated module.

Having no OCAM available to provide the objects to the object requiring modules, we decided to make use of NFS to have access to the Non Real Time objects located in the remote server and a dedicated software developed in the project to have access to the video sequences located in the server at a sufficient throughput for immediate decoding and display in the terminal.

As far as the M&H Level was concerned, the only solution which could be agreed for implementation in the time frame of the project was the MHEG solution, fitting with the recommended solution for the commercial service.

As far as the data syntaxes were concerned, we had to consider what was available as commercial products and from the project partners, trying to distinguish between temporary choices for the laboratory demonstrator and recommended choices for the commercial service.

The media syntax choices we made for the laboratory demonstrator implementation were dependent on the availability of decoding solutions for the terminal platform and of coding solutions for the production platform. Here are the syntax choices we made for the different media types :

Text : Service providers and end-users require to have a rich text syntax with the following characteristics :
- multilingual
- multifont
- multisize
- multistyle (bold, italic, underlined, etc.)
- multicolour

The only solution we were able to provide in the time frame of the project was based on WMF format, but we were unable to recommend such a solution for the commercial service.

Graphics : No standardised solution was available from the partners. We decided to use WMF format and BMP Format.

Still pictures : There was no need to discuss on this. JPEG is the adequate standard for both the Laboratory Demonstrator and the Commercial Service.

Moving pictures : MPEG1 was an emerging standard with products already announced. We decided to use MPEG1 in the project.

For the commercial service MPEG1 is of course a good candidate but with the arrival of MPEG2, the choice is more opened, will we have both or only MPEG2 at the end.

Media Type	Solution implemented in IMS1 Laboratory Demonstrator project	Recommended solution for Commercial Service
Text	WMF, BMP	?
Graphics	WMF, BMP	CGM ?, PHIGS ?
Still pictures	JPEG	JPEG
Video sequences	MPEG1	MPEG1 ?, MPEG2
Audio	MPEG1 Audio	MPEG1 Audio
Audio-visual sequences	MPEG1	MPEG1 ?, MPEG2

Table 1. Media Syntaxes for Laboratory Demonstrator and Commercial Service

The applications to be demonstrated have been selected in different application domains to be representative of potential commercial applications :
- Video-on-Demand Application developed in Portugal,
- Hotel chain application developed in Spain,
- Visual Encyclopaedia application developed in France,
- Mail Order application developed in Germany,
- Yellow pages application developed in Sweden.

No production tool has been developed inside the project. The application developers have monitored developments made in this area and selected what was suitable for their needs. Following the standardisation activity around MHEG, an MHEG editor called ANIMA has been developed by ADV, a French software house. It was the only tool available for the production of MHEG Objects.

Other tools for the content production have been investigated. For JPEG still pictures a number of products were available and it was easy to select one of them.

For text production, we can use either the ANIMA MHEG Editor for simple text or Microsoft Word for rich text with a conversion in WMF format for interchange between server and terminal.

For graphic production, a number of production tools are available and any graphics production package which provides a WMF format output is suitable for that.

For MPEG1 production, it was not so easy. At the beginning no real time MPEG1 encoder was available and it was difficult to imagine that one could be available before the end of the project. But in fact, this went far faster than expected and it was possible to make use of such a product for the production of IMS1 applications.

For the application development, we had no script language, so we had to use the C++ Programming Language for that, which of course can not be recommended as the commercial solution. For the commercial service we are waiting for the scripting language standard which is under study in the ISO MHEG Expert Group.

In parallel with the development of the IMS1 Laboratory Demonstrator described above, work on the OCAM, necessary for the commercial service configuration, has been started. But the project time frame did not allow the integration of the OCAM in the Laboratory Demonstrator configuration, only an OCAM Demonstrator was developed.

Conclusion.

The Laboratory Demonstrator configuration, although not directly applicable for a commercial service, is a first step toward this goal. The development of a Laboratory Demonstrator was necessary to understand the technical problems still to be solved to run a pilot with real end-users and to share a common view for the introduction of the commercial service.

Through the development of such a Laboratory Demonstrator, the network operators will have gained expertise in the development of applications, in the use of the selected standards and in the knowledge of the network requirements from video retrieval services.

The next step (pilot phase) will be carried out outside Eurescom because it requires the participation of other active players in the multimedia retrieval services different from network operators such as service providers, content owners, user and network equipment manufacturers.

This is why we have set up the MARS Project (Multimedia Audio-visual retrieval Service) under the RACE umbrella. It is scheduled to set up a European service pilot by the end of 1995.

Hypermedia Information Retrieval System
Using MHEG Coded Representation
in a Networked Environment

Jong-Jin Sung, Mi-Young Huh, Hyoung-Jun Kim, and Jin-Ho Hahm

Protocol Engineering Center, ETRI
P.O.Box 106, Yusong, Taejon, 305-600, Korea

Abstract. A prototype hypermedia information retrieval system in a networked environment, is developed. For the growing importance of information sharing and interchange, the system uses the international standard, MHEG, for its multimedia information encoding and representation. The system is built in a client/server configuration having client/server communication between a server system and multiple client systems over N-ISDN. The server system stores MHEG encoded information and the client system retrieves, processes, and displays the information. For its time critical characteristics of multimedia data, the system provides functionalities to handle synchronization of transmitting data.

1. Introduction

With realization of the high speed networks such as FDDI, DQDB, and B-ISDN and real-time multimedia information processing, moves to integrate both technologies to create advanced teleservices are actively going on[1]. These services cover a wide range of different fields including education, training, advertisement, entertainment, publication and so on. Requirements of these services commonly reflect following points, i.e., real-time processing capability, availability on a networked environment, sharing of information among different systems, and friendly GUI. Among these requirements, sharing of information is having greater importance. Multimedia systems from different vendors and for different purposes can not share information with each other, while multimedia information itself is a valuable asset.

We have developed a hypermedia information retrieval system called HIRS as a model system for multimedia teleservices. Focusing on the importance of the information sharing, HIRS uses MHEG(Multimedia and Hypermedia information coding Expert Group) coded representation in dealing with data representation and processing. MHEG is a standard coded representation making multimedia information interchangeable and sharable[2,3].

HIRS is in a client/server configuration. The information encoded in MHEG representation is stored in a remote server system while users of client systems can retrieve the information through the client/server communication protocol. The client/server communication is built over a N-ISDN, and provides some features for the time-critical multimedia data transmission.

In this paper, overview of the system configuration and key functionalities of HIRS are described.

2. MHEG Technology

MHEG, standardized by ISO/IEC JTC1/SC29/WG12, defines the representation and encoding of multimedia and hypermedia objects to be interchanged as a whole within or across applications or services, by any means of interchange[2]. These objects, encoded using ASN.1 or SGML will provide a common base for other ITU-T Recommendations and ISO standards, and for the many multimedia and hypermedia applications which will be developed in the forthcoming year in a wide range of domains.

Our system follows this standard for the encoding of information using ASN.1[4,5]. Although the standard is not in the state of IS(International Standard) at this moment, we decided to use the MHEG representation defined in CD(Committee Draft) 13522-1 in our first prototype system. With the results obtained from the development, we will actively contribute to the standard and play an important role in the international standardization activity.

3. System Configuration

HIRS consists of a server system and multiple client systems which interchange information through client/server communication over N-ISDN. There are three major system components, i.e., client systems, a server system, and communication parts. Overall system configuration is illustrated in the figure 1.

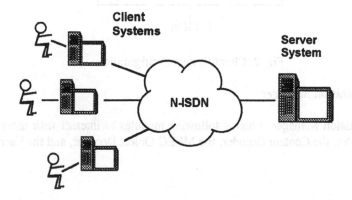

Fig. 1. HIRS Configuration

The client system, interacting with users, accepts input from them, and based on this input, sends requests to the server system to acquire wanted data. After receiving the results from the server, it displays them to the users. On the remote side, the server system, receiving requests from the client systems, finds requested data from stored objects and sends them as replies to the client systems. Communication parts connect the client systems with the server system and interchange requests and replies using a designed client/server communication protocol over N-ISDN.

3.1 Client System

The client system, as illustrated in the figure 2, comprises the Presentation Manager and the MHEG Processing Manager. The Presentation Manager provides a user interface and manages display of information to the user and MHEG Processing Manager analyzes and schedules MHEG encoded objects to be delivered in an appropriate way to the Presentation Manager.

The Client Communication Manager, placed between the client system and the N-ISDN, sends requests to the server and receives replies from it.

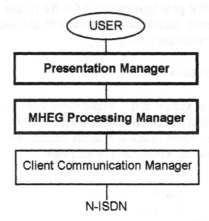

Fig. 2. Client System Configuration

3.1.1 Presentation Manager

The Presentation Manager contains following modules to interact with users : The Event Decoder, the Content Decoder, the MHEG Object Decoder, and the User Interface.

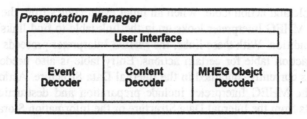

Fig. 3. Presentation Manager

The Event Decoder analyzes user interactions such as menu selection, button click, and keyboard typing and transfers them to the MHEG Processing Manager.

The Content Decoder and *the MHEG Object Decoder* analyze and transfer MHEG information to be presented to the User Interface. Among various MHEG encoded objects, *the Content Decoder* deals with content objects such as text, image, graphics, audio, and video objects. So this module has a couple of sub modules to handle different types of the media contents. *The MHEG Object Decoder* deals with all the other MHEG encoded objects used for presentation except the content objects. This module takes care of temporal and spatial arrangement of the content objects including synchronization.

The User Interface module displays an initial interface screen when the system starts and directly interact with users.

3.1.2 MHEG Processing Manager

The MHEG Processing Manager, as illustrated in the figure 4, consists of the MHEG Parser/Formatter, the Internal Data Structure, and the MHEG Interpreter.

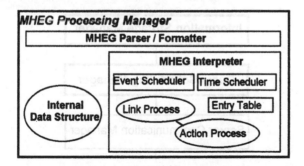

Fig. 4. MHEG Processing Manager

The MHEG Parser/Formatter is in charge of encoding and decoding of MHEG objects coming from the server system and then constructs *Internal Data Structure*.

The MHEG Interpreter plays a main role in the process of the MHEG objects. It interprets the data in *the Internal Data Structure* to perform event scheduling, link processing, and other actions for the objects. For these actions, the MHEG Interpreter builds link table and action table. When an input event comes from the Presentation Manager, the MHEG Interpreter looks up in the link table to find links with matching trigger condition. With these links, the MHEG Interpreter proceeds further processes on the action table for certain actions. Entry table is also needed to manage MHEG objects currently contained in the Internal Data Structure. Actions to be carried out by the MHEG Interpreter include preparation and destruction of the requested objects from the Internal Data Structure or the Information Storage Manager. Synchronization and ordering of time-critical objects are also the role of the MHEG Interpreter. The Event Scheduler and the Time Scheduler of the MHEG Interpreter take these scheduling roles.

The MHEG Parser/Formatter, the MHEG Interpreter, and the Link Processor in the MHEG Interpreter are designed to run on their own processes, and these processes could communicate each other through IPC(Inter Process Communication). But current status of the HIRS does not support these techniques because of non-preemptive multitasking of Windows 3.1 on which the first prototype HIRS is built. However, the 2nd prototype will be on Windows NT and support these.

3.2 Server System

The server system, as shown in the figure 5, is composed of the Information Storage Manager and the Query/Scheduling Manager. The Information Storage Manager manages storage of MHEG encoded objects. The Query/Scheduling Manager queries the Information Storage Manager on requested MHEG encoded objects and schedules the replied objects to be sent in effective order to the client system.

The Server Communication Manager, placed in the front end of the server system, takes a responsibility of receiving and sending data from and to the client systems.

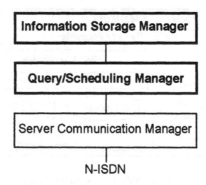

Fig. 5. Server System Configuration

3.2.1 Information Storage Manager

The Information Storage Manager stores MHEG encoded data and media contents. MHEG encoded data is stored in the internal database and media contents in files. The Information Storage Manager uses an object-oriented database system as an internal database. For easy access to this database, we made simple APIs in front of the Information Storage Manager.

Data structure of the objects to be stored in the internal database is shown in the figure 7. A stored object has two attributes, i.e., *MHEG Identifier* and *MHEG Encoded Stream*. MHEG object data delivered from the client system is stored in the field *MHEG Encoded Stream* with its ASN.1 notation without any change, while MHEG identifier stored in the field *MHEG Identifier* is used as a key reference value for a stored MHEG object for direct search.

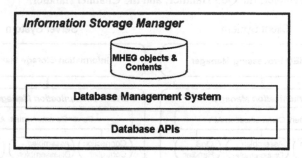

Fig. 6. Information Storage Manager

Content media can be either embedded in the MHEG content object or linked to the MHEG content object by an external reference which includes file name of the content media. In our first prototype HIRS, all the content media except text are stored in files and linked by references from the MHEG content objects.

MHEG Identifier	MHEG Encoded Stream (ASN.1)

Fig. 7. Structure of Stored Objects

3.2.2 Query/Scheduling Manager

The Query/Scheduling Manager accepts requests from the client system via Communication Managers and queries the Information Storage Manager on requested MHEG encoded objects and media contents. If the requested data is a MHEG encoded object this module queries through database APIs and if the requested data is a media content it asks the Information Storage Manager for wanted media file.

Scheduling of the objects to be sent to the client system is also an important role of this module. The scheduling information itself is made by the MHEG Processing Manager of the client and is delivered to this module to be used for effective data transmission. The use of this scheduling scheme is described later in the section 4.3.

3.3 Communication Managers

Communication Managers connect client systems with a server system and interchange requests and replies using a designed client/server communication protocol over N-ISDN. A Communication Manager on the client system is called Client Communication Manager and one on the server system is Server Communication Manager. The server system of our HIRS is a connection-oriented and concurrent server. For data transmission, both the client system and the server system use S-Interface cards. A connection setup is established on a D channel and data transmission is on B channels

Component modules of the Client (and the Server) Communication Manager are the Client (or the Server) Communication API, the Connection Controller, the Assembler/Disassembler, the QoS Handler, and the Channel Handler.

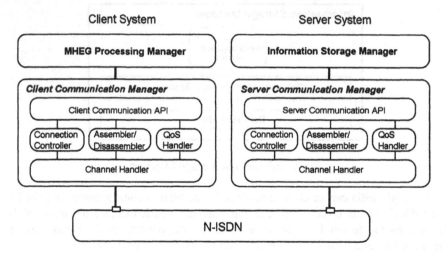

Fig. 8. Communication Managers

The Channel Handler creates and manages logical channels. HIRS uses N-ISDN 2B channels for one client-server data transmission. On this channel, MHEG objects are delivered between a client and a server. Channel Handler can create and manage multiple logical channels on this 2B channel to carry multiple MHEG objects concurrently. Especially for different media contents to be transmitted simultaneously, support of the multiple logical channels is important.

The Connection Controller takes care of the connection establishment and release. The Server Communication Manager provides concurrent connection control mechanism, so the server system can connect multiple client systems at the same time and provide services concurrently. The Server Communication Manager listens until a connection establishment request from a Client Communication Manager arrives and forks a Connection Controller process to establish a connection, then it listens for another connection request. Each data from multiple client systems comes through its own Connection Controller of the Server Communication Manager, and is packed into a queue for entering the Query/Scheduling Manager. On each connection, there can be multiple channels. These channels are controlled by the Channel Handler.

The Assembler/Disassembler handles assembling and disassembling of the MHEG objects to be carried over through N-ISDN S-Interface cards.

The QoS Handler is supposed to deal with such characteristics as throughput, delay, jitter, flow control, and rate control. But the first prototype HIRS does not cover these and postpones them until the next version.

Figure 9 shows *communication APIs* for the Client and the Server Communication Managers. These APIs are for connection establishment, connection release, data send, and data receive on both client and server side. These client/server communication APIs are mapped into the N-ISDN APIs.

	Client Communication API	Server Communication API
Connection Establishment	Connect	Listen Accept
Data Transmission	Send Receive	Send Receive
Connection Release	Close	Close

Fig. 9. Communication APIs

4. Prototyping

4.1 An Example Scenario

To explain overall operations of our first prototype HIRS, we describe a simple example scenario depicted in the figure 10.

Screen 1 is the initial menu screen to accept user's selections. When button A1 is selected, Screen 2 is invoked and displayed. Screen 2 shows parallel presentation of object B1 and B2. From the Screen 1, when button A2 is selected, the Screen 3 is invoked to display sequential presentation which presents C1, C2, and C3 in orderly manner. Button Ax is for termination of the application and Bx and Cx are for going back to the previous screen, Screen 1.

4.2 Overall Operations

Above example scenario is previously encoded in MHEG representation and stored in the Information Storage Manager of the server system. And initially, there is no MHEG object stored in the client system.

Encoding of the above example scenario results in three MHEG composite objects and many related component objects that are referenced by or embedded in the three

MHEG composite objects. The three composite objects are the root objects of Screen 1, 2, and 3.

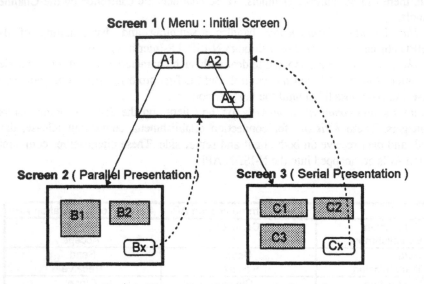

Fig. 10. Illustration of the Example Scenario

When HIRS is started by a user, MHEG Processing Manager of the client system requests the initial MHEG object which is the root composite object of the Screen 1. This request is first applied to the local storage devices. But, since there is no MHEG object in the client system, the request is redirected to the Information Storage Manager of the server system. After receiving the root object of the Screen 1 from the server, the MHEG Processing Manager interprets the object information, constructs internal data structures and processes links (including Container Startup Link, Presentation Startup Link, and so on) and action rules. While interpreting and processing the root composite object, the MHEG Processing Manager finds out related necessary MHEG objects to present Screen 1, and fetches these objects from the server system.

Screen 1 is a menu screen having selection buttons A1 and A2 for Screen 2 and 3. When one of the buttons is selected, the status of the button selection becomes a trigger condition for another action(s), thus the MHEG Processing Manager looks up internal link table to find a matching link and processes it. In the case of the button A1, according to the matching link, the Screen 2 is invoked. Rest of the operation is similar to the case of the fetch and display of the menu screen, Screen 1, until the fetch of content objects B1 and B2.

B1 and B2 are presented in parallel when Screen 2 is displayed. It requires the Server Communication Manager to send two objects simultaneously in order to achieve effective parallel presentation on the client side. Further discussion about this issue is described in the next section.

When button A2 is selected, Screen 3 is displayed in which C1 is firstly presented. At the moment of the presentation of C1 is over, that of C2 is started. And after C2 is over then finally C3. In this sequence, the object C1, C2, and C3 are delivered from the server to the client in timely arrangement by the Client/Server Communication Managers. Also, the timely transmission of the objects from the server to the client is discussed in the next section.

4.3 Synchronized Data Transfer

Multimedia information encoded in MHEG representation contains not only media contents but also synchronization schemes. The synchronization scheme describes the way how related media contents can be activated(displayed) together. Multiple objects(i.e., media contents), to be synchronized in certain period of time, had better be sent together in parallel to the client system. Otherwise, none could be activated until the whole related objects arrive at the client system. Synchronized data transfer means that sending multiple objects in an orchestrated scheme, so that more timely display of the data on the client system can be achieved[6-11].

To realize the synchronized data transfer, the Server Communication Manager needs to know the control information such as the presentation schedules and the sizes of the requested objects. Here, the presentation schedule means starting point and duration of every object. In our prototype system, this information is constructed by the Time Scheduler of the MHEG Processing Manager of the client system during the interpretation of the fetched MHEG objects. And this information is delivered to the Server Communication Manager and used for proper bandwidth alignment and transmission scheduling. Control packets are used for Server Communication Manager to inform the Client Communication Managers to rearrange the data receiving conditions.

4.4 Prototyping Environment

The first prototype HIRS system is implemented on 486PCs and MS-Windows 3.1. Programming languages are C, C++, and Windows SDK, and the database system for the Information Storage Manager is made of a developer library toolkit for object-oriented database programming. ISDN S-Interface cards used in our system are previously made and tested in another project of our research institute.

Because of the non-preemptive multitasking and the lack of proper interprocess communication of Windows 3.1, some of the design features are implemented in restricted ways. But currently, development of the second prototype system is going on over Windows NT which covers previous restrictions, supporting multi-threading mechanism and useful interprocess communications and so on.

5. Conclusion

HIRS, a hypermedia information retrieval system in a networked environment, is developed. For the growing importance of information sharing and interchange, the system uses the international standard, MHEG, for its multimedia information encoding and representation. The system is built in a client/server configuration having client/server communication protocol between a server system and multiple client systems over N-ISDN.

Overview of the system configuration and key functionalities of HIRS are described. For its time critical characteristics of multimedia data, HIRS provides such functionalities as to handle synchronization of transmitting data.

References

1. W. L. Kemper: Delivering Multimedia in a Networked Environment Multimedia Communications'93. Banff. Canada. (1993)

2. ISO/IEC. Committee Draft 13522-1 Coded Representation of Multimedia and Hypermedia Information Objects (MHEG) Part 1. June 15, (1993)

3. R. Price: MHEG: An Introduction to the Future International Standard for Hypermedia Object Interchange. Proc. of ACM Multimedia93, 121-128 (1993)

4. ISO/IEC. IS 8824 Specification of Abstract Syntax Notation One (ASN.1). Second Edition. (1990)

5. ISO/IEC. IS 8825 Specification of Basic Encoding Rules for Abstract Syntax Notation One (ASN.1). Second Edition. (1990)

6. C. Nicolaou: An Architecture for Real-Time Multimedia Communication Systems. IEEE J. on Selected Areas in Communications. Vol. 8, No. 3, April (1990)

7. B. Prabhakaram and S. V. Raghavan: Synchronization Models for Multimedia Presentation with User Participation. Proc. of ACM Multimedia93, 157-166 (1993)

8. T. D. C. Little: A framework for synchronous delivery of time-dependent multimedia data. Multimedia Systems, 1:87-94 (1993)

9. D. J. Gemmell: Multimedia Network File Servers: Multi-channel Delay Sensitive Data Retrieval. Proc. of ACM Multimedia93, 243-250 (1993)

10. T. D. C. Little and A. Ghafoor: Scheduling of bandwidth-constrained multimedia traffic. Computer Communications, Vol. 15, No. 6. July/Aug. (1992)

11. K. Lee and et. al.: Temporal Specification and Synchronization for Multimedia Database Queries. Proc. of International Symposium in Next Generation Database System and Their Applications. 198-204. Fukuoka. Japan (1993)

Transparent ATM LAN Interconnection over ISDN

Patrick Droz and Jean-Yves Le Boudec

IBM Research Division, Zurich Research Laboratory, 8803 Rüschlikon, Switzerland

Abstract. Narrowband Integrated Service Digital Networks (N-ISDNs) are emerging as key services for multimedia over public networks; at the same time ATM-based hubs [1] are starting to form backbones for providing multimedia services in customer premises networks (CPNs). In this context, it is necessary to be able to interconnect CPNs not only by means of broadband services (ATM or SMDS), but also by means of N-ISDN. (This is of course different from the dual, traditional problem of supporting N-ISDN connections over ATM).

In this paper we address the requirement of transparent interconnection of ATM local area networks over public switched N-ISDN services [2, 3, 4, 5]. The method presented not only reduces the cost of using the ISDN service, it also provides a high level of transparency to the end systems. It has the following features:

- ATM LAN-attached systems are not aware that ISDN circuits may be used; from their point of view, ATM connections are carried end-to-end.
- ISDN circuits (B-channels) are set up on demand, based on both the requested bit rate and the actual traffic occurrence of the established ATM connection.

This method is adaptive in the sense that the B-channels are allocated according to the actual traffic occurrence, thus reducing link costs drastically compared to leased lines. It supports multiple destinations at the same time by assigning B-channels to the currently needed destinations. This paper covers the full range of topics involved, i.e. adapter design, protocol architecture as well as transparent integration into the ATM architecture. Our method is a first example of use of bandwidth on demand by CPNs. It can be extended to the use of other switched services on the public network side.

1 Overview of the Method

The components of the method are illustrated in Fig. 1, where the nonshaded areas represent components assumed to be available in an ATM node [6]:

- the ATM node performs switching and control functions for a pure ATM environment. In particular, the control point services perform internal resource management, routing and topology exchange, address resolution and connection control for ATM connections.

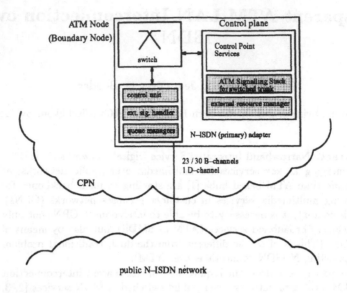

Fig. 1. Node structure

- the ISDN adapter provides access to an ISDN service. Only switched circuits (B-channels) are assumed in this document. In particular, the ISDN adapter provides a set of signalling functions used to control the establishment and release of B-channels. It uses a D-channel for this purpose. A primary interface is assumed (23 or 30 B-channels, depending on the service, plus one D-channel).

The new components are shaded in Fig. 1. They are:

- The external resource manager (ERM) function allows ISDN circuits to appear as regular ATM trunks to the rest of the control point services. It acts as the interface between the pure ATM architecture and the non-ATM architecture. It initializes and controls all activities to make ISDN circuits appear as regular ATM trunks. To achieve these functions, the ERM is integrated into the interface between the inbound and outbound ATM protocol stacks.
- The control unit (CU) is the module that coordinates the pure ISDN part. Its responsibility is to make concatenated B-channels to one destination appear as one trunk. It also coordinates the assignment of B-channels to different destinations. The CU has an interface to the ERM to obtain the required information.
- For each connected destination there is a queue manager (QM) that monitors and controls the set of assigned B-channels. It measures the traffic and triggers the allocation and deallocation of ISDN B-channels to the actual needs. The QMs are generated dynamically by the CU whenever a new destination has to be connected. The QMs have an interface to the CU.

- The external signalling handler (ESH) handles the ISDN signalling activity. It controls the setting up and clearing of ISDN circuits. The ESH has an interface to the CU.
- The ATM signalling stack for switched trunks controls connections that use the virtual trunks set up over ISDN circuits. It is a slight modification of the Q.93B stacks present in the control point services. The modification to Q.93B is to have a more efficient integration of the N-ISDN service. This includes different time-out values for certain Q.93B timers. The modification could be avoided by allowing a less streamlined implementation.

In the following, the ATM node with the additional functions described above is called a boundary node. The method works as follows:

- During normal control point service activity, connection requests are presented to the boundary node (called in this context the source boundary node). The method by which the control point services determine that a specific boundary node is involved in the connection is beyond the scope of this document, and so is the method by which the control point services determine that the boundary node should make use of the ISDN service and connect to a peer boundary node. The address resolution that results in the determination of the peer boundary node (the target boundary node) is also assumed to be performed by the control point services.
- When the ERM in the source boundary node detects a connection setup message for a remote destination that can be reached by the N-ISDN service, it triggers the CU to provide connectivity. The ERM sends the entire setup message including the desired destination and the QOS parameters to the CU. This is described in Section 3.
- The CU checks first whether the desired destination is already connected. If the destination is not yet connected, a new QM is created with the required number of B-channels. If there are not enough B-channels available, the new connection will be refused. This is described in Section 4.
- If connectivity is already provided, only the additional B-channels are assigned to the appropriate QM. This is described in Section 5.
- In the case of a new destination, the QM triggers the ESH to establish one B-channel to the desired destination so that later on, the ATM signalling can be transported transparently across it. It also allocates enough buffers to support the additional connection. As soon as connectivity is provided, it will be reported back to the CU and further back to the ERM. From this point in time, the connection between the two boundary nodes appears to the control plane as a trunk. This is described in Section 7.
- In parallel to the ISDN signalling, the "ATM signalling stack for switched trunk" is initialized. This is described in Section 8. This component will eventually determine whether the connection can be established by the usual means of Q.93B.
- As a permanent background function, the queue manager adapts the number of B-channels between the source and the target boundary node. It tracks

the actual traffic in order to minimize the number of B-channels used. This
is described in Section 6.

2 Protocol Architecture

First, we present the protocol architecture to merge ATM and ISDN into one
architecture without necessitating many changes to either of these two protocol
families. In Fig. 2, the protocol stacks are depicted for a simple configuration.
The important part is the two nodes in the middle, which are called the boundary
nodes.

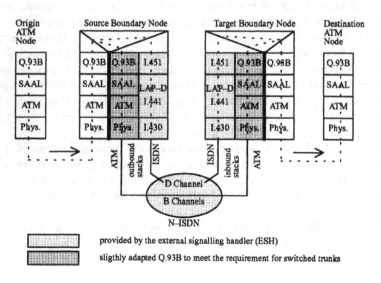

Fig. 2. Signalling protocol stacks

Let us assume that the connection setup runs from left to right. Then the
boundary node on the left is the source boundary node. Similar to the internal hop-by-hop connection setup with Q.93B, there exists on each intermediate
node an inbound and outbound protocol stack. The basic idea to integrate the
N-ISDN service is to have two protocol stacks instead of only one: an almost
unmodified Q.93B protocol stack, which is called here an ATM signalling stack
for a switched trunk, and an N-ISDN signalling stack. On the source boundary
node, two outbound stacks are built, whereas on the target boundary node, there
are two inbound protocol stacks.

The Q.93B protocol stacks for switched trunks are basically the same as
for nonswitched trunks. In a simple implementation they could be exactly the

same. But in order to achieve a more streamlined concept, minor adaptations should be made to assimilate the properties of the switched service. This involves the adjustment of time-out values. From Fig. 2 it can be seen that the ATM signalling is transported end-to-end transparently over the B-channels, while the ISDN signalling is done over the D-channel.

Figure 3 depicts the data protocol stack. Here it can be seen that ATM cells flow on the B-channel. Therefore the only difference between internal and external links is the physical (lowest layer) interface employed.

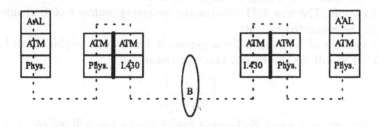

Fig. 3. Data protocol stack

In the following section the individual steps will be described in more detail. Figure 1 will be used for explanation.

3 Initial Processing by the External Resource Manager

The ERM provides the interface between the pure ATM architecture and the ISDN architecture. It can be regarded as the glue that merges the two architectures. It converts ATM setup messages into ISDN setup messages and initiates the building of the ISDN outbound stack. The following steps are performed by the ERM.

- Upon detection of an ATM setup message for a destination that can be reached by the N-ISDN service, it determines the capacity needed for the connection based on the parameters included in the ATM setup message. In this paper the only QOS parameter considered is the peak cell rate. It is assumed that NRB traffic is controlled by means of a specific mechanism not described here. One possible control mechanism is backpressure [7, 8].
- The requested bit rate and the desired destination address are then sent to the CU for further processing. In addition the "ATM outbound stack for switched trunk" is initialized. The buildup of this stack runs in parallel to the ISDN-related activities.

The ERM can be integrated into the interface between the inbound and the outbound stack. During the connection setup, all necessary information flows from the inbound to the outbound stack. The ERM can extract all the needed information.

4 Initialization of Dynamic Trunks over ISDN Circuits

Upon receipt of a setup message from ERM, the CU checks first whether the target boundary node is already connected. In case no connectivity is available, the following scenario will take place.

- The CU creates a new QM, which will handle the connections to the new destination. The new QM obtains the necessary number of B-channels to carry the requested bit rate.
- The number of B-channels for a requested bandwidth of pbit/s can be calculated as follows (for 64,000 kbit/s B-channels):

$$\left\lceil \frac{p}{64,000} \right\rceil \ .$$

- If there are not enough B-channels available, the CU will indicate this back to the ERM. The ERM then initiates the clearing of the connection.
- Then the queue manager allocates the necessary number of buffers and sends a request to the ESH to establish one B-channel to the new destination. The ESH implements the signalling protocol suite of the ISDN D-channel. As soon as one B-channel is established, the ATM signalling can be transported to the target boundary node transparently.

The CU on the target boundary node is also responsible for accepting new connections. The inbound signalling protocol stack for N-ISDN is generated by the ESH. After completion of the stack, ESH sends the request to the CU. The control unit then creates a new QM with one B-channel. From this point in time, ATM setup messages can be received by the target boundary node transparently.

As soon as the Q.93B setup message arrives at the target boundary node, the ERM will be informed with the desired QOS parameters. ERM then sends the transformed request to the CU. At this point it is possible to decide definitely whether the new connection can be accepted. In order to accept the connection, the source and target boundary node need a certain number of B-channels for this connection. If not enough B-channels are available, the CU will report this back to the ERM, which then creates an ATM clear connection request. If the B-channels are available, they will be assigned to the appropriate QM.

By exploiting user-to-user signalling on the D-channel, the QOS parameters can be transmitted to the destination boundary node to improve the performance of deciding whether a connection can be accepted.

For security reasons it is necessary to go through some authentification process. This can be done by user-to-user signalling on the D-channel. The authentification mechanism is beyond the scope of this paper.

In case of busy lines (no B-channels available at the target boundary node), the message flow goes from the ESH to the CU and back to the ERM.

5 Modification of Dynamic Trunks at Connection Setup

If the CU sees that the desired destination is already connected, the following steps will be executed.

- The CU determines the additional number of B-channels needed for the new connection. If the necessary number of channels is available, they will be assigned to the appropriate QM.
- If there are not enough B-channels available, the same action as described above will be taken (clear connection request).

6 Adaptation of B-Channels to the Traffic Occurrence

The architecture is designed to minimize the link costs across the public ISDN network. Links are allocated and deallocated dynamically according to the traffic load. The connections to one destination are combined in one QM. The bundling of the B-channels to achieve higher bandwidths than 64 kbit/s is hidden from the control plane. The control point sees only one channel with the aggregated capacity.

As B-channels to the same destination can be routed on different paths, the end-to-end delay can vary among the channels. To overcome this problem, some buffers are allocated to each B-channel. The bundled channels to a destination are then used in a strict round-robin fashion.

At startup the CU has a pool of B-channels which can be assigned to different destinations. During operation the B-channels walk through different states which are illustrated and described below. After releasing a connection, the B-channels are returned to the pool.

A link to a specific destination always consists of one or more B-channels. A distinction between the first and the additional B-channels is made in order to improve the end-to-end delay characteristic. During the connection setup the first B-channel is always allocated and activated (physically established). This is necessary for Q.93B messages to be carried to the peer boundary node. The first B-channel per destination passes through the states shown in Fig. 4. Note: if there are more than just one ATM connection to the same destination, they share the first B-channel.

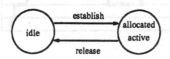

Fig. 4. States for the first B-channel

- *Allocated* means that the link is now reserved for this destination. The channel remains allocated until the connection is released.
- *Active* means that the link is actually established, i.e. data can flow across it.

The additional B-channels are handled more dynamically. They are released if there is not enough traffic for them. But the number of B-channels needed to support the requested bandwidth is reserved! This ensures that the ISDN adapter never becomes overbooked. The administration of the B-channels is done through queues. The queue length and the growing rate are the two factors that trigger the allocation and deallocation of the additional B-channels. The architecture will be presented in Section 9. These additional B-channels go through the states shown in Fig. 5.

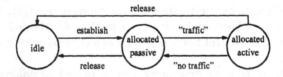

Fig. 5. States of an additional B-channel

The two events "traffic" and "no traffic" are generated by the QM according to the actual traffic. At the beginning, the additional B-channels are merely allocated but not actually established. If there is enough traffic they will be physically allocated and used to carry the traffic. If the traffic load drops below a certain rate, the channel will be physically released, but remains reserved. There is a predefined sequence in which the channels are allocated and deallocated. Figure 6 illustrates the behavior of a virtual channel consisting of no more than five B-channels.

Fig. 6. Example of a maximum of five active B-channels

Time runs from left to right. The B-channels that lay under the stressed line are active; the channels above are passive. The channels are put stepwise into the active state or released to the passive state. This stems from the fact that D-channel signalling is sequential. For the entire duration of the connection, the five channels remain reserved. With this allocation and deallocation strategy, costs for external links can be reduced.

When a connection is cleared, the additional B-channels will be returned to the CU. The first B-channel remains as long as there are connections to that particular destination. If the last connection to a destination is cleared, the first B-channel will also be returned to the CU. As soon as a B-channel is back in the pool, it is available again for other connections.

7 External Signalling Handler

The ESH implements the standard signalling protocol suite for N-ISDN D-channels. In Fig. 2, the various layers of the protocol were shown.

- In the uppermost layer there is the call control protocol I.451. This is the ISDN counterpart to Q.93B in ATM, which means that this component deals with the establishment and release of B-channels.
- The LAP-D (I.441) is the SAAL counterpart. It provides a reliable transport service to its users.
- I.430 implemented on the lowest layer. This part deals with the physical requirements of N-ISDN (voltage levels, signal encoding, timing and so on).

Up to now the asymmetry between the source and the target boundary node has not been mentioned. It is necessary to define some rules so that race conditions cannot occur. The problem of race conditions is that both sides can initiate connection setups for completely new connections or for additional B-channels. In addition there is a propagation delay for D-channel signalling as well.

By using the call-back mechanism it is possible to make one of the two involved boundary nodes the manager of the B-channels. The boundary node with the smaller E.164 address can be chosen as the B-channel manager. When the called party detects that allocated and passive B-channels have to be activated, it can trigger the other side through the call-back mechanism. The task of bringing a link from the active to the passive state is more difficult because the caller has to monitor the incoming traffic as well (reverse traffic). In a more sophisticated implementation, user-to-user signalling on the D-channel can negotiate the release of channels.

If two parties try to reach each other simultaneously, the one with the smaller E.164 number can be chosen to be the caller and the other one is the called party. As the calling party number is an information element in the ISDN setup message, the decision whether to accept or refuse a call is straightforward.

Let us make one remark on the concept of making a distinction between the first and the additional B-channels. With this concept the end-to-end delay can be improved. In addition it is expected that there are more than just one ATM connection to a destination and that in this case all connections share the first channel. An alternative would be to add additional criteria, i.e. the queue must be empty for at least 20 seconds or so.

8 Connection Control by the ATM Signalling Stack

As mentioned above, the ATM signalling stack for switched trunks is Q.93B with minor adaptations for switched trunks. The modifications are to improve the cooperation between N-ISDN and Q.93B. However, it would be possible to use an unmodified Q.93B stack.

In order to have a more streamlined integration, some time-out values from Q.93B can be adjusted to the N-ISDN needs. This is because the connection setup through N-ISDN is slower, and therefore more time has to be allowed to establish a connection. In addition, the interface between Q.93B and ERM can be improved if Q.93B and ERM communicate directly with each other.

9 Buffer Management

For each connected destination, a QM is assigned. The QM is responsible for bundling the B-channels and for the initiation of state changes of the additional B-channels. It also has to provide the correct number of buffers to store the incoming traffic until the additional B-channels become active.

The following parameters determine the number of allocated and active B-channels for one destination (the example is for 64 kbit/s B-channels):

- a measured arrival rate in kbit/s,
- d departure rate (to the ISDN) in kbit/s ($m \times 64$ kbit/s),
- y peak bit rate (known from the setup message) in kbit/s,
- t time to activate a B-channel in seconds.

With these parameters the following numbers can be calculated:

- allocated links:

$$n = \left\lceil \frac{y}{64} \right\rceil \quad ,$$

- allocated and active links:

$$m = \left\lceil \frac{a}{64} \right\rceil \quad .$$

The buffer size must be big enough to bridge the time it takes to establish an additional B-channel in case the arrival rate increases. The worst case is when the arrival rate changes from 0 to the declared peak cell rate. Then the required storage size is

$$\text{size} = n \times t \times y \quad .$$

This is the upper and save limit because the signalling on the D-channel can be done in a pipelined manner. It is not necessary to wait until one B-channel is activated before signalling for the next one.

The parameter a must be determined by the QM itself. In a simple estimation model, the QM samples the queue length from time to time. With the time interval and the newly arrived data, a can be calculated as follows:

$$a = \frac{\text{arrived data}}{\varDelta t} .$$

The time interval has to be made reasonably small to achieve good results. More powerful estimation models are beyond the scope of this paper.

10 Support of Non-Reserved Bandwidth (NRB)

It was mentioned above that the only traffic type supported is peak cell rate and non-reserved bandwidth (NRB). In this section we present the management of NRB.

In the case of NRB, it is not possible to reserve the number of required B-channels in advance. But the concept for the first B-channel to a destination still works (see Section 6). For additional B-channels the following policies are used:

- As long as there are B-channels available, they can be used by NRB traffic. They will be assigned to a destination on demand, and go directly into the allocated and active state.
- As soon as the estimation model determines that one of the B-channels can be put into the passive state, the B-channel will be released and returned to the CU so that it is available for other connections.
- Connections with reserved bandwidth have precedence over NRB connections. This means that additional B-channels can be requested back by the CU to pass them to reserved bandwidth connections.
- The maximum number of B-channels for NRB connections between two boundary nodes is given by the boundary node with the smallest available number of B-channels.

If necessary, additional rules can be applied, i.e. the maximum number of B-channels for NRB connections could be set to a certain threshold to reduce costs. Or the priority of NRB connections could be improved against the RB (reserved bandwidth) connections by making the additional B-channels non-preemptive. The rules could even be dependent on the destination or time of day.

In order to guarantee proper operation of the backpressure mechanism, it is necessary to be able to store the amount of data which can be transported during a round-trip delay time. In the worst case a B-channel is routed over a satellite which introduces a round-trip propagation delay of 0.5 s. Thus per 64 kbit/s B-channel, 4 kbytes ($0.5 \times 64{,}000/8 = 4$ kbyte) of buffer space is required.

11 Design of ISDN Adapter

In this section we present one possible implementation. Note that the placement of the individual components can be varied. Let us assume hereafter that the components are arranged as shown in Fig. 1.

The presented architecture of the adapter shows the necessary basic components, but does not imply that each of these components must be provided in hardware. It is very well possible to implement some of these components in software by using a signal processor as the hardware architecture. In an ISDN primary access node, all the 30 (23) B-channels and the D-channel are time-division-multiplexed into one physical channel. Each individual channel has its predefined time slot. The adapter is responsible for meeting the technical requirements that are standardized for N-ISDN. The unused B-channels are filled with special empty patterns. In Fig. 7, the high-level architecture of the N-ISDN adapter is depicted.

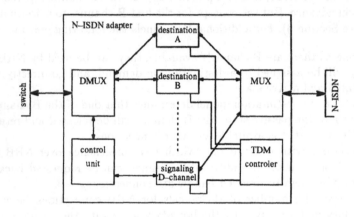

Fig. 7. High-level adapter design

All data, ATM signalling as well as user data, enters and leaves the adapter through the switch. There is a demultiplexer on the adapter that routes the incoming packets according to VPI/VCI labels. That is, it separates the packets for the different B-channels and the D-channel.

For the D-channel signalling, a predefined label is necessary. From the DMUX the cells are passed to the queue managers for each currently connected destination. These queue managers are responsible for bundling the B-channels to the same destination. They also trigger the allocation and deallocation of the additional B-channels. The control unit gets its data from the ERM through the switch on a predefined VPI?VCI label.

The TDM controller is responsible for meeting the corrent timing for the ISDN attachment. It triggers the read-out for the appropriate buffers. The TDM controller is also managed by the CU according to the assignment of the B-channels. The MUX brings the traffic from all the B-channels and the D-channels into one pipe. The output from the MUX meets the physical standard of the N-ISDN (voltage level, signal encoding and timing).

As the traffic is full-duplex, the architecture of the adapter is full-duplex as well. But the incoming traffic can by routed directly to the switch. So the incoming traffic is only buffered to reconstruct the ATM cells. The cells are then passed directly to the switch. This could be done without separating the traffic from the different destinations. But by separating the traffic it is easier to monitor the upstream traffic. This is necessary to decide whether an additional B-channel can be put in the passive state.

On the adapter there is a certain set of information about the connections needed for management purposes. Per QM the following information is needed:

- All allocated B-channels (i.e. B3, B9, B17 and B30),
- All the currently active channels (i.e. B3 and B9),
- All transient links (i.e. B17 going into the active state),
- A list of all connections including their VPI/VCI labels, their requested bandwidth and the assigned B-channels.

This information is administrated by the control unit, which is also responsible for the distribution among the various components on the adapter. The CU controls for instance the DMUX that the cells are directed to the appropriate queue managers. Because the DMUX routes according to VPI/VCI labels, it must obtain them from the CU.

12 Extension to ATM VP service

The presented architecture works also for virtual paths over N-ISDN if they can be set up the same way as virtual channels. This means basically that they can be set up by Q.93B. But an SNMP subagent could also be provided so that an operator could set up the VPs. Because N-ISDN service is used in a transparent way, the VPs can be implemented as on an internal trunk.

An SNMP subagent could be integrated in the CU. This would also be necessary to read out the current configuration and state of the adapter. But it could also be used explicitly to set up or take down connections.

13 Summary

We have presented an adaptive and transparent ATM LAN interconnection mechanism over a narrowband ISDN service. The method adjusts the number of allocated B-channels to the actual needs, thus reducing costs for WAN links drastically. The paper presented the full range of pertinent topics, including adapter design, protocol architecture and transparent integration.

References

1. **ATM User Network Interface Specification 3.0**, December 1993.
2. **Overview of ATM Networks: Functions and Procedures**, *E.D. Sykas, K.M. Vlakos and M. J. Hillyard,* Computer Communications, Vol 14, No. 10, December 1991.
3. **The Asynchronous Transfer Mode; A Tutorial**, *Jean-Yves Le Boudec,* Comp. Netw. ISDN **24** (1992) pp 279-309.
4. **ISDN, An Introduction**, *William Stallings,* Macmillan Publishing Company 1989, ISBN 0-02-415471-7.
5. **Internetworking with ATM WAN's**, *John David Cavanaugh, Timothy J. Solo,* Minnesota Supercomputer Center, Inc. December 14, 1992.
6. **FALCON: A Switch-Based ATM LAN**, *W.E. Denzel, J.-Y. Le Boudec, E. Port and H.L. Truong,* IBM Research Report RZ 2449, April 1993.
7. **A GFC Protocol for Congestion Avoidance in the ATM Connectionless Service**, *J. Cherbonnier and J.Y. Le Boudec,* Proc. EFOC/LAN'92, Paris, France, June 1992, pp 305-309.
8. **Backpressure Flow Control at the UNI to Support the Best-Effort Service on ATM**, *J. Cherbonnier, D. Orsatti and J. Calvignac,* ATM Forum contribution 93-1005, Stockholm, November 1993.

CPU Utilization of Multimedia Processes:
HeiPOET – The Heidelberg Predictor of Execution Times Measurement Tool

Hartmut Wittig
Lars C. Wolf
Carsten Vogt

IBM European Networking Center, Vangerowstraße 18, D-69115 Heidelberg
Mail: { wittig, lwolf }@vnet.ibm.com, vogt@fh-koeln.de

Abstract: Due to the time characteristics of audio and video data, the processing of multimedia applications has to be done using real-time mechanisms. Scheduling algorithms used within such systems require information about the processing time requirements, the CPU utilization of the applications, to perform schedulability tests. Since multimedia applications are often constructed by combining processing modules (often called stream handlers), processing time determination for these modules is required. The multitude of these modules and the large variety of computer systems calls for a measurement tool. In this paper we define the term CPU utilization for multimedia processing and describe the CPU utilization measurement tool HeiPOET. The presented measurements show that the tool provides measurement results with good accuracy.

1 Introduction

Multimedia applications have characteristic quality-of-service (QoS) requirements that must be satisfied by the underlying computer system, i.e., its local and its network resources. The CPU, being the principal local resource to be managed, requires functions for workload policing, capacity reservation, and process scheduling to enable the processing of real-time multimedia data. In general, application generated CPU workloads can be distinguished into the following service classes:

- non real-time processes
- control processes handling real-time and non-real-time tasks
- real-time processes with soft deadlines
- real-time processes with hard deadlines

The first class contains processes without deadlines, for example a compiler process. Because these applications have no real-time requirements they may run with the lowest priority among these classes. In most CPU reservation and scheduling schemes, at least a small amount of CPU capacity will be reserved for such applications to avoid starvation. This part of CPU capacity can be managed using classical strategies such as round robin or multi-level feedback.

Control processes regulate the access to certain resources. For example, an important control process for the CPU is the scheduler. A control process can also be used to reserve network resources in the setup phase of multimedia communication. On the other hand the same control process can be involved in the transmission of real-time messages, e.g., acknowledgment messages to trigger retransmissions of lost or corrupted data units [6]. The control processes should be handled in a deterministic manner with high priority.

Real-time applications with hard deadlines (e.g., manufacturing control systems) are processed in dedicated systems. For these systems exists a great variety of scheduling and reservation mechanisms ([5]). Though it is difficult to integrate hard real-time systems with multimedia applications, it can be done in principle ([13]).

Most multimedia processes can be considered as soft-real-time processes. If a data unit is too late or lost, this is not necessarily noticed by the human viewer. To avoid disruptions a characteristic quality of the media presentation is required. To describe these processing requirements of multimedia applications, a characteristic QoS parameter set must be defined. For example, the playback of a video stream requires a picture loss rate below a certain percentage, and it is necessary that each logical data unit (e.g., video frame) is displayed at a certain point in time. In a multimedia system resource reservation mechanisms and QoS enforcement strategies are used to guarantee QoS requirements of multimedia applications. This paper focuses on multimedia applications which process continuous media with soft deadlines.

CPU scheduling of multimedia processes is based on the classical real-time scheduling algorithms, Rate Monotonic (RM) [11] and Earliest Deadline First (EDF) [11]. RM scheduling is a static scheme which is defined in a context of tasks that require periodic CPU processing. Therefore, RM scheduling is well-suited for applications processing continuous media. According to their processing periods, processes have different, fixed priorities. Processes with small periods are executed with the highest priority. Low priorities are assigned to processes with large periods. At any time, the dispatcher of the CPU chooses the highest-priority process to run next. EDF scheduling is a dynamic real-time scheduling algorithm. The task with the earliest deadline has the highest priority. Because the timing constraints of all processes change, the priorities of tasks are dynamically adapted. To test whether a new task can be scheduled using RM or EDF scheduling, the following condition must be met:

$$\sum_{\forall i \,\in\, Tasks} R_i \times P_i \leq B$$

Index i runs through set of all multimedia tasks, R_i denotes the maximum processing rate of task i, P_i is the processing time per-period, and B is the schedulability bound. For RM scheduling the schedulability bound is determined by $B = \ln(2) \approx 0.69$. The schedulability bound of EDF scheduling is $B = 1$.

Each multimedia application has to specify the workload it will generate. This workload specification consists of the processing rate R and execution time per period P. The resource management system performs the schedulability test to decide whether this new application can be accepted. If enough CPU capacity is available to execute the new application without disturbing existing applications, the schedulability test returns successfully. As seen from the above formula, the processing time is an important parameter for the schedulability test. However, the determination of these values is still a largely unresolved issue in the area of CPU scheduling. The main problems measuring CPU utilization of multimedia applications are:

- Great variety of multimedia applications and interfaces
- Exact definition of the term "CPU utilization" of multimedia applications
- Handling of Variable-Bit-Rate Streams

This paper focuses on the problem of how to determine CPU utilization of multimedia applications. The HeiPOET (Heidelberg Predictor of Execution Times) measurement tool provides a computer-based prediction of CPU utilization and a refinement of measured values during the execution time of multimedia applications. The highlights of the HeiPOET tool are:

- Automation of CPU utilization measurements
- Specification language for the description of various parameters
- High precision of measurement results by eliminating operating system interruptions
- Computation of statistical metrics to check the reliability of measurements
- Re-usability of measurement results by storing measurements in a database

In the next section, we describe the model chosen for CPU utilization. Then design principles of HeiPOET are explained. Measurements presented in Section 4 show the validity of the architecture before we give our conclusions.

2 CPU Utilization of Multimedia Processes

In this section a definition of the term "CPU utilization" of multimedia processes and a characterization of stream handlers as the heart of multimedia applications are given.

We define the CPU utilization of an application process as follows:

The CPU utilization of an application process is the overall duration in which the CPU is occupied in order to perform this application task.

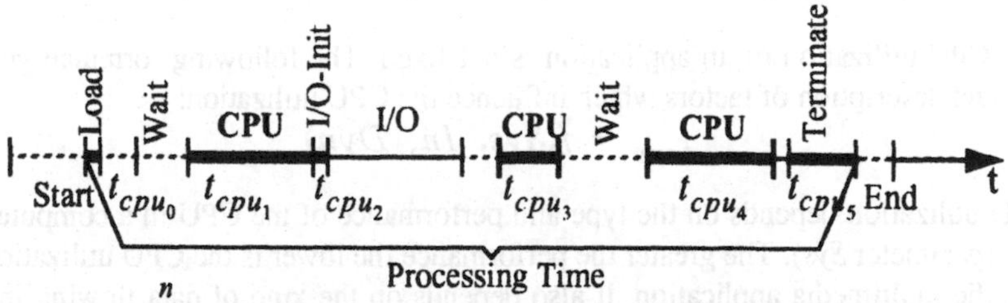

$$t_{cpu} = \sum_{i=0}^{n} t_{cpu_i} \qquad \begin{array}{l} t_{cpu_i}..\text{Duration of i-th processing cycle} \\ t_{cpu}.. \text{ Overall CPU utilization of the multimedia module} \end{array}$$

Fig. 1. Definition of CPU utilization

In Figure 1 it can be seen that CPU utilization consists of two different parts. In the first place, it includes the pure application code execution time on the CPU. Additionally, there are operating system activities to make the execution of multimedia applications possible, e.g., context switches, initialization and termination of I/O operations. An example is a multimedia application which plays an MPEG video stream coming from the hard disk. Most parts of the CPU are utilized for the software decompression algorithm, yet the read operation also needs CPU time, at least to set up an asynchronous disk I/O operation to read the multimedia data.

The operating system activities are characterized by the multimedia application process model (see Figure 2) derived from the classical UNIX process model ([2]). To determine the CPU utilization of a periodic multimedia application, three states are of specific interest: the states "Running", "Asleep" and "Ready to Run". If a process is in the state "Running" its machine code will be executed. The time an application is in this state is the prominent part of its CPU utilization. The transitions between these states represent the operating system activities which also require CPU capacity. Thus, initialization, termination, context switch, wakeup and sleep operations (①-⑥) are the operating system parts in the definition of the CPU utilization of a multimedia application. In the life cycle of process initialization (①) and termination (⑥) operations are system activities which are performed only one time, namely at the beginning and the end of the multimedia processing. Because the initialization and the termination of a multimedia application are not time-critical, and periodic scheduling algorithms do not consider aperiodic parts in the schedulability test. In the context of this paper, these operations are of no interest for the measurement of CPU utilization.

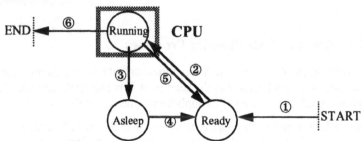

Fig. 2. State transition diagram of multimedia processes

The CPU utilization of an application is not fixed. The following formulae gives an abstract description of factors which influence the CPU utilization:

$$t_{cpu} = f(Sys,\ In,\ Dyn)$$

CPU utilization depends on the type and performance of the CPU in a computer system (parameter Sys). The greater the performance the lower is the CPU utilization of a specific multimedia application. It also depends on the kind of data flowing into the multimedia application, e.g., control information like the picture rate of an MPEG stream, or the compressed data stream containing the MPEG video (parameter In). A third influence to the CPU utilization are dynamic properties of the operating system, e.g., the scheduling algorithm and timing (parameter Dyn). For example, using preemptive scheduling algorithms causes more context switches than non-preemptive scheduling algorithms.

Often multimedia systems are structured in modules, objects or devices (e.g.,

[9],[3],[1],[14]). Each module is responsible for a logical operation on the continuous-media stream, e.g., read operation from a multimedia file system, mixing of audio and video streams. These modules are called stream handlers (SH) (e.g., [9],[14]). In general, an SH is a software entity processing continuous-media data streams. The general structure of an SH is shown in Figure 3.

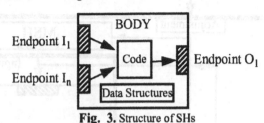

Fig. 3. Structure of SHs

An SH consists of a body and at least one endpoint. The SH body includes data structures for control information and the executable code. SHs transfer multimedia data via endpoints[1]. Data flows unidirectional between SHs. Endpoints for sending data are called output endpoints, endpoints for receiving data are input endpoints.

A multimedia application can be modelled as an acyclic directed graph of SHs beginning with source SHs and ending with sink SHs. The SH graph is constructed at the application start-up time and can also be changed dynamically at application run time.

The CPU utilization of a multimedia application consisting of SHs is the sum of the CPU utilization of each SH in the SH graph. Hence, for the CPU utilization measurement the application can be divided into its basic parts, the SHs, which are then measured individually. The advantage of this approach is its modularity, which reduces the measurement effort. SHs are well structured and easy to handle for measurements. Measuring the CPU utilization based on SHs allows to handle dynamic changes in the multimedia application. E.g., if a new SH is added to the existing graph at application run time, the CPU utilization of the new SH can obviously be added to the CPU utilization of the SH graph.

3 Design Principles and Architecture

HeiPOET is a measurement tool for CPU utilization of multimedia SHs. The architecture of HeiPOET is based on related work in the field of measuring response times in hard real-time systems (see [7, 8, 10]), however, since response time is not of primary interest for multimedia applications, the results of this work from hard real-time systems is not usable in multimedia systems. The design of the HeiPOET measurement tool is shown in Figure 4.

---

1. An exception are source and sink SHs: the source SH receives data from device drivers and the sink SH sends data to device drivers.

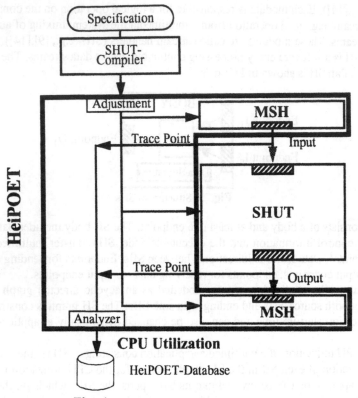

Fig. 4. Design of the HeiPOET measurement tool

HeiPOET is structured into three major parts:

- SH specification language and compiler: A language describes a set of SH initialization values to measure the SH under various input parameters and conditions. All control attributes of SHs can be modified using this specification language. Thereby, the SHs can be adjusted on the fly and CPU utilization can be measured using various parameter settings of the SH. A description of the SH specification language is given in Appendix A.
- Measurement stream handlers: Specific SHs to send/receive test data to/from the stream handler under test (SHUT) are called measurement stream handlers (MSH). They are used to form the environment in which the SHUT is embedded in the measurement phase. Because of the variety of SHUT the MSHs must be flexible. According to the elements in the specification language the SHUTs, the input MSH and the output MSH are adjusted. Major parameters to be adapted are the number of input and output endpoints of the SHUT, the number of incoming and outgoing packets of each endpoint per processing period, and the size of incoming packets.
- Analyzer: The analyzer reads the recorded measurements, analyzes the measure-

ments, computes statistical metrics, and stores the results in the HeiPOET data-base.

The HeiPOET tool works as follows: HeiPOET reads a SH specification file containing the description of a set of SH initialization values (see Appendix B). The MSHs and the SHUT are initialized with the first set of specified initialization values. The sequence of operations in the HeiPOET measurement phase is shown in Figure 5.

Input MSH   **SHUT**   **Output MSH**

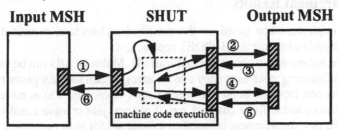

Fig. 5. HeiPOET measurement phase

Depending on the actual scenario, the input MSH sends a data frame to the SHUT. When the control flow changes from the MSH to the SHUT, a timestamp is recorded using a system trace mechanism (①), and the machine code of the SHUT is executed. When the SHUT finishes the processing for the current data packet (i.e., the processed packet is sent to the other MSH), a second timestamp is saved (②). The output MSH receives the data and the control flow immediately returns to the SHUT. This is the beginning of the next SHUT activation, a new timestamp is recorded (③). After sending data packets through other endpoints to the output MSH and saving the timestamps of these operations (④,⑤) the control flow returns from the SHUT to the input MSH (⑥). The duration between the timestamps ① and ⑥ is a complete processing period of the SHUT. The SHUT is in the state "Running" between timestamps ① - ②, ③ - ④, and ⑤ - ⑥. This time can directly be added to the CPU utilization of the SHUT.

Interrupts and other system operations (e.g. sleeping, paging, interrupts) in the execution phase of the SHUT are also recorded by the trace facility (the whole kernel of the used AIX™ operating system is instrumented with calls to trace functions). Thereby, it is possible to detect, to measure, and to subtract all interruptions occurring during the measurement phase of the SHUT (e.g., by higher priority processes or interrupt handling routines). This technique guarantees measurement results with high accuracy. Analyzing the resulting trace with the saved timestamps, HeiPOET can compute the CPU utilization of the SHUT. To increase the precision of the measurement, the measurements are repeated for each set of SHUT parameters for a certain number of times. Statistical metrics, such as mean CPU utilization, standard deviation, and the 0.95 confidence interval, are computed to get information about the reliability of the CPU utilization measurements.

These steps are repeated until the CPU utilization for each set of parameters in the SH specification file has been measured. All results are saved in the HeiPOET database.

Since the CPU utilization of a SH depends on the performance of a computer, it is necessary to measure the CPU utilization on each computer system. E.g., the measurements can be done at SH installation time and when the system configuration changes. As part of the workload specification of an applications SH graph, the CPU utilization values of the SHs are retrieved from the HeiPOET database.

4 Measurement Results

Using an implementation based on the described architecture, measurements have been performed to validate the HeiPOET approach.

The first measurement object is a multicast SH. Multicast SHs can be used in multimedia conferencing systems to copy one incoming stream to many partners. A multicast SH has one input endpoint and many output endpoints. The number of output endpoints can dynamically be changed, e.g., if partners join or leave a multimedia conference. The first measurement is based on a multicast SH with two output endpoints. In one processing period, the multicast SH duplicates a packet (10 KByte) coming from the input MSH and forwards them through two output endpoints to the output MSH. The CPU utilization measured with the HeiPOET tool[2] is shown in Figure 6.

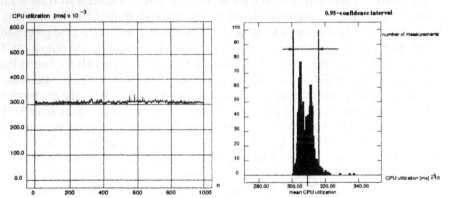

Fig. 6. CPU utilization of a multicast SH with two endpoints

The diagram on the left side shows the measurement series. Though the test machine was heavily loaded, it can be seen that the CPU utilization is approximately constant. The diagram on the right side of Figure 6 gives statistical metrics to quantify the confidence of the measured results. The mean CPU utilization of the multicast SH is 310µs. The standard deviation in the measurement series is 3µs, this means less than one percent of the mean CPU utilization. The 0.95 confidence interval is [301µs, 316µs].

Additional measurements on various computer models are presented in Table 1. These measurements are based on a mixer SH which mixes two incoming streams and produces one outgoing stream (e.g., to mix digital audio streams coming from microphones). It can be seen from Table 1 that the standard deviation is very small.

---

2. The measurements were performed on an IBM RISC System/6000 Model 530™ running AIX 3; the system was used by other tasks and users during the measurement.

In all cases the standard deviation has never exceeded three percent of the mean CPU utilization. It can be resumed that the HeiPOET tool works with a very high precision. In practice, HeiPOET is sufficient to measure the CPU utilization of SHs.

Table 1: CPU utilization of a mixer SH

Computer (RISC System/6000)	mean CPU utilization [ms]	standard deviation [ms]
Model 360	8.0	0.02
Model 340	12.2	0.04
Model 530	16.1	0.03

Finally, a multicast SH has been measured with a varying number of output endpoints. Using the SH specification language measurements are performed with a number from two to ten output endpoints (see Appendix B for the SH specification file). Figure 7 shows the results. Each point in the diagram represents 1000 single measurements. It can be seen that there is a linear relation between the number of output endpoints and the CPU utilization of the multicast SH. Because of the complexity $O(n)$ of the multicast operation (where n is the number of output endpoints), these measurements can be taken as an additional proof for the correct work of the HeiPOET tool.

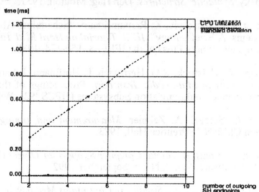

Fig. 7. Multicast SH with two to ten endpoints

5 Conclusions

HeiPOET was developed to solve the problem of CPU utilization measurement of multimedia applications. Based on the CPU scheduling algorithms and the SH model, the architecture of HeiPOET was designed. Using the SH specification language, HeiPOET is suitable to measure SHs with various numbers of input and output endpoints. The modular SH model is used to split a multimedia application into single stream handlers, measure CPU utilization of each SH, and combine these measurement results to the overall CPU utilization of the multimedia application. The measurements have shown that the HeiPOET tool is able to produce results with sufficient precision; mean CPU utilization and other statistical metrics show the trustworthiness of the measurements. These values are stored in the HeiPOET database, and can be used by CPU resource managers to test the schedulability of the multimedia tasks.

Further measurement will focus on SHs processing variable bit rate streams, like MPEG encoding and decoding SHs.

References

[1]   M. Altenhofen, J. Dittrich, R. Hammerschmidt, T. Käppner, C. Kruschel, A. Kückes, T. Steinig: *The BERKOM Multimedia Collaboration Service.* Proceedings of the ACM Multimedia '93 Conference, Anaheim, 1993.

[2]   M.J. Bach: *The Design of the UNIX Operating System.* Englewood Cliffs N.J., Prentice Hall, 1986.

[3]   G. Blair, G. Coulson, P. Auzimour, L. Hazard, F. Horn, J.B. Stefani: *An Integrated Platform and Computational Model for Open Distributed Multimedia Applications.* Proceedings 3rd International Workshop on Network and Operating System Support for Digital Audio and Video, San Diego, 1992.

[4]   N. Chomsky, *Syntactic Structures.* Den Hag, Mouton, 1957.

[5]   C.S. Cheng, J.A. Stankovic, K. Ramamritham, "Scheduling Algorithms for Hard Real-Time Systems: A Brief Survey", *IEEE Tutorial on Hard Real-Time Systems,* Washington D.C., Computer Society Press of the IEEE, S. 150-174, 1988.

[6]   L. Delgrossi, C. Halstrick, R.G. Herrtwich, F. Hoffmann, J. Sandvoss, B. Twachtmann: *Reliability Issues in Multimedia Transport.* Proceedings of the Second Workshop on High Performance Communication Subsystems (HPCS '93), Williamsburg, Sept. 1993.

[7]   D. Ferrari, G. Serazzi, A. Zeigner: *Measurement and Tuning of Computer Systems.* Englewood Cliffs N.J., Prentice Hall, 1983.

[8]   P. Gopinath, T. Bihari, R. Gupta: *Compiler Support for Object Oriented Real-Time Software.* IEEE Software, Vol. 9, No. 5, September 1992.

[9]   R.G. Herrtwich: *Timed Data Streams in Continuous Media Systems.* TR-90-026, International Computer Science Institute Berkeley, May 1990.

[10]   K.B. Kenny, K.J. Lin: *Measuring and Analyzing Real-Time Performance.* IEEE Software, Vol. 8, No. 5, September 1991.

[11]   C.L. Liu, J.W. Layland: *Scheduling Algorithms for Multiprogramming in a Hard Real-Time Environment.* Journal of ACM, Vol. 20, No. 1, January 1973.

[12]   S. Shenker, D.D. Clark, L.Zhang: *A Scheduling Service Model and a Scheduling Architecture for an Integrated Services Packet Network.* Internet Draft, 1993.

[13]   L. Wolf, R.G. Herrtwich: *The System Architecture of the Heidelberg Transport System.* ACM Operating Systems Review, Vol.28, No. 2, April 1994.

Appendix A:
Syntax of the SHUT Specification Language

The syntax of the SHUT specification language can be described with a context-free grammar G consisting of set of terminal symbols T, non-terminal symbols N, transformation rules R and a starting symbol s [4]:

$$G = (T, N, R, s)$$

The SHUT specification language follows this grammar:
T = { IN , OUT , ATTRIBUTES , ENDPOINT, LOOP , ; , , , = , TO , STEP
      SUB , 0..9 , a-z , A-Z }
N = { SHUT_specification,
      identification, input_descriptor, output_descriptor, attribute_descriptor,
      SHUT_name, SHUT_type, endpoints, endpoinds_ea,
      attribute_list, simple_list, attributes, attributs, attribute_type,
      simple_type, enumeration_type, loop_type, comment }
s = SHUT_specification

The rules R of the SHUT specification language are described in Backus-Naur form:
<SHUT_specification> ::=
      <identification> <input_descriptor> <output_descriptor> <attribute_descriptor>
<identification> ::=
      IDENT <SHUT_name> <SHUT_type>
<input_descriptor> ::=
      IN <endpoints>
<output_descriptor> ::=
      OUT <endpoints>
<attribute_descriptor> ::=
      ATTRIBUTES <attribute_list>
<SHUT_name> ::=
      NAME = <string> ;
<SHUT_type> ::=
      TYP = (SYNC | ASYNC <integer>) ;
<endpoints> ::=
      ε | <endpoint> <endpoint_ea>;
<endpoint_ea> ::=
      ENDPOINT <integer> = <integer>;
<attribute_list> ::=
      <simple_list> | <attribute_list> <simple_list>)
<simple_list> ::=
      ε | LOOP { <attributes> }
<attributes> ::=
      attribute | attributes attribute
<attribute> ::=
      <integer> SUB <integer> = <attribute_type> ;
<attribute_type> ::=
      <simple_type> | <enumeration_type> | <loop_type>
<simple_type> ::=
      <integer> | <string>
<enumeration_type> ::=
      <simple_type> | <enumeration_type> , <simple_type>
<loop_type> ::=
      <integer> TO <integer> STEP <integer>

Comments are allowed to be inserted everywhere into the SHUT specification .
<comment> ::=
      (# <string> \n) | (# <string> #)

Appendix B:
SHUT Specification of a Multicast Stream Handler

An example of a SHUT specification file of a multicast stream handler (MSH) is presented.

```
IN
      ENDPOINT 1 = 1 # one incoming packet per processing period
                       for input endpoint 1 #
OUT
      ENDPOINT 1 = 1 # one outgoing packet per processing period
                       for output endpoint 1 #
      ENDPOINT 2 = 1 # one outgoing packet per processing period
                       for output endpoint 2 #
      ...
      ENDPOINT 10 = 1# one outgoing packet per processing period
                       for output endpoint 10 #
ATTRIBUTES
  LOOP
      1002 SUB 1 = 2 TO 10 STEP 1
                           # attribute 1002 of the MSH specifies the number
                               of MSH output endpoints (2 .. 10) #
      1024 SUB 0 = 10
                           # attribute 1024 of the MSH specifies the size
                               of incoming packets #
  ...
```

In the MSH specification there are three different descriptors: data input, data output, and attribute descriptor.

The input characteristics of the MSH are described in the input descriptor. The input descriptor contains the number of incoming packets in one processing period of the MSH.

The output descriptor specifies the number of outgoing packets in one processing period of the MSH. According to the multicast tree of the multimedia application, the number of output endpoints varies. It is specified that there is one outgoing packet per output endpoint in one processing period of the MSH.

One of the main goals of the HeiPOET tool is the automation of the CPU utilization measurement of stream handlers with varying parameters. In the attribute descriptor various parameter settings of the MSH can be described using the loop type (see Appendix A). The loop type in the example specifies that the CPU utilization must be measured for an MSH with 2 to 10 output endpoints. The size of the incoming data packets is 10 KByte.

Extending the Rate-Monotonic Scheduling Algorithm to Get Shorter Delays

Ingo Barth

University of Stuttgart
Institute of Parallel and Distributed High-Performance Systems (IPVR)
Breitwiesenstraße 20-22, D-70565 Stuttgart, Germany
barth@informatik.uni-stuttgart.de

Abstract. A modification for the rate monotonic scheduling algorithm is presented. This modification allows a shorter delay and jitter for a class of tasks. This will result in a shorter end-to-end delay for distributed and pipelined data stream handling. Beside the advantage of the shorter delay and smaller jitter, the disadvantages of the modifications are discussed in quality and quantity. Measurement results of a split-level scheduler using the modified algorithm and a simulation of the algorithm are compared.

1 Introduction

The integration of continuous media, e.g. audio and video, is a result of the growing performance of computer systems. In [2] the relationship between the requirements for handling multimedia data and the development of computing power has been shown. It is argued, that there are sufficient but scarce resources to handle video data. Scarce resources demand for the management of resources.

Within the resource management of multimedia communication systems the CPU management is a demanding task, because the CPU normally controls all activities, including, for instance protocol handling. As the information of continuous media is time-dependent, resource management is required to guarantee a limited jitter and delay for multimedia data handling.

Scheduling algorithms like the earliest-deadline-first algorithm EDF and the rate-monotonic algorithm RM enable a scheduler to give such guarantees, where RM is normally used because of the lower system overhead. As the RM algorithm can only guarantee a minimal delay of one period for one execution step of a stream, the delay introduced by scheduling the CPU to handle the execution grows fast in a pipelined execution system.

In distributed multimedia applications, where a multimedia stream can span over multiple systems, the delay increases, since every system is at least one step in the pipeline. Video data with 25 frames/sec have a delay of 40 msec per execution step. This can easily lead to a total delay of half a second or more in a conferencing scenario. A delay of half a second is known from telephone calls over satellite links and disturbs interactive communication. We will present a scheduling algorithm, based on the standard rate-monotonic algorithm, that can assure different levels of quality for jitter and delay, especially for shorter delays.

The remainder of the paper is organized as follows. In the next section we present our task model used for scheduling continuous media. In Section 3 we give a brief overview of related work, followed by the description of the modified algorithm in Section 4.

Practical experience and test cases are presented in Section 5, followed by a brief summary. In Appendix A we present some measurement and simulation results.

2 Scheduling Model

We use a scheduling model based on periodic tasks with the definitions shown in Fig. 1. Most definitions correspond to definitions in different models proposed in the literature.

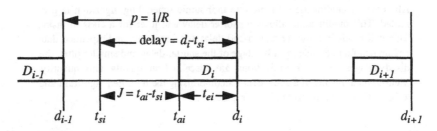

Fig. 1. Scheduling model

The scheduled unit used in this paper is a task, which may be a process or a thread in a real implementation. A task processes a stream of data units D_i which have to be processed before a given point of time, the deadline d_i for the data unit. The data units must be processed periodically with a rate R resulting in the period $p = 1/R$. The execution time t_{ei} is the time it takes to process data unit D_i without any interruptions. Because the execution time for different data units may differ, we use the maximal execution time $t_e = \max t_{ei}$ to reserve the resource. The activation time t_{ai} for a data unit is defined to be the time when the task has to be scheduled to hold the deadline exactly assuming there are no other tasks ($t_{ai} + t_{ei} = d_i$). This would result in a minimal jitter.

In our scheduling model the scheduler gets the activation times for the tasks and schedules them according to the task's rate. The scheduling time t_{si} is the moment when the scheduler makes the task runable for the data unit D_i. The scheduling time t_{si} defines the maximal jitter ($J = t_{ai} - t_{si}$) and at the same time the minimal delay (d_i-t_{si}), as the data unit has to be available before the scheduling time has arrived and may be processed within the given boundaries.

3 Related Work

Scheduling in multimedia systems was discussed using scheduling algorithms from real-time systems and considering the multimedia assumptions like periodicity of data handling and the unpredictability of the system. In practice two scheduling algorithms are considered in the literature, when discussing scheduling in multimedia systems. These algorithms are the earliest-deadline-first algorithm EDF ([5], [9]) and the rate-monotonic algorithm RM ([6]).

The EDF algorithm looks for the task with the earliest deadline and schedules this one at every moment. The algorithm allows a processor utilization of 100% and is optimal for preemptive scheduling with dynamic priorities on a non-parallel processing unit. For EDF a sorted list of tasks with their deadlines must be built to schedule the tasks. For every new data unit the deadline must be inserted into this list.

The RM algorithm schedules the runable task with the highest rate at every moment. A sorted list of tasks with their rates must be built to schedule the tasks. This list must be updated each time a new task is added or a task is deleted. The RM algorithm allows the processor to be utilized up to 69% to guarantee no deadline violation. This algorithm is optimal for preemptive scheduling with static priorities.

As most multimedia systems are added on top of existing operating systems ([7], [3]) a split-level scheduling is used [1]. Split-level scheduling means, that the scheduler of the operating system is used without changes for multimedia and a special scheduler is implemented to manipulate the behaviour of the operating system scheduler. Most operating system schedulers use priorities to determine the scheduling behaviour. To guide the operating system scheduler to handle multimedia data, the split-level scheduler has to change the priorities of the tasks according to their states. A split-level scheduler using the EDF algorithm must assign the highest priority to the task with shortest deadline. As priority changes are not cheap operations (about 150μs on SparcStation 10/40 under SunOS 5.3), the drawback for EDF is the high frequency of dynamic changes of priorities, while RM priority changes are only needed when new tasks are added. The drawback for RM is that we cannot use 100% of the processing power for periodic real-time tasks. As we want to use the scheduler in a multimedia environment, we assume that we must execute non-real-time tasks too and therefore we do not need utilization of 100%.

4 Extending the Rate-Monotonic Algorithm

The minimal delay and the jitter to be guaranteed by RM is based on the period p of the task and the execution time t_e of data units. As RM is non-blocking, the processing of data unit D_i must be finished before data unit D_{i+1} is scheduled. This means, that within one period the processing of a data unit must be finished. The deadline of data unit D_{i-1} is the scheduling time t_{si} for data unit D_i. The jitter J is the difference between the scheduling time and the activation time of a data unit

$$J = t_{ai} - t_{si} = (D_i - t_e) - D_{i-1} = (D_i - D_{i-1}) - t_e = p - t_e.$$

The scheduler can guarantee a delay of p. This is the minimal delay for ressource reservation protocols to be negotiated.

For a pipelined execution (see Fig. 2) the minimal delay of all processing units must be summed up. In a system built on distributed functional units, the delay may grow very high. Therefore we searched for a scheduling method to get shorter delays. A shorter delay can be achieved by a higher rate. This means, when we insert additional dummy data units into the data stream we get a higher rate and therefore a shorter delay. We don't need to insert these dummy data units, because the scheduler does not derive the activation time directly from the rate. The activation time is derived from the time stamp of the data unit and relayed to the scheduler. This means, we can leave out activation times from the stream without confusing the scheduler. A drawback of this higher rate is, that we must reserve more processing time for the task and the processor utilization will fall.

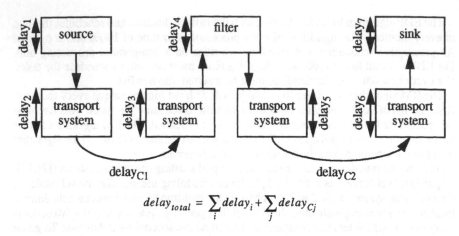

$$delay_{total} = \sum_i delay_i + \sum_j delay_{Cj}$$

Fig. 2. Tasks in a distributed, pipelined execution

Based on this virtual higher rate we extend the RM-algorithm to handle two classes of tasks. Task set T_1 will contain all tasks with a shorter delay and task set T_2 those with normal delay. If we limit the amount of processing time used for T_1 to $n\%$ of the totally available processing time, we will get a virtual rate of $R_v = 100*R/n$ or a virtual period of $p_v = p*n/100$. If we would reserve R_v*e processing time for the task in T_1 we could handle the tasks of T_1 and T_2 according to RM and schedule them. A lot of processing time would be unusable for real-time processing. As we know that the minimal processor utilization is a value for the worst case and typically 88% instead of 69% can be used [7], we split the priority levels into two ranges, one with the higher priorities for T_1 and the other for T_2. This means, that all tasks in T_1 have a higher priority than tasks in T_2 without regarding the rate. Within every task set the priority is assigned according to the RM scheme.

Using these task sets leads to shorter delays for tasks in T_1 with jitter $J = p_v - e$ and a delay of p_v. In task set T_2, we can get priority inversion with tasks in T_1 according to RM and therefore the assumptions for RM are no longer fulfilled. This means, we can get deadline violations in T_2. A violation can only occur, when a priority inversion has occurred or the virtual processor utilization (using R_v as the rate in T_1) is more than 69%. To limit the deadline violation in T_2, we define the minimal rate for tasks in T_1 to R_{min}. For all tasks in T_2 with a lower rate than $R_{vmin} = 100*R_{min}/n$ no priority inversion can occur and therefore a deadline violation because of the priority inversion is not possible.

The worst case for a violation of the deadline is given, if all tasks in T_1 are executed without any interruption by tasks from T_2. The maximum time t_{max} for such an execution is given for one task in T_1 with the minimal rate R_{min} and the total amount of processing time usable for T_1

$$U_{1max} = U_{max}*n/100 = 69\%*n/100 \text{ with } t_{max} = U_{1max}/R_{min}.$$

U_{max} is the maximal processor utilization to be managed with RM: $U_{max} = 0.69$.

We find a second source for deadline violations with this algorithm. We simulate a higher rate by using higher priorities but we do not split the execution time for tasks of set T_1 according to this higher rate. This results in a behaviour as if multiple data units of one data stream are computed in a burst. Because of this burstiness we get deadline violations in Table 3 (Appendix A) at 38 Hz for tasks in task set T_2 although priority inversion does not occur before 50 Hz. Fig. 3 shows a deadline violation because of the burstiness.

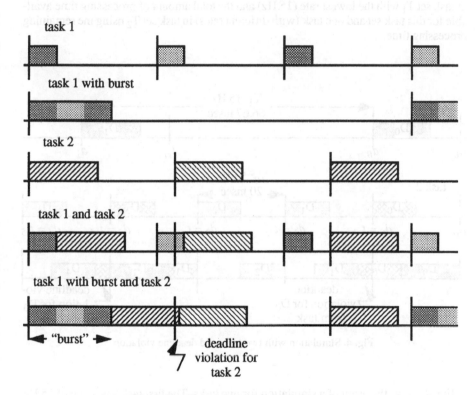

Fig. 3. Deadline violation because of burstiness

5   Implementation and Measurement

To test the theoretical results, we have implemented a scheduler [4] that uses this modified algorithm. This scheduler is integrated into the *CINEMA* system [8] and is used to do the scheduling for multimedia data processing.

Critical states for the algorithm can be tested in a test environment, which will be described in this section together with the test results. The critical states for RM occur, when all tasks should be scheduled at the same moment. In this case tasks with low priority have to wait the longest time and can get critical for deadline violations. In our test environment we can define multiple tasks with different rates for both task sets T_1 and T_2 and with different execution times. The activation times t_{ai} for the tasks are calculated in a form, that the scheduling times T_{si} for tasks are the same at the beginning. Additionally, we can use the total amount of processing time available. The limit that we use for minimal rate R_{min} in T_1 is defined to be 15Hz and the total amount of processing time is set to $n = 30$ (real utilization: $U_{1max} = U_{max}*30\% = 69\%*30\% = 20.7\%$). These values are selected because 15 Hz is the lower bound for video picture rate and the total amount of processing time defines the improvement for the delay ($delay = p_v = p*n*100$). In Appendix A Tables 1 - 4 show the result of a simulation using one task in task set T_1 with the lowest rate (15 Hz) and the total amount of processing time available for this task set and one task (with different rates) in task set T_2 using the remaining processing time.

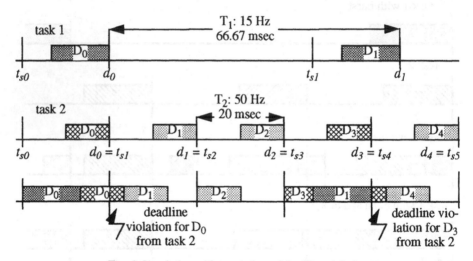

Fig. 4. Simulation with two tasks and deadline violation

Fig. 4 shows the result of a simulation for two tasks. The first task has a rate of 15 Hz, uses 30% of the available execution time and is in task set T_1. The second task has a rate of 50 Hz, uses 70% of the execution time and is in task set T_2. The simulation shows, that a deadline violation of 3.5 msec has occurred. Fig. 5 shows another two task simulation, but in this test the second task has a rate of 30 Hz and no deadline violation occurs.

We have simulated the scheduling algorithms for multiple tasks with different rates. We found no deadline violation using 4 and 5 tasks in the range from 50 -5 Hz and different assignments to the task sets T_1 and T_2. With more tasks with different rates (and rates in T_2 lower than 15 Hz) the worst case will rarely occur and violation will be smaller because of the smaller execution times for the higher rates. For further measurement results see Appendix A.

As a high rate for a task results in high management overhead (e.g. for task switching) we think that in todays systems rates above 50 Hz will rarely be used and will not occupy the total amount of processing time used by T_2. Violations within the range below 10 msec won't be recognized for video data because the picture frequency nor-

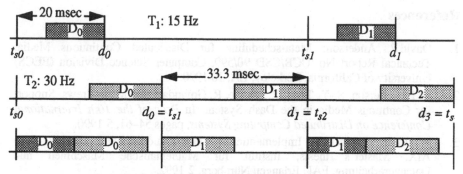

Fig. 5. Simulation with two tasks without deadline violation

mally is lower than 100 Hz. If multiple violations of the deadline for one data unit occur, they are not summed up as the activation time is derived from the time stamp of the data unit. This means, that a violation can be compensated for the data unit by following executions.

6 Summary

Multimedia systems must include resource management to handle different and dynamically changeable multimedia configurations. The basic resource for data handling is the CPU, that is managed by the scheduler. Scheduling algorithms for multimedia data handling, like the rate-monotonic-algorithm, are often used in multimedia systems, but they have the disadvantage of introducing a minimal delay of the size of one period of the stream for every execution step in a pipelined processing system. In a distributed system where different stream handling functions are placed on different systems this delay sums up to a value that can disturb interactive communication.

We introduced a modified rate-monotonic algorithm that uses different task sets, where each task set is scheduled according to rate-monotonic and has its own priority range. The improvement of the algorithm for tasks with shorter delays depends on the limit of processing time assigned to tasks of this class. Using n% of the available processing time for these tasks, we can guarantee a delay of $n*p/100$ when p is the period of the task. The drawback for this algorithm is, that deadline violations for tasks with the normal delay can occur. As the presented measuring results are based on worst case constructions we can assume, that in a real system the deadline violations are smaller than presented here.

Future work will be to integrate the scheduler into the *CINEMA* system together with a ressource reservation protocol to handle real multimedia data and to get experience with this algorithm in a running system. The real influence of the possible deadline violations must be examined with real presentations of multimedia data.

The applicability of this algorithm for multi-processor systems will be looked at in the future. As rate-monotonic scheduling is not optimal for multi-processor systems we will test the algorithm in a running system looking at the influence on the presentation.

References

1. David P. Anderson: Meta-scheduling for Distributed Continuous Media. Technical Report No. UCB/CSD 90/599, Computer Science Division (EECS) University of California, Berkeley, CA, 10 1990.
2. D. P. Anderson, S.-Y. Tzou, R. Wahbe, R. Govindan, and M. Andrews: Support for Continous Media in the Dash System. In *Proc of the 10th International Conference on Distributed Computing Systems*, pages 54–61, 5 1990.
3. W. Burke: Entwurf und Implementierung eines Pseudo-Echtzeit-Schedulers für AIX. Master's thesis, Institut für Mathematische Maschinen und Datenverarbeitung, FAU Erlangen Nürnberg, 2 1992.
4. Walter Fiederer: Entwicklung eines Ressourcen-Managers für die CPU. Master's thesis, IPVR, Universität Stuttgart, 2 1994.
5. Ralf Guido Herrtwich: Betriebsmittelvergabe unter Echtzeitgesichtspunkten. *Informatik-Spektrum*, 14(3):123–136, 6 1991.
6. C.L. Liu and James W. Layland: Scheduling Algorithms for Multiprogramming in a Hard-real-time Environment. *Journal of the ACM*, 20(1):46–61, 1 1973.
7. Andreas Mauthe, Werner Schulz, and Ralf Steinmetz: Inside the Heidelberg Multimedia Operating System Support: Real-time Processing of Continuous Media in OS/2. Technical Report 43.9214, ENC European Networking Center, 9 1992.
8. Kurt Rothermel, Ingo Barth, and Tobias Helbig: CINEMA - An Architecture for Configurable Distributed Multimedia Applications. Submitted for publication.
9. Ralf Steinmetz: *Multimedia Technologie*. Springer Verlag, 1993.

Appendix A

We have tested the scheduling of different combinations of tasks using the presented algorithm. The following tables show the results from executing different tasks with different rates, execution times and task set assignments on a SparcStation 10/40 under SunOS 5.3 and from a simulation of the algorithm. The simulator has a resolution of $1\mu sec$ and eliminates all external influences as measurement inaccuracy or limitations from the scheduler of the operating system.

The tables show the task set, the rate and the execution time for one data unit. The listed delay value is the guarantee given by the scheduler. We tested the scheduler by defining a common scheduling time for all tasks (which is repeated depending on the tasks' periods) and measuring the difference between the scheduling time and the time the task has finished the execution ($t_{dif} = t_{si} - t_{fi}$). In the last column we calculate the difference between the deadline and the finishing time ($d_i - t_{fi}$). The tables show the limit values over a period of 4 seconds. A negative value in the last column represents a deadline violation.

Tables 1-4 show the simulation results for two tasks in different task sets using the maximum available processing time. For task set T_1 the lowest rate was selected to determine the first occurrence of a deadline violation.

Tables 5-8 show the same sets of tasks simulated (Table 5 and Table 7) and executed in real system using a split-level scheduler (Table 6 and Table 8). The differences found between simulation and real execution are based on errors made when measuring the

total execution time. As the implementation of the split-level scheduler can only schedule tasks with rates that are divisors of 100 Hz without anomalies, we have selected
tasks with rates according to this limitation. The anomaly is based on the usage of the
timers of the thread package, which are looked at every 10 msec.

All examples use the total amount of processing time usable for real-time processing.
This is 0.69 s within one second which is the limit for RM.

Table 1. 2 tasks (15 Hz and 30 Hz and a processor utilization of 69% simulated) without a deadline violation

Task set	Rate/Hz	Exec. time/msec	Delay/msec	t_{dif}/msec	d_i-t_{fi}/msec
1	15	13.8	29.7	13.8	15.9
2	30	16.1	33.3	29.9	3.4

Table 2. 2 tasks (15 Hz and 37 Hz and a processor utilization of 69% simulated) without a deadline violation

Task set	Rate/Hz	Exec. time/msec	Delay/msec	t_{dif}/msec	d_i-t_{fi}/msec
1	15	13.8	29.7	13.8	15.9
2	37	13.1	27.0	26.9	0.2

Table 3. 2 tasks (15 Hz and 38 Hz and a processor utilization of 69% simulated) with a deadline violation

Task set	Rate/Hz	Exec. time/msec	Delay/msec	t_{dif}/msec	d_i-t_{fi}/msec
1	15	13.8	29.7	13.8	15.9
2	38	12.7	26.3	26.5	-0.2

Table 4. 2 tasks (15 Hz and 50 Hz and a processor utilization of 69% simulated) with a deadline violation

Task set	Rate/Hz	Exec. time/msec	Delay/msec	t_{dif}/msec	d_i-t_{fi}/msec
1	15	13.8	29.7	13.8	15.9
2	50	9.7	20.0	23.5	-3.5

Table 5. 7 tasks in the range from 50 - 5 Hz and a processor utilization of 69% simulated

Task set	Rate/Hz	Exec. time/msec	Delay/msec	t_{dif}/msec	d_i-t_{fi}/msec
1	25	4.1	14.9	4.1	10.8
1	20	5.2	18.6	9.3	9.3
2	50	2.1	20.0	11.4	8.6
2	25	4.1	40.0	15.5	24.5
2	20	5.2	50.0	20.7	29.3
2	10	10.3	100.0	31.1	69.0
2	5	13.8	200.0	44.9	155.2

Table 6. 7 tasks in the range from 50 - 5 Hz and a processor utilization of 69% using a split-level scheduler

Task set	Rate/Hz	Exec. time/msec	Delay/msec	t_{dif}/msec	d_i-t_{fi}/msec
1	25	4.1	14.9	4.1	10.8
1	20	5.2	18.6	9.3	9.3
2	50	2.1	20	11.6	8.4
2	25	4.1	40	15.7	24.3
2	20	5.2	50	23	27
2	10	10.4	100	33.8	66.2
2	5	13.8	200	66.4	133.6

Table 7. 4 tasks in the range from 50 - 20 Hz and a processor utilization of 69% simulated

Task set	Rate/Hz	Exec. time/msec	Delay/msec	t_{dif}/msec	d_i-t_{fi}/msec
1	20	10.3	22.2	10.3	11.9
2	50	4.1	20.0	14.5	5.5
2	25	8.3	40.0	22.8	17.2
2	20	3.5	50.0	26.2	23.8

Table 8. 4 tasks in the range from 50 - 20 Hz and a processor utilization of 69% using a split-level scheduler

Task set	Rate/Hz	Exec. time/msec	Delay/msec	t_{dif}/msec	d_i-t_{fi}/msec
1	20	10.4	22.3	10.5	11.8
2	50	4.1	20	14.6	5.4
2	25	8.3	40	28.0	12.0
2	20	3.4	50	31.5	18.5

A Multimedia Application Adaptation Layer (MAAL) Protocol

Wei Liu

Computer Science Department
University of Maryland Baltimore County
Baltimore, Maryland 21228, USA

Abstract. The goal of this research is to resolve the difference between variable-sized packets (or frames) in multimedia application and fixed and small size cells in Broadband Integrated-Services Digital Networks (B-ISDN) with Asynchronous Transfer Mode (ATM). In particular, I propose a new operation mode of multimedia application adaptation layer (MAAL) as a supplement of existing four AAL types. The new AAL protocol supports streams of co-related cells in order to reduce the processing overhead and simplify the traffic scheduling at the application level. Furthermore it re-synchronizes the multimedia traffics through the Asynchronous Transfer Mode network interface.

1 Introduction

Recent development of Broadband Integrated-Services Digital Network (B-ISDN) has revolutionized the architecture of computer communication networks. In the future key high speed transport facility, the information transfer unit is the 53 byte Asynchronous Transfer Mode (ATM) cell [7]. This fixed and small size data element (called cell) structure is favored by the voice-carrier community, while the data communication industry has long been practicing with the variable-sized packets or frames for data traffic. For the future multimedia networking application, it becomes an important issue to effectively resolve the difference between the variable-sized application data unit and the fixed-size and small ATM cell in the integrated services environment.

With fixed cell overhead, the performance of ATM network does not scale up when the video frame is being transmitted which requires thousands of consecutive cells. While the cell parameters are attractive for priorities, multiplexing, and robustness in multimedia traffic mixes, the small cell size presents a problem. Small size implies large number of cells for a single application data unit, and under current transmission technology, the cell traffic-jam forces large overheads for interrupt service and bus transactions on the host and switch processor. To eliminate or reduce the overheads, a new video application adaptation mode is needed to accommodate the new networking applications. Meanwhile, the data packets of traditional networking application also need to be segmented into ATM cells when the traffic is passed through a public B-ISDN ATM backbone network. Again, it motivates an efficient adaptation operation. This issue becomes critical in order to deliver high speed end-to-end performance in the near future interconnection environment.

This paper will focus on one important layer interface issue: the encapsulation of application traffic units (such as video frames or data packets) into the cell stream and the efficiency of the encapsulation. In particular, we are going to develop a new operation mode for the Application Adaptation Layer (AAL) based on a new cell-stream framing structure.

This research is closely related to the previous works. In [9], I have derived a quantitative relationship between the number of user packets and the ATM cell rate for traffic control purpose, but the encapsulation issue is still open. None of the four existing AAL modes [7, 15], is designed for integrated multimedia communication and application in the sense that one AAL is designed for a single traffic type. AAL1 and AAL2 support constant and variable rate for real time traffic respectively. AAL3/4 is for connection-oriented and connectionless data service. AAL5 provides connectionless LAN emulation in ATM LANs. What we need is a new high performance AAL for integrated services.

Some previous research addressed the intermedia synchronization for events and streams [4, 8, 12]. Especially, Minzer[12] discussed a call model for multimedia services, which breaks a call object into more elementary components. His discussion only limited to the service level from the users' perspective. But it is important to divide a user level connection into subconnections from the networking's view, which is discussed in this paper.

Other related research has been reported on many implementation issues. Feldmeier[5] suggested that complete information be provided for explicit labelling of a trunk of data, which tends to introduce larger headers in order to contain enough information for independent packet processing. This approach may be justified for large and variable-length packets as in certain network testbeds. It is not suitable for small ATM cells, which are normally inter-dependent. Moors and Cantoni[14] addressed the nondeterministic behavior of the receiver when implementing reassembly, buffer allocation, scheduling and connection management for ATM interface.

This paper addresses the requirements of multimedia application adaptation and a possible approach in section 2. Then section 3 shows the design of a Multimedia Application Adaptation Layer (MAAL) protocol based on the proposed approach. Finally, in section 4 we discuss the impact of these results and point out future research.

2 Cell Encapsulation of Packets/Frames

The key of our study is to provide a framing technique for a sequence of cells which carry the integrated service traffic. This "frame" is treated as an independent subconnection such that interleaving among the subconnections will be restricted according to the synchronization and scheduling constraints (e.g., lipsync). These constraints and the multi-cell-framing structure could be provided by a new multimedia application adaptation protocol, with four requirements that should be satisfied:

- Preserving small ATM cell, which is the B-ISDN standard and is required for scheduling real time traffic;
- Providing a framing structure, which encapsulates a sequence of cells or subconnections, for application data (e.g., video frames, data messages) but without changing the ATM cell structure;
- Adding synchronization information and scheduling constraint into the frame for connection management, integrated multiplexing and performance guarantee;
- Eliminating duplication of identical headers in a consecutive sequence of cells in order to achieve high performance.

Because transmitting variable-size packets may introduce a delay and delay jitter intolerable for real time voice/video packets, a small size packet, called ATM cell, is used as the structure of a transmission unit in the B-ISDN standard. A large packet may have to be divided into small ATM cells in order to achieve asynchronous transport. The cell structure tends to introduce more headers for longer messages, thus increasing the processing overhead.

If a packet to be transmitted fits within the ATM cell payload, then it is encapsulated directly. Otherwise the packet must be segmented by the source and transferred in multiple cells: the first being the Beginning Of Message (BOM), the last the End Of Message (EOM), and any intermediate cells Continuation Of Message (COM) cells. In order to achieve very high speed processing of integrated services for multimedia applications, we propose a new connection scheme to encapsulate a sequence of ATM cells.

In view of the high speed transmission, a single message is a "long" cell stream in terms of cells. We can treat the whole message/cell-stream as a "connection". Inside a BOM cell, we provide completely self-describing information to the following COM cells which require a single processing. The contents of the BOM and EOM cells are the control fields for the "frame" which contains a sequence of consecutive ATM cells. By doing so, we allow multiplexing at the "cell" level and processing at the "connection" level, and thus meeting the above requirements.

The cell connection is different from the user level connection in that the former uses the cell as a unit while the later uses bits for the whole session. A cell connection implements multiplexing/demultiplexing, direct memory access, and a single context switching if applicable; while a user level connection negotiates and reserves networking resources. Throughout this paper the "connection" will refer to a cell-level connection, and the user level will be explicitly identified.

One important advantage of introducing a framing structure for the cell connections is to facilitate adding synchronization information and scheduling constraint. Problems with completely independent cell scheduling and switching are the loss of co-related cell information and difficulty of re-synchronization. Usually ATM cells from different connections may be arbitrarily interleaved, which imposes quite difficulty to synchronize the video frame traffic. The new BOM and EOM signaling message will alleviate the processing overhead of multimedia traffic by restricting the traffic interleaving. We call it controlled-interleaving.

One way of reducing the header overhead and speed-up processing is to avoid duplication of identical headers. This can be achieved by encoding the common control information for a consecutive sequence of cells into some marking cells. A natural choice of those marking cells would be using the BOM and EOM cells. Because of the critical role of the marking cells, the BOM and EOM should no longer be a part of the data/video cell stream. Instead, we recommend that they become new elements of the B-ISDN signaling system. As a result, a new adaptation/transmission mode is allowed as discussed in the next section.

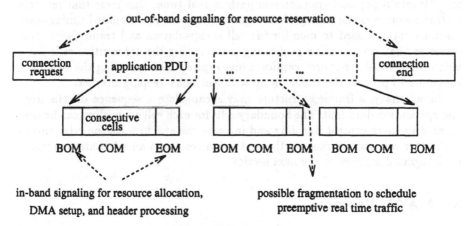

Fig. 1. Cell connections

2.1 Signaling Support

To implement the new cell level connection or subconnection concept, we also extend the signaling capability (see Figure 1). First we define the in-band and out-of-band signaling. In-band signaling is a mean to transfer control signal together with the voice/data signal, while out-of-band signaling is a technique to separate control traffic and user traffic [11].

Today's ATM technology only provides out-of-band facility with logical channels, which is inherited from Q.93B [1] and Signaling System No.7 [10]. But with the same transmission and switching technology for both user traffic and signaling traffic, in-band multiplexing and signaling may be also feasible.

In-band signaling can be implemented by designating certain ATM cells to carry control signals only. For examples, the BOM and EOM cells can server as signaling cells. With the in-band BOM and EOM signals, we can extend the signaling capacity to achieve real time multiplexing and demultiplexing of our cell-connection concept.

Before, a stream of bits is flowing in the network. To determine the boundary of the packet, flags and bit-stuffing techniques are used. Now a stream of cells is flowing in the network. The application data unit (e.g., video frame) consists of a sequence of cells. The connection boundary is marked by the signaling control messages such as BOM and EOM, which are boundary cells to satisfy the second requirement listed before. The In-band BOM and EOM cells also carry the synchronization and scheduling information to implement the third requirement.

Call control functions remain the same, but the connection control function is modified to include the new cell stream boundary capability. The new signaling function establishes, releases, and reconfigures the VP (Virtual Path) and maps the VP into a physical transmission path in real time. One issue that requires further research is the timing of the signaling process. An extended timing state machine may be used to monitor the cell encapsulation and transmission processes so that we can allow continuing transmitting the consecutive cells for a video frame until a resource caption timer expires. This enabling the switching controller to process the traffic on fly with less book-keeping overhead.

In summary, a framing structure may encapsulate a sequence of cells from one application data unit. The boundary cells for each cell connection can be designated to carry control signaling and indicate the synchronization relationship with other cell connections. Further extension results in an adaptation protocol for integrated services in the next section.

3 MAAL Protocol

Unlike the other AALs, where the adaptation information is encoded in every ATM cell, our MAAL only encodes the control information in the BOM and EOM cells (see Figure 2). Here we are only interested in the protocol functions and services. The number of bits in each field is subject for further design.

The ID number has local significance for a connection oriented transmission. An ID is needed as the connection may be fragmented into multiple subconnections or cell-connections may be merged into a new cell connection as co-related cells accumulated after passing many number of hops (similar phenomenon was observed in internet [13]).

The Time-Stamp field is used for real time traffic. If the whole video frame arrives too late to be played out, the whole sequence of cells will be dropped in a switch.

The Cross-Connection Reference is a collections of IDs for related cell connections. It is essential for scheduling multiple connections to achieve lip-sync.

The performance requirements in the QoS (Quality of Service) are copied from the initial user-level connection setup information. If the users are allowed to change the requirements during the session (e.g., asking for more bits for a video channel), the increment will be reflected in the next cell-level connection

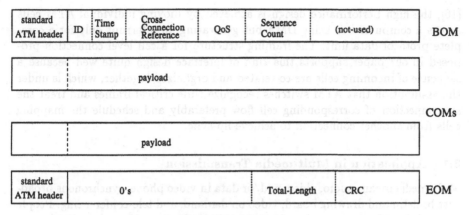

Fig. 2. MAAL design

and is indicated by the next BOM in the MAAL QoS. The detail QoS (Quality of Service) fields can be found in the ATM UNI-Specification [1] and is omitted here.

The change of QoS allows dynamic filtering of the traffic as in the RSVP [17] except that the change is initiated by the forwarding side. This feature is important for multicast distribution.

The Sequence Count is incremented by the number of the following COM cells. Unlike the Sequence-Count in existing AALs, which is modulo 8, here we suggest a much larger range so that it can serve two purposes. First is to check missing cell-connections. Secondly, by computing the difference between two BOMs, the buffer size can be derived for the following cells.

Finally, in EOM cell an optional checksum is used for the whole cell-level connection, as the header of the COM cells may not need to be examined.

Our long term research goal is to demonstrate the proposed MAAL protocol and apply it to implement a framework for multimedia traffic synchronization in B-ISDN network. Intermedia synchronization is the temporal coordination of multiple types of related traffic (voice, video). The corresponding elements in the concurrent traffic streams must be played out at identical times. Note that the concurrent traffic streams may represent different classes of data in the network, have different packet size, and experience different delay. The delay variance complicates the multimedia traffic scheduling [8].

Our signaling cell allows a timestamp to be carried for the traffic cell stream. The cross connection reference fields indicate the timing relationship between two cell level connections. A simple example of the communication relationship is the association of an audio with a video connection. Further research is needed to determine how to implementation of any complex communication relationship by using the cell connection id, cross connection reference, and timestamps.

Although this MAAL protocol is proposed for switches of multimedia traffic, it helps to achieve high performance host-network-interface as well. For examples in ATM interface hardware architecture by Davie [3] and Traw with Smith

[16], the high performance design is achieved by moving individual ATM cells across a computer bus using DMA but generating an interrupt only for a complete protocol data unit. The framing structure for a cell level connection proposed in our paper supports this kind of interface design quite well because a sequence of incoming cells are co-related and organized together, which is under the assumption that ATM switches recognize time critical frames and treat the subconnection of corresponding cell flow preferably and schedule the mapping cells from another connection to achieve lip-sync.

3.1 Application in Multimedia Transmission

Multimedia means audio, video, and/or data in video phone, synchronous broadcast bulletin and drawing board, video on demand, and tele-conferencing. People may ask for different scenarios of very complicated call configurations, but in general calls are implemented with separate connections being added separately. The communication relationship (e.g., synchronization) between connections should be handled accordingly by the communication network servers.

It is obviously impractical to check the timing requirement for every ATM cell at the transport layer. On the other hand, the application level software may only implement the abstract logical topology and general QoS requirements [2] and may not be able to access the transport scheduler. The multimedia application adaptation layer provides a well defined framing structure for scheduling a session of cell connections.

As the initiating party invokes a multimedia call, the server process at user-network interface sends out the connection requests with a connection description field. This connection description and any other subsequent negotiated and agreed information are recorded in MAAL servers at the user network interface. The MAAL servers are responsible for generating and consuming the control information in the BOM and EOM cells. For large scale ATM networks, we propose additional MAAL servers inside some of the switches (see Figure 3).

For ordinary ATM switches without the MAAL processing capability, the BOM and EOM in-band signaling cells are forwarded as normal ATM cells. The MAAL server process may run at the user network interface or at some special ATM switches with MAAL capability. The extra overhead of processing the BOM and EOM control cells in the special switches can be offset by eliminating processing the COM cell headers. This can be achieved by setting up a by-pass path to forward the COM cells without going through the switch fabric since all common header information is contained inside the COM cell. One possible implementation is to setup a DMA transfer from the incoming port of the BOM cell to the buffer until the EOM is received.

The timestamp in the BOM is checked with the QoS requirement, and the cross-connection reference information and the connection id are used to access the MAAL server and any scheduler for the synchronization process. Meanwhile, the cell stream is forwarded to the output port by a DMA transfer. Further research is needed to quantify the characteristics of consecutive cell streams and any continuous processing overhead.

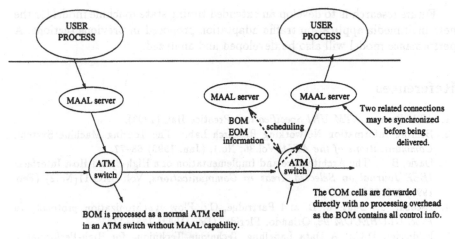

Fig. 3. Application of MAAL

4 Summary and Discussion

There are four classes of existing ATM transport services, which are defined from certain combinations of three parameters: bit rate, delay sensitivity, and connectioness. In this paper, we add a new parameter: the communication relationship among cell streams.

The key process in this research is to break the application connections into subconnections by following a proposed adaptation process. We can replace the overhead of the Application Adaptation Layer processing by the signaling processing. By doing so, we not only enable to enforce the synchronization requirements for multimedia application, but also reduce the processing overhead by introducing a well defined cell stream structure for each video frame and voice/data packet. The benefits of the proposed encapsulation and connection boundary concepts are:

1. The boundary cell provides a framing structure and imposes synchronization and multiplexing requirements.
2. A special switch which implements this concept could skip processing the header of the intermediate cells in a cell stream, thus reducing processing delay and improving performance.
3. In a local area network (especially in ATM-LAN) environment, if the cells are not going to pass the public switches, the header in the intermediate cells can be completely ignored, and the payload capacity is increased proportionally. However, a cell bridge is needed in the internetwork boundary.

Future research is to develop an extended timing state machine model for the new multimedia application traffic adaptation proposed in previous sections. A performance model will also be developed and analyzed.

References

1. ATM-Forum: *ATM UNI Specification*. Prentice Hall (1993).
2. Bellcore Information Networking Research Lab.: The Touring Machine System. *Communications of the ACM*, Vol.36, No.1, (Jan. 1993) 68–77.
3. Davie, B.S.: The Architecture and Implementation of a High-Speed Host Interface. *IEEE Journal on Selected Areas in Communications*, Vol. SAC-11(No.2) (Feb. 1993) 228–239.
4. Escobar, J., Deutsch, D. and Partridge, C.: Flow synchronization protocol. In *Proc. GLOBECOM'92*, Orlando, Florida, (Dec. 1992) 1381–1387.
5. Feldmeier, D.C.: A Data Labelling Technique Technique for High-Performance Protocol Processing. In *Proc. of ACM SIGCOMM'93*, San Francisco, (Sept. 1993) 170–181.
6. N.S.F.: Research Priorities in Networking and Communications Research. *Workshop Report to* the NSF Division of Networking and Communications Research and Infrastructure, (1992).
7. Kawarasaki, M. and Jabbari, B.: B-ISDN Architecture and Protocol. *IEEE Journal on Selected Areas in Communications*, Vol. SAC-9(9) (Dec. 1991) 1405–1415.
8. Little, T.D.C. and Ghafoor A.: Multimedia synchronization protocols for broadband integrated services. *IEEE Trans. Communications*, Vol.9, (1991) 1368-1382.
9. Liu, W.: ATM Traffic Control Interface Model. In *Proc. of IC^3N '93, International Conference on Computer Communications and Networks*, San Diego, California, (June 1993) 161–165.
10. Liu, W., Krieger, U., and Akyildiz, I.F.: An Admission Control Model Through Out-of-band Signaling Management. In *Proc. of IEEE INFOCOM'92*, Florence, Italy, (May 1992) 987–995.
11. Miller, P.A. and Turcu, P.N.: Generic Signaling Protocol: Architecture, Model, and Service. *IEEE Trans. on Communications*, Vol. COM-40(No.5) (1992) 957–979.
12. Minzer, S.: A Signaling Protocol for Complex Multimedia Services. *IEEE Journal on Selected Areas in Communications*, Vol. SAC-9(No.9) (Dec. 1991) 1383–1394.
13. Mogul, J.C.: Observing TCP dynamics in real networks. In *Proc. of ACM SIGCOMM'92*, Baltimore, MD, (Aug. 1992) 305–317.
14. Moors, T. and Cantoni A.: ATM Receiver Implementation Issues. *IEEE Journal on Selected Areas in Communications*, Vol. SAC-11(No.2) (Feb. 1993) 254–263.
15. Partridge, C.: *Gigabit Networking*. Addison Wesley, (1994).
16. Traw, C.B.S. and Smith, J.M.: Hardware/Software Organization of a High-Performance ATM Host Interface. *IEEE Journal on Selected Areas in Communications*, Vol. SAC-11(No.2) (Feb. 1993) 240–253.
17. Zhang, L., Deering, S., Estrin, D., Shenker, S., and Zappala, D.: RSVP: a new resource reservation protocol. *IEEE Network Magazine*, 9(No.5) (Sept. 1993) 8–18.

Adaptation Layer and Group Communication Server for Reliable Multipoint Services in ATM Networks

Georg Carle

Institute of Telematics
University of Karlsruhe, D-76128 Karlsruhe, Germany
Telephone: ++49 / 721 / 608-4027, Fax: ++49 / 721 / 388097
E-Mail: carle@telematik.informatik.uni-karlsruhe.de

Abstract. Upcoming applications have demanding communication needs. One requirement is the provision of a reliable high performance multipoint communication service. In order to meet high performance requirements and in order to allow an efficient use of network resources, powerful error control mechanisms are required. This paper presents a novel concept for support of multipoint communication in ATM networks. It is based on a new adaptation layer type, called the Reliable Multicast ATM Adaptation Layer (RMC-AAL), and on a new network element, called the Group Communication Server (GCS). A set of error control mechanisms tailored for multipoint communication are integrated into RMC-AAL and GCS. Error control is based on ARQ and FEC schemes, allowing to select the mechanism that is most suitable for the application requirements in a specific communication scenario. The functionality of adaptation layer and group communication server are described, and a basic implementation architecture is presented. Performance results obtained by means of simulation and analysis are given.

1 Introduction

In the evolution of high speed networking, two developments will be of growing importance. One issue is the fast growing deployment of ATM networks, both in local and in wide area networks. The other issue is the increasing importance of group communication scenarios. Upcoming applications, for example in the areas of computer-supported co-operative work (CSCW), distributed applications and virtual shared memory systems require point-to-multipoint (Multicast, 1:N) as well as multipoint-to-multipoint (Multipeer, M:N) communication [1]. For a growing number of applications such as multimedia collaboration systems, the provision of a multicast service with a specific quality of service (QoS) in terms of throughput, delay and reliability is crucial.

If multipoint communication is not supported by the network or by the end-to-end protocols, multiple point-to-point connections must be used for distribution of identical information to the members of a group. The support of multicasting is beneficial in various ways: It saves bandwidth, reduces processing effort for the end systems, reduces the mean delay for the receivers and simplifies addressing and connection management.

Various issues need to be addressed in order to provide group communication services in ATM networks [2, 3]. Switches need to incorporate a copy function for support of 1:N virtual channels (VCs). Signaling must be capable of managing multipoint connections, and group management functions need to be provided for administration of members joining and leaving a group. Procedures for routing and call admission control (CAC) need to be adapted for multicast communication. Another key problem that must be solved to provide a reliable multipoint service is the recovery from cell losses due to congestion in the switches.

If a reliable service in ATM networks is based on traditional transport protocols like TCP, severe performance degradations may be observed [4]. Additional problems occur for the provision of a reliable multipoint service, where transmitters need to deal with many receivers and where cell losses occur more frequently.

This paper focuses on suitable error control mechanisms for correction of cell losses. After presenting alternatives for the provision of a basic multipoint service in ATM networks, the problem of potential cell loss is explained in more detail. Then, an overview on existing error control mechanisms and on protocols that apply these mechanisms is given. The conceptual framework for the provision of a reliable multipoint service is presented, comprising of suitable mechanisms, required components, and basic implementation architectures. Results of a perfomance evaluation by analysis and simulation are given which allow the provision of guidelines for appropriate selection of the mechanisms.

2 Multipoint Communication in ATM Networks

2.1 Multipoint Bearer Service in ATM Networks

Applications may require the following types of multipoint communication: one-to-many, many-to-one and many-to-many. There are a number of ways to support these communication types in ATM networks [5]. Virtual paths and virtual channels may be of the types point-to-point and point-to-multipoint. Many ATM switch designs are already prepared to copy incoming cells to multiple output ports, providing a basic support for multicast communication in ATM networks.

Support of multipoint connections in signaling protocols is currently under development. In the draft recommendation of the signaling protocol for B-ISDN [6], support of multipoint connections is not yet included. In the User-Network Interface (UNI) specification version 3.0 of the ATM Forum [7], phase 1 signaling is specified which allows the management of point-to-multipoint connections. Multipoint-to-multipoint connections are not supported by phase 1 signaling, but two techniques are proposed for multipeer communication.

According to the first proposal, each node in a group that wishes to communicate has to establish a point-to-multipoint connection to all of the other nodes of the group. N point-to-multipoint connections are required for a group with N members. This solution does not scale well for large groups. For large, long-lived groups, numerous virtual channels need to be maintained. If one receiver joins or leaves a group, every multicast tree must be modified.

According to the second proposal, each node has to establish a point-to-point connection to a 'Multicast Server'. A point-to-multipoint connection from the Multicast Server to every member of the group is used to transmit messages to the members of the group. This requires N point-to-point connections and one point-to-multipoint connection, improving the scalability significantly. If this approach is selected, mechanisms must be applied in order to distinguish cells of different senders [8, 3]. One possibility is to distinguish the cells based on an identifier in the payload of the cell. The Message Identifier (MID) of AAL3/4 [9] may be used for this purpose. In this case, MIDs must be negotiated, and a MID demultiplexing function must be integrated into every receiver. AAL5 [9] allows a simpler implementation of the adaptation layer, but it does not provide a field for demultiplexing cells. If cells of different frames are mixed, the receiver is only able to detect the collision by checksum violation and to discard the affected frames. In order to avoid these collisions, the multiplexing of different VCs onto a single VC needs to be done in a way that every receiver receives all cells of one frame before receiving cells of another frame. Such a mechanism may operate either in reassembly mode or in cut-through mode. In reassembly mode, forwarding of an incoming AAL5 frame starts after the reception of the last cell of this frame. In cut-through mode, already the first incoming cell of a frame may be forwarded if no other frame of the group is in the process of forwarding.

2.2 Cell Loss in ATM Networks

Two factors must be considered which cause ATM networks to discard cells: transmission bit errors in the cell header field due to noise, and buffer overflow in multiplexing or cross connecting equipment. While fibre optic transmission technology allows to keep the bit error probability very low, the most frequent cause for cell loss is buffer overflow. In ATM networks, statistical multiplexing provides a high degree of resource sharing. Short periods of congestion may occur due to statistical correlations among variable bit rate traffic sources, resulting in buffer overflow. The probability for cell loss may vary over a wide range, depending on the strategy for usage parameter control (UPC) and call admission control which is applied. If very low cell loss probabilities are to be guaranteed even for highly bursty sources, only part of the network resources may be utilised. Utilisation may be increased on the risk of higher cell loss rates. Cell losses due to buffer overflow occur during situations of congestion, caused by superpositon of traffic bursts. Therefore, they do not occur randomly distributed, but in bursts and show a highly correlated characteristic [10, 11]. If a reliable service has to be provided, mechanisms are required which are able to handle this type of error efficiently. For ATM multicast connections, the problem of cell losses is even more crucial than for unicast connections. Collisions of the multicast VC with different unicast VCs may occur independently at every output port of a switch. For multicast switches with dedicated copy networks, additional collisions may occur for correlated arrivals of bursts in different multicast VCs [12].

2.3 Error Control Mechanisms

For applications that cannot tolerate the cell losses of the ATM bearer service, error control mechanisms are required. Error control consists of two basic steps: error detection and error recovery. For error recovery, two mechanisms are available: Automatic Repeat ReQuest (ARQ) and Forward Error Correction (FEC). Error control is difficult in networks that offer high bandwidth over long distances. High data rates in combination with a long propagation delay result in high bandwidth-delay products, causeing problems for the following reasons:

- End-to-end control actions require a minimum of one round-trip-delay, and retransmissions require large buffers and may introduce high delays;
- Efficient error control with timer-based loss detection is difficult, because delay variations do not allow very accurate timer setting, causing deterioration of the service quality;
- Processing of error control needs to be performed at very high speeds, if no bottle-neck is to be introduced.

ARQ Methods. For go-back-N ARQ protocols, transmitter and receiver implementations may be very simple, and no buffering is required for the receiver. For selective repeat protocols, transmitter and receiver implementations are more complex, and a large buffer is required by the receiver. Processing overhead of ARQ methods is proportional to the number of data and acknowledgement packets that are processed. For point-to-point communication, ARQ mechanisms are well understood, and a number of protocols for data link layer and transport layer, employing these mechanisms, are known. For multicast communication, there are still many open questions concerning acknowledgement and retransmission strategy, achievable performance and implementation. Large groups require that the transmitter stores and manages a large amount of status information of the receivers. The number of retransmissions is growing for larger group sizes, decreasing the achievable performance. Additionally, the transmitter must be capable of processing a large number of control information. If reliable communication is required to every multicast receiver, a substantial part of the transmitter complexity is growing proportionally to the group size. To overcome this problem, a scheme that provides reliable delivery of messages to K out of N receivers may be applied (K-reliable service).

FEC Methods. FEC methods promise a number of advantages for multicast communication in high-speed networks. The delay for error recovery is independent of the distance, and large bandwidth-delay products do not lead to high buffer requirements. In contrast to ARQ mechanisms, FEC is not affected by the number of receivers. However, FEC has two main disadvantages. It is computationally demanding, and it requires constantly additional bandwidth, limiting the achievable efficiency. Additionally, the problem needs to be addressed that cell losses frequently occur in bursts.

A number of proposals exist on how to use FEC for ATM networks. In [13], the generation of one (res. three) redundant cells in a block of k cells, based on XOR-operation, is proposed. This coding scheme is capable of correcting one (res. two) cell losses in the block, while sequence numbers in the cells are used for loss detection. In [11], a scheme is proposed where a sequence of cells is arranged as a two-dimensional array and where XOR-operations are performed to generate one redundant cell per row and one per column. For h columns, a burst error of up to h consecutive cell losses may be corrected. Loss detection is performed using the redundant cells of the rows. No cell sequence numbers are required, permitting application of the scheme also for VP and VC connections independent of the adaptation layer protocol. The use of a special Reed-Solomon Code that is called RSE (Reed-Solomon erasure code) was proposed in [14, 15]. The coder produces h redundant cells from a block of k information cells, and the decoder is capable of correcting up to h cell losses. Cell sequence numbers are required for loss detection. **Hybrid Error Control.** Hybrid error control schemes combine ARQ and FEC. Type I hybrid ARQ schemes use FEC only for error correction and a separate system for error detection. In type II hybrid ARQ schemes, coding is used for error detection and for error correction. A code that is only used for error correction is able to correct more missing information than a code that is also used for error detection. In a type I hybrid scheme, the redundancy may be fully utilised for regeneration of missing cells. Applying this scheme reduces the mean number of required retransmissions, which allows to reduce mean delay and jitter.

2.4 Protocols for Error Recovery

Adaptation Layer Protocols. In the B-ISDN protocol reference model it is planned to integrate error control mechanisms into the Service Specific Convergence Sublayer (SSCS) of the adaptation layer [16, 17]. This is called assured mode service [9]. Up to now, only two SSCS-Protocols that offer error control mechanisms are specified in the B-ISDN recommendations. The Service Specific Connection Oriented Protocol (SSCOP) is subject of standardisation for a SSCS that offers assured mode service for signaling. The protocol provides end-to-end flow control and retransmission of lost or corrupted data frames by operating in either go-back-N or selective retransmission mode. However, SSCOP does not support assured mode multicast connections. For AAL1, a SSCS with FEC is proposed [9], based on a Reed-Solomon-Code that uses 4 redundant cells for 124 information cells allowing the regeneration of up to four missing cells.

Transport Protocols. Transport protocols that are suitable for a connectionless network layer, as for example TCP, TP4 and XTP, provide more functionality than the functionality that is required for a SSCS-Protocol. These transport protocols need to handle network packets that are received out of sequence without performing error recovery. A SSCS protocol for a reliable service may be simpler, as it may use sequence number gaps for error detection. The TP++ Transport Protocol [18] is designed for a heterogeneous internetwork with large bandwidth-delay product and is suitable for ATM networks. TP++ uses a type I hybrid ARQ scheme and is at present the only transport protocol for high speed networks with FEC. It is only capable of unicast communication. Up to now, no hybrid ARQ protocol was presented for multicast communication in ATM networks.

Protocol Implementation. While transmission capacity was growing enormously over the last years, protocol processing and system functions in the transport component turned out to be a performance bottleneck. High performance communication subsystems, based on parallel protocol processing [19], and hybrid architectures with hardware components for time-critical operations [20, 21] are required for provision of a service with high throughput and low latency. For highest performance, complete VLSI implementations of transport subsystems are planned [22]. The performance bottleneck of the transport component that can be observed for point-to-point-communication is even more crucial for reliable multipoint connections. For a growing number of receivers, processing of a growing number of control packets and management of a large amount of status information needs to be performed.

Selection of Protocol Mechanisms and Protocol Configuration. In order to offer a wide range of services to the applications for various network parameters, several concepts of flexible communication subsystems are under development. The parallel transport system PATROCLOS [21] is a parallel implementation of a high performance transport system, offering a wide range of protocol mechanisms that may be selected according to the needs of an application. The Flexible Communication SubSystem (FCSS) [23] is a configurable, function-based transport system. It utilises a de-layered communication architecture that performs the complete transport component functionality for a specific data stream. It provides flexibility and dynamics of QoS selection and control, supporting the application-specific configuration of the protocol machines based on automatic selection of protocol mechanisms out of a protocol resource pool. Additionally, a set of predefined service classes is provided. The selection of appropriate protocol mechanisms in combination with the reservation of resources in network and end systems allows to enhance delay and loss characteristic of the ATM bearer service in order to provide applications with a specific service quality [24].

3 Conceptual Framework for Reliable Multipoint Communication in ATM Networks

A conceptual framework was developed that allows the use of error control mechanisms best suited for a specific multipoint communication scenario at locations that allow highest performance. Figure 1 presents the ATM network scenario with multicast mechanisms in the adaptation layer of ATM end systems and in dedicated servers. For large groups, the servers may be used hierarchically.

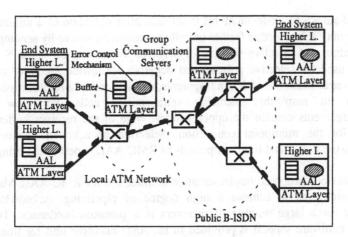

Figure 1:Support for reliable multipoint communication in servers and end systems

3.1 Reliable Multicast ATM Adaptation Layer (RMC-AAL)

The integration of error control mechanisms into the Adaptation Layer needs to be done in a way that high throughput and low latency are guaranteed. In order to offer a reliable and efficient high performance multicast service, the concept of a Reliable Multicast ATM Adaptation Layer (RMC-AAL) was developed. Its ideas are based on the proposal of a configurable extended adaptation layer [24], on the parallel transport system PATROCLOS and on the flexible communication subsystem FCSS.

RMC-AAL extends basic functions of AAL5 by selectable error control mechanisms. Error recovery is based on three schemes: pure ARQ, type I hybrid ARQ and pure FEC. A K-reliable and a fully reliable service are offered. Retransmissions may be sent by multicast or by unicast in selective repeat or go-back-N mode. Complete frames or frame fragments be retransmitted. It can be selected if retransmissions are sent by multicast or individually. When FEC is used, h redundant cells for $l \cdot h$ information cells are generated based on XOR-operations and matrix interleaving. Frames are distinguished using the 'end-of-message' identifier of AAL5 in the payload type field of the cell header. Frames are identified by a sequence number (with frame sequence numbers of 24 bit) and carry the payload length (16 bit) in the frame header. Cell sequence numbers (6 bit) are provided for detection of missing cells. Two options are available for additional frame based error detection. The payload of a frame may be protected by the cyclic redundancy check CRC-32 of AAL5 for a minimum Hamming distance of four when applied to a payload with up to 11454 bytes. If the mode for retransmission of frame fragments is selected, the payload may be protected by a weighted sum code of 32 bit (WSC-2 of [18]). This alternative approach requires a more complex processing unit, but allows to evaluate the code for payload protection in any order. For links with a high bit error probability, the per-cell cyclic redundancy check CRC-10 of AAL 3/4 for a minimum Hamming distance of four may be applied. Receivers send acknowledgements periodically, after reception of a frame in which an 'immediate acknowledgement' bit is set, and after detection of a missing frame. Receivers may acknowledge frames or cells cumulative positive by sending a lower window edge, and selective positive or negative by sending bit maps. Similar bit maps are used for selective positive and negative acknowledgments of individual cells. For retransmissions of frame fragments, the first cell of a retransmission frame carries a bit map that identifies retransmitted cells. For flow control, acknowledgements contain the upper window edge of the receiver buffer section reserved for the multipoint connection. Selection of acknowledgement mode, retransmission mode, and time-out periods of RMC-AAL is performed using control frames.

Figure 2 shows a proposed implementation architecture for RMC-AAL. Main focus of the design was to achieve a high degree of pipelining. Acknowledgement processing for a large number of receivers is a potential bottleneck. Therefore, dedicated hardware support is provided in the *ARQ manager* unit for filtering and processing of acknowledgements, and for managing the status information of the group and of individual receivers. A component for window processing generates

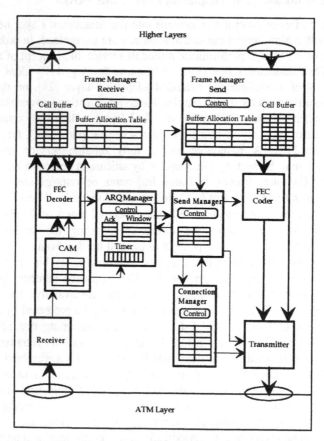

Figure 2: Architecture for the Reliable Multicast ATM Adaptation Layer

multicast flow control information required by the send manager. Generation of acknowledgements is also performed in the *ARQ manager* unit. The *send manger* unit schedules between ordinary transmissions, retransmissions and acknowledgements. The *connection manager* unit schedules between different connections and is also responsible for rate control and spacing. Additional hardware components are required for CRC, FEC, buffer management, list and timer management. For cell demultiplexing at the receiving side, a content addressable memory (CAM) is used to map the large VPI/VCI address space on smaller internal identifiers. Control of the units is provided by a microprogrammable machine, as it was proposed in [25] for the implementation of a programmable AAL interface.

3.2 Group Communication Server (GCS)

The presented reliable multicast adaptation layer represents an important step towards a high performance reliable multicast service. Further improvements of performance and efficiency may be achieved by the deployment of dedicated servers in the network that provide support for group communication. In many cases of multicasting, the achievable throughput degrades fast for growing group sizes. A significant advantage can be achieved if a hierarchical approach is chosen for multicast error control. The proposed Group Communication Server (GCS) integrates a range of mechanisms that can be grouped into the following tasks:

- Provision of a high-quality multipoint service with efficient use of network resources;
- Provision of processing support for multicast transmitters;
- Support of heterogeneous hierarchical multicasting;
- Multiplexing support for groups with multiple transmitters.

For the first task, performing error control in the server permits to increase network efficiency and to reduce delays introduced by retransmissions. Allowing retransmissions originating from the server avoids unnecessary retransmissions over common branches of a multicast tree. The integration of FEC mechanisms into the GCS allows regeneration of lost cells and reinsertion of additional redundancy for adjusting the FEC coding scheme according to the needs of subsequent hops.

For the second task, the GCS releases the burden of a transmitter that deals with a large number of receivers, providing scalability. Instead of communicating with all receivers of a group simultaneously, it is possible for a sender to communicate with a small number of GCSs, where each of them provides reliable delivery to a subset of the receivers. Integrating support for reliable high performance multipoint communication into a server allows better use of such dedicated resources.

For the third task, a GCS may use the potential of diversifying outgoing data streams, allowing conversion of different error schemes and support of different qualities of service for individual servers or subgroups. A group communication server may offer the full range of error control mechanisms of the reliable multicast adaptation layer. For end systems, it is not required to implement the full functionality of RMC-AAL. It will be sufficient to have access to a local GCS for participation in a high performance multipoint communication over long distances. The error control mechanisms of individual end systems have only negligible influence onto the overall performance, as simple error control mechanisms are sufficient for communication with a local GCS. If a priority field is used in the frame format, the server is able to distinguish packets of different importance. One example application would be hierarchically coded video. For information with different importance, different FEC codes may be applied inside one VC, or specific frames may be suppressed for certain outgoing links.

For the fourth task, the GCS provides support for multiplexing of AAL5 frames onto a single point-to-multipoint connection. It may be selected by signaling if the GCS operates in reassembly or in cut-through mode.

The implementation architecture of the group communication server may be very similar to the implementation architecture of the RMC-AAL. A main difference is that the incoming data is not passed to the upper layers, but transmitted instead shortly after reception. Therefore, a single frame management unit for cell buffering is sufficient.

3.3 Signaling

For the management of multipoint connections based on RMC-AAL and GCSs, an extended signaling protocol was developed which is based on the signaling protocols of ITU [6] and ATM Forum [7]. It allows the negotiation and selection of the set of error control mechanisms used for a specific multipoint connection. Dynamic change of call participation is supported. Information of group membership is stored in a central database, administered by a group management server.

4 Performance Evaluation

It is important to know which error control scheme is best suited for a given situation. Analytical methods were applied and simulations were performed in order to evaluate the achievable performance of the proposed error control schemes for the envisaged multicast scenarios. For modelling the correlation properties of lost cells, a two state Markov model (Gilbert Model) may be applied. Based on the worst case observations of [11], a probability of 0.3 was used for a cell discard following a cell discard. This is equivalent to cell losses with a mean burst lenght of 1.428 cells. Using this error model, four different error control schemes were simulated in a point-to-multipoint scenario with four receivers. A multicast tree with one common link and four individual links was assumed, and the same error model was applied to all links. A data rate of 100 Mbit/s, a distance of 100 km, and a frame length of 50 cells was used. The first scheme applied selective retransmissions of frames, the second scheme allowed selective retransmission of missing cells. In the third scheme, FEC with 5 redundant cells was combined with selective retransmission of frames, while the same FEC with selective retransmission of missing cells was combined in the fourth scheme. Figure 3 shows the efficiency (relation of usable cells to total number of transmitted cells) of the four schemes for different cell loss probabilities. Figure 4 shows mean delays that were observed. Maximum efficiency may be achieved by the ARQ scheme with retransmissions of individual cells. In

this scheme, only discarded cells affect efficiency. If only complete frames are retransmitted, a part of the efficiency is wasted by cells that were already successfully transmitted. For the two FEC schemes, the redundancy of 10% limits the achievable efficiency to 0.9. This disadvantage is traded off by the fact that the delay remains constant over a wide range of cell loss probabilities. Figure 4 also shows a constant delay of 0.4 ms caused by FEC. For the distance and data rate of the simulation, this constant delay is a significant part of the round trip time. Therefore, the mean delay of the ARQ schemes is lower than the mean delay of the hybrid schemes up to a cell loss rate of 10^{-4}. However, for equivalent mean delays the ARQ scheme causes already a large jitter. For longer distances and larger groups, FEC will show an even higher advantage. In order to select an appropriate error control mechanism, the following question is of high interest: up to which limit of cell loss probability results a framebased ARQ scheme in higher efficiency than a framebased hybrid ARQ scheme? An interpolation of the simulation results shows a cell loss probability q_s of approximately $\log(q_s) = -3.4$. Applying analytical methods, a formula was derived for the general efficiency equilibrium: If N denotes the number of receivers, n denotes the number of cells in a packet, and h denotes the number of redundant cells of a FEC scheme, the cell loss probability q_s for which a hybrid ARQ scheme achieves the same efficiency than a simple ARQ scheme was evaluated to:
$$q_s = \frac{h/n}{(n-h)(N+1)} .$$
Applying this result to the parameters of the simulation results in $\log(q_s) = -3.352$, which indicates a high correspondence of analysis and simulation.

Analytical methods were applied in order to evaluate the achievable performance of RMC-AAL in selective repeat (SR) and go-back-N (GBN) mode and to evaluate the potential gain by deployment of GCSs. Figure 5 shows the efficiency of the two retransmission modes in three different scenarios. Scenario 1 represents a basic 1:N multicast without GCS. Scenario 2 represents 1:N multicasting with a GCS that performs retransmissions as multicast. In scenario 3, the GCS uses individual VCs for retransmission. The analysis is based on the following assumptions: protocol processing times may be neglected, acknowledgements are transmitted over a reliable connection, and buffers are sufficiently large. A group of 100 receivers and a data rate of 622 Mbit/s are assumed. Two cases are distinguished. The upper diagram of figure 5 shows the efficiency for an overall distance of 1000 km (distance of 500 km from GCS to the receivers), and the lower diagram shows an overall distance of 505 km (distance of 5 km from GCS to the receivers). The analysis shows that in all cases, the efficiency is increased significantly by the GCS. Highest efficiency may be achieved for scenario 3 and selective repeat. Scenario 2 improves significantly for a shorter distance between GCS and the receivers. Go-back-N retransmissions show acceptable performance only for moderate bandwidth-delay products. Regarding efficiency, scenario 3 and selective repeat should be selected. However, this solution requires the highest implementation complexity for end systems and GCS.

135

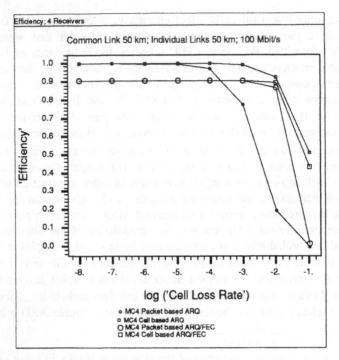

Figure 3: Simulation results for efficiency

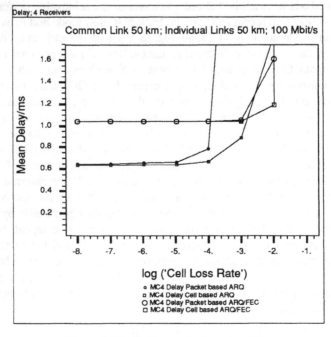

Figure 4: Simulation results for mean delay

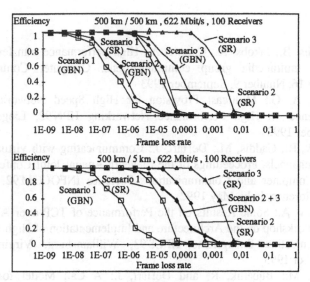

Figure 5: Efficiency analysis for go-back-N and selective repeat in scenarios with and
without group communication server

5 Conclusions

It was pointed out that existing strategies do not allow the provision of an efficient
and reliable high performance multipoint service in ATM networks. A new concept
was presented which has the potential to fulfil the requirements of upcoming
distributed applications. It is based on the integration of multicast ARQ and FEC
error control schemes into a new adaptation layer type called Reliable Multicast
ATM Adaptation Layer (RMC-AAL) and into a new network element called Group
Communication Server (GCS). The functionality of these elements is presented, and
an implementation architecture is proposed. A first performance evaluation is given
which shows the potential benefits of hybrid error control schemes onto service
quality of multipoint connections, and potential improvements if GCSs are
integrated into the network.

Subject of ongoing work is a more detailed evaluation of the achievable
performance, including investigation of the influence of processing times and of
limited buffers. Implementation complexity will be evaluated to allow a better
comparison of the alternative approaches. This should allow to derive guidelines for
the deployment of GCSs and for the selection of the error control scheme best suited
for a given situation.

Acknowledgement

The author would like to thank Martina Zitterbart and Torsten Braun for valuable
discussions. Special thanks are to Comdisco Systems for providing the simulation tool and to
Axel Westenweller for implementing the simulation model. The support by the
Graduiertenkolleg "Controllability of Complex Systems" (DFG Vo287/5-2) is also gratefully
acknowledged.

References

1. Heinrichs, B.; Jakobs, K.; Carone, A.: High performance transfer services to support multimedia group communications. Computer Communications, Volume 16, Number 9, September 1993
2. Waters, A. G.: Multicast Provision for High Speed Networks. 4th IFIP Conference on High Performance Networking HPN'92, Liège, Belgium, December 1992
3. Bubenik, R.; Gaddis, M.; DeHart, J.: Communicating with virtual paths and virtual channels. Proceedings of the Eleventh Annual Joint Conference of the IEEE Computer and Communications Societies INFOCOM'92, pp. 1035 - 1042, Florence, Italy, May 1992
4. Romanov, A.: Some Results on the Performance of TCP over ATM. Second IEEE Workshop on the Architecture and Implementation of High Performance Communication Subsystems HPCS'93, Williamsburg, Virginia, U.S.A., September 1993
5. Gaddis, M.; Bubenik, R. and DeHart, J.: A Call Model for Multipoint Communication in Switched Networks. Proceedings of International Conference on Communications ICC '92, pp. 609 - 615, June 1992
6. ITU-TS Draft Recommendation Q.2931 (former Q.93B): B-ISDN User-network Interface Layer 3 Specification for Basic Call/Bearer Control. Geneva, 1993
7. ATM Forum, UNI Specification Document Version 3.0. PTR Prentice Hall, Englewood Cliffs, NJ, U.S.A., 1993
8. Wei, L.; Liaw, F.; Estrin, D.; Romanow, A.; Lyon, T.: Analysis of a Resequencer Model for Multicast over ATM Networks. Third International Workshop on Network and Operating Systems Support for Digital Audio and Video, San Diego, U.S.A., November 1992
9. ITU-TS Draft Recommendation I.363: BISDN ATM Adaptation Layer (AAL) Specification. Geneva, 1993
10. Brochin, F., Pradel, E.: A Call Traffic Model for Integrated Services Digital Networks. Proceedings of IEEE International Conference on Communications ICC'93, Geneva, Switzerland, May 1993
11. Ohta, H., Kitami, T.: A Cell Loss Recovery Method Using FEC in ATM Networks. IEEE Journal on Selected Areas in Communications, Vol. 9, No. 9, December 1991, S.1471-1483
12. Shimamoto, S., Zhong, W., Onozato, Y.,Kaniyil, J.: Recursive Copy Networks for Large Multicast ATM-Switches. IECE Trans. Commun., Vol. E75-B, No. 11, pp. 1208-1219, November 1992
13. Shacham, N.; McKenny, P.: Packet recovery in high-speed networks using coding. in Proceedings of IEEE INFOCOM '90, San Francisco, California, June 1990, pp. 124-131
14. McAuley, A.: Reliable Broadband Communication Using a Burst Erasure Correcting Code. Presented at ACM SIGCOMM '90, Philadelphia, PA, U.S.A., September 1990

15. Biersack, E. W.: Performance Evaluation of Forward Error Correction in an ATM Environment. IEEE Journal on Selected Areas in Communication, Volume 11, Number 4, pp. 631-640, May 1993

16. ITU-TS Recommendation I.321, BISDN Protocol Reference Model and its Applicaton. Geneva, 1991

17. ITU-TS Recommendation I.362, BISDN ATM Adaptation Layer (AAL) Functional Description. Geneva, 1992

18. Feldmaier, D.: An Overview of the TP++ Transport Protocol Project. Chapter 8 in High Performance Networks - Frontiers and Experience. Ahmed Tantawy (Ed.), Kluwer Academic Publishers, 1993

19. Zitterbart, M.; Tantawy, A.N.; Stiller, B.; Braun, T.: On Transport Systems For ATM Networks. Proceedings of IEEE Tricomm, Raleigh, North Carolina, April 1993

20. Carle, G.; Siegel, M.: Design and Assessment of a Parallel High Performance Transport System. in Proceedings of European Informatics Congress - Computing Systems Architectures Euro-ARCH '93 (October 1993, Munich, Germany); P. P. Spies (Ed.), Springer Verlag Berlin Heidelberg, 1993

21. Braun, T.; Zitterbart, M.: Parallel Transport System Design. 4th IFIP Conference on High Performance Networking HPN'92, Liège, Belgium, December 1992

22. Schiller, J.; Braun, T.: VLSI-Implementation Architecture for Parallel Transport Protocols. IEEE Workshop on VLSI in Communications, Stanford Sierra Camp, Lake Tahoe, California, U.S.A., September 1993

23. Zitterbart, M.; Stiller, B.; Tantawy, A.: A Model for Flexible High Performance Communication Subsystems. IEEE Journal on Selected Areas in Communications, Volume 11, Number 4, pp. 507 - 517, May 1993

24. Carle, G.; Röthig, J.: BISDN Adaptation Layer and Logical Link Control with Resource Reservation for a Flexible Transport System. Proceedings of European Fibre Optics Communications and Networking Conference EFOC&N'93, The Hague, Netherlands, June 30 - July 2, 1993

25. Johnston, C.; Young, K.; Walsh, K.; Cheung, N.: A programmable ATM/AAL interface for gigabit network applications. Proceedings of IEEE Globecom '92, Orlando, Florida, U.S.A., December 1992

TIP: A Transport and Internetworking Package for ATM

Stefan Böcking, Per Vindeby

Siemens AG, Corporate Research and Development
Otto Hahn Ring 6, D-81730 Munich, Germany

Abstract. Common transport protocols do not satisfy the requirements of upcoming applications (e.g. multimedia) and may not benefit from networks with new properties (e.g. ATM). The innovative transport system TIP provides an architecture which allows extendability and optimization of protocols to satisfy new requirements. A protocol is constructed from atomic protocol building blocks and may be tailored according to application requirements and network properties as provided by ATM. Since TIP basically uses a connection-oriented network service and QoS parameters, an efficient mapping onto ATM is possible.

1. Introduction

With ATM coming up, providing high bandwidth and low error rate, new multimedia applications requiring high throughput and guaranteed performance seem to become achievable. Not only in a local test configuration but also in a real networking environment.

high throughput service
low service latency
multipoint conversations
guaranteed performance qualities

new application requirements

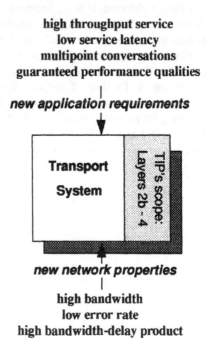

new network properties

high bandwidth
low error rate
high bandwidth-delay product

Fig. 1. New Application Requirements and Network Properties

At transport system level, today's common LAN protocols are designed to handle connectionless networks with relative low bandwidth and can not support the new applications in a satisfying manner. If the sophisticated features of different networks are hidden by only using the smallest common service of all networks (as with the internet protocol IP or ISO 8473 [8]), applications may not benefit from the new network services.

Nevertheless, there is a general desire to employ today's protocols such as TCP/IP on top of ATM [a] to support traditional applications. This is especially true for the LAN environment. The general approach is to define a connectionless service on top of the ATM service which in some way or another makes ATM behave like a traditional connectionless LAN. The *ATM LAN emulation* being defined by the ATM Forum providing a MAC (Medium Access Control) service is an example for this [6]. Such kind of solution may enable these legacy protocols to operate over ATM, but all the properties of ATM are ignored. Due to bad matching of protocol procedures in ATM and common transport systems (e.g. protocol redundancy) valuable resources may even be wasted. [7]

Not only traditional applications, but also the services required by innovative applications must be satisfied. These applications strongly ask for new services as provided by ATM. For this reason, a new transport protocol is needed which should satisfy all the above requirements. A new protocol and architecture called *TIP (Transport and Internetworking Package)* will be discussed in the following. TIP is based on a sophisticated network service. Networks not providing this service have to be enhanced, but the transport layer and hence the application will benefit from networks such as ATM providing a high level of service.

2. TIP's Architecture

TIP consists of the subnet convergence layer *Netglue*, the transport and internetworking protocol called *Tempo*, and *Channels*, the protocol run-time environment.

Netglue is only responsible for enhancing the features of the underlying network to a well defined network service. Netglue may enhance the service of any type of subnet, but in this contribution we will focus on ATM.

Tempo contains all the essential protocol mechanisms for end-to-end control (error and flow control, security, etc.). In principle, TIP could be a part of an AAL, but since Tempo shall work for any network type and provides end-to-end control over multiple heterogenous subnets it is located above ATM.

Channels is a run-time environment dedicated for real-time protocol processing. It supports the dynamic configuration of Tempo and Netglue and simplifies their integration into existing operating systems or onto dedicated hardware platforms. For further information on Channels see [15].

---

a. The term ATM is used to describe the common service of ATM including signaling [4, 5] and AAL (user plane)

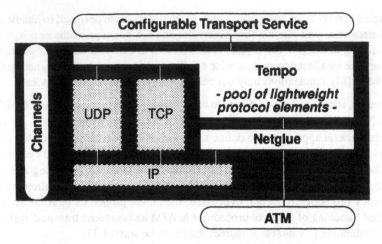

Fig. 2. TIP's Architecture

By having a modular architecture, TIP can be configured to provide just the service requested by the user, such as high throughput, low service latency, security level, multipoint conversation or guaranteed performance [14]. In the same way it can be optimized to the services of the network and consequently avoid unnecessary protocol overhead and redundancy. For example, if segmentation is not needed, it does not have to be configured and therefore does not cause any overhead during communication. Thus, the main advantages of this configurable approach is a scalable functionality. The performed protocol functions may be optimized to users requirements and network services satisfying high throughput and low service latency [13].

The internal structure of TIP consists of a pool of independent protocol building blocks which may be composed depending on the *Quality of Service* (*QoS*) requested by the user at connection establishment. Some basic building blocks are error control, flow control and security. For each basic building block, different mechanisms may be selected, such as *FEC* (*Forward Error Correction*) or retransmission for error control. Standard configurations may be pre-configured by default to allow a faster connection establishment. Since Netglue is also composed of the elementary protocol building blocks, it is actually not an isolated protocol layer.

New protocol building blocks can be introduced supporting new transport services without changing the existing environment. The protocol properties of TIP can easily be extended with the evolution of ATM.

Furthermore, TIP supports the smooth migration of today's protocols and applications on different levels. The TCP/IP protocol stack, for example, can be integrated as one or multiple building blocks enabling transparent communication with other pure TCP/IP systems. If the IP internet is to be used only as a transit network, Tempo may access the IP protocol via a dedicated Netglue. By only using a subset of the powerful TIP *API* (*Application Programming Interface*) as a standard socket interface, for example, today's applications may use TIP without modifications.

Just as ATM, TIP has a separate control and user plane. The control plane is responsible for management of TIPs associations (establishment, release, etc.) and the user plane basically only handles the data transfer. Hence, a direct and efficient mapping of TIP control information to ATM signaling is possible.

Between control and user plane entities within TIP, local information can be exchanged. For example, if ATM notifies a congestion via the control or management plane, this information is forwarded to TIP's user plane in order to reduce the transmission rate.

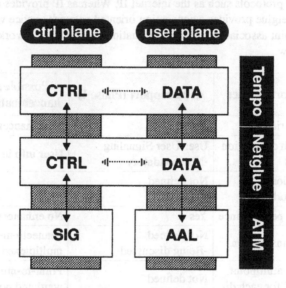

Fig. 3. Mapping of TIP's Control and User Plane to ATM

Some common LAN applications such as FTP uses a connection-oriented (CO) transport service as provided by TCP [16]. Since IP is connectionless (CL), TCP has to map CO to CL. But ATM is basically connection-oriented, so some protocol layer is required (e.g. LAN emulation) to re-map from CL back to CO. This causes a lot of undesired overhead. Using TIP no such overhead occurs because TIP may have a throughout connection-oriented protocol stack.

Other common applications like RPC uses a datagram transport service (CL) as offered by UDP. When using TIP, these applications may use a new *transaction*-oriented transport service. It consists of the exchange of request and reply messages. This service is based on an implicit connection establishment and release.[a] It may be mapped by Netglue onto a connectionless service or a *fast reservation protocol* which is likely to be provided by ATM [1, 10].

Multiplexing and de-multiplexing are time-consuming and generate protocol overhead. TIP on ATM does not need any multiplexing of connections, since one transport connec-

---

a.   A transaction-oriented service is an extension to the datagram and connection-oriented service defined by ISO.

tion may be mapped onto exactly one ATM virtual channel (multiplexing is done by ATM).

3. Netglue

Netglue is placed on top of ATM to enhance the ATM service to the network service as required by Tempo. TIP's network service differs from the network service provided by today's network protocols such as the internet IP. Whereas IP provides a low level datagram service, Netglue provides a connection oriented network service with bidirectional point-to-multipoint associations. More information about TIP's network service is given in the table below.

TIP's network service	ATM support (Phase 1)	Possible service enhancements for Phase 1
Connection-oriented	Yes	No enhancement needed
User info with connection establishment	User-User Signaling -Not mandatory	User info in user plane
QoS negotiation (User - Network)	Not defined -Phase 2 (?)	-
'Guaranteed' performance	Yes	No enhancement needed
Best effort transmission	Not defined -Being discussed	connectionless server, multiplexed VCs,...
Bidirectional multipoint, different QoS for each direction	Not defined -Phase 2 (?)	Point-to-multipoint forward and point-to-point backward
Adding/deleting users in bidirectional multipoint connections	Not defined -Phase 2 (?)	Add/delete leaf and establish/release backward VC
Priority	cell loss priority -User parameter at AAL API?	No enhancement needed
Backward congestion indication	backward explicit congestion notification -Being discussed	Receiver returning forward congestion indication (if supported) to sender.
Outband signaling	Yes	No enhancement needed

Tab. 1. TIP's Network Service and the required ATM Service Enhancements in Netglue

The present stage of ATM does not provide the network service required by the transport layer Tempo. Some ATM features are not defined as mandatory (User-User Signaling) and some features are still only being discussed but may be included in the follow-up specifications of the ATM Forum ('Phase 2'). When only assuming the ATM service as defined in 'Phase 1', the network service required by Tempo is provided by enhancements in Net-glue. The enhancement will be based on today's ATM services.

The final goal is to have an improved ATM service making a Netglue protocol for ATM obsolete. However, some local functions, such as selecting the appropriate AAL type based on the QoS parameters provided by the user, may still be needed.

Services which are available within ATM but not listed in the table above will be used in order to optimize performance by preventing alike functions to be performed redundant. For example, if AAL type 5 is used the transport layer does not need to do bit-error detection since this is already performed by the AAL. However, for reliable communication Tempo must do the error correction.

4. Tempo

Tempo is a collection of protocol mechanisms which can be composed specific to the application, to form an optimal protocol. The selected and designed protocol mechanisms are *lightweight*, i.e. the instruction path and the number of network packet exchanges are minimized for error-free communication, the mechanisms are simplified for hardware implementation and are functionally decomposed for parallel processing.

The transport service of Tempo is also configurable [14]. It allows to express different transport models such as connection, datagram or transaction. Furthermore, various communication structures such as point-to-point, point-to-multipoint or multipoint-to-multipoint are supported.The QoS of each data stream can be adjusted to the application needs.

4.1 Service

A transport service is applied by transport users in order to exchange information independent of their network location and types of subnets such as ATM or Ethernet. User information is transferred by the transport protocol as transport service data units (*TSDUs*).

The TIP addressing scheme uses *address patterns* which describe any simple or complex communication between an unlimited number of transport users which is called a *transport association*. Since each TIP-endsystem (node) only sets up the outgoing transport streams and the complete association is set up node by node as specified in the address pattern, multipoint-to-multipoint associations with different QoS parameters for each stream are supported using the restricted capabilities of ATM alone. This highly flexible addressing scheme also enables adding and deleting of nodes in an existing association; e.g. adding a partner in a tele-conference session.

Most of the existing addressing schemes, such as the one used in TCP/IP and ISO/TP, only provide pairs of source and sink addresses to define transport associations, thus they are basically restricted to point-to-point conversations. Even though the sink address may contain a multicast address, these protocols may not benefit directly from the multipoint capability of ATM, since ATM requires that the initiator of a multipoint connection knows the individual address of all sinks.

Tempo will provide means against hacking and eavesdropping which threatens the availability of the system and the confidentiality and integrity of the transferred data. Furthermore, authentication and access control may be configured.

Transport stream. Transport associations themselves only provide the logical binding between users. The way of communication within an association is defined in terms of *transport streams.* A stream is a sequence of TSDUs which unidirectionally flow from one source to one or more sinks. The sinks group of a stream can be either well-known or anonymous. A stream to multiple sinks can be mapped to an ATM point-to-multipoint connection reducing network load. Each stream in an association may have different transport qualities. The following figure shows a multicast transport stream.

Fig. 4. A Multicast Transport stream

The following figure illustrates a bidirectional point-to-multipoint transport association.

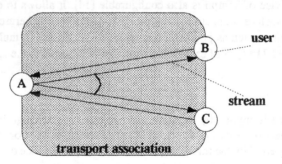

Fig. 5. A bidirectional Point-to-Multipoint Transport Association

Transport quality. Transport quality defines a collection of attribute/value pairs which is related to a transport stream. Streams of the same association may have differently assigned transport qualities. The transport attributes generally cover performance, reliability, priority, costs, synchronization and security properties. Our present service definition comprises performance, reliability and security qualities as shown below.

Quality	Parameter
Reliability	TSDU damage
	TSDU loss
Performance	guarantee level
	TSDU delay
	TSDU interval
	TSDU workahead
	TSDU size
Security	TSDU confidentiality
	TSDU integrity
	Authentication
	Access control
Costs	
Synchronization	further study
Priority	

Tab. 2. TIP's Quality of Service Parameter

The level of the performance guarantee may be specified as *best effort, soft guaranteed* or *hard guaranteed*. Streams can basically be classified according to their guarantee level:

Stream type	Stream quality
asynchronous stream	Stream whose performance qualities are provided best as possible (*best effort*). Issued quality values are used as a hint by the protocol.
continuous stream	Stream whose performance qualities are guaranteed best as possible (*soft guaranteed*) or also under worst case condition (*hard guaranteed*).

Tab. 3. Performance Guarantee Levels

4.2 Protocol

A transport protocol is performed by a set of *transport entities*. They use the underlying network service in order to exchange protocol information conveyed in transport protocol data units (*TPDUs*). Their content, format and coding, and the transport entities behavior are defined by the protocol.

Tempo assumes a connection-oriented network service which supports point-to-multipoint connections with an unreliable and bidirectional data transfer phase. A certain bandwidth of the network connection can be required for each direction. Tempo's protocol design has been simplified by these assumptions in the following way: Tempo needs no connection-management mechanism such as the three-way handshake of TCP or the timer-based mechanism combined with a key mechanism of XTP [17]. Thus, Tempo has simplified TPDUs, a reduced number of TPDU exchanges and a lowered protocol processing expense. Furthermore, Tempo needs no address information conveyed in each TPDU during the data transfer phase which is necessary in connectionless networks. Thus, Tempo's protocol information in a TPDU could be reduced to 16 bytes compared to a minimum of 40 bytes of TCP/IP PDUs and 44 bytes of XTP PDUs. Additionally, Tempo uses the network property to reserve ATM network bandwidth to provide continuous transport streams at application level. This ATM feature is lost by using XTP 3.6 or TCP/IP.

Supporting Tempo's transport service on ATM networks, protocol mechanisms such as segmentation, error control, flow control and congestion avoidance are required.

4.2.1 TPDU Format

Protocol entities exchange information in terms of transport protocol data units (TPDUs). A TPDU usually has three segments: header, data and trailer. All three segments are used to convey protocol data. However, user data (TSDUs) are only transmitted in the data segment.

TPDU		
header	data	trailer

Fig. 6. TPDU Format

Two TPDU types are defined: (1) request TPDUs which flow from stream source to stream sink carrying data such as TSDUs and (2) response TPDUs which flow from stream sink to stream source carrying data such as acknowledgments. The following figure shows the data flow at transport user and provider level and the terms used in the subsequent description of the protocol description.

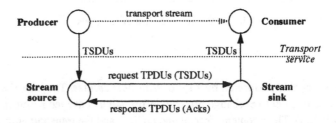

Fig. 7. Data Flow at Transport User and Provider

Both types have a common header format. Header fields have a constant size, building a header of 16 bytes. The trailer format and size depend on the protocol composition. The trailer is mainly used to carry check sums and integrity check values. A tailored header according to the composed protocol mechanisms which may even shorten the header is for further study. The currently defined header format is as follows:

header					
4	4	4	1	1	2(bytes)
time	len	seq	type	option	header check

Fig. 8. TPDU Header Format

A TPDU may fit into a single ATM cell. The following figure shows that up to 24 bytes of user data may be transferred in one ATM cell using AAL type 5. Tempo's TPDU trailer may be omitted because AAL type 5 already performs checksumming on the AAL-SDU [2].

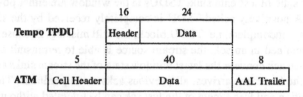

Fig. 9. Up to 24 Bytes of User Data may fit into one AAL type 5 Cell

4.2.2 TSDU Segmentation

TSDUs are transmitted in the data segment of one or more TPDUs as *TSDU blocks*. The last block of a TSDU is tagged by the stream source so that the stream sink can recognize the TSDU boundary. TSDU segmentation is only necessary if the maximum required TSDU size including TPDU header and trailer is greater than the provided NSDU (Network Service Data Unit) size.

Using ATM AAL type 5 the maximum NSDU size is 64KB. For most applications, this will make segmentation at transport level superfluous. However, large NSDUs may produce bursty traffic in ATM and may increase cell loss probability. The impact on Tempo's protocol mechanisms is under investigation.

4.2.3 Error Control

TPDUs may become lost in a network because of physical interference or congestion in nodes and subnets. An occurrence of gaps in a TPDU sequence may be ignored, notified or corrected by Tempo. Loss is detected by sequence numbers. If lost TPDUs need to be corrected, Tempo uses a selective acknowledgment scheme.

Sequence numbering. The stream source continuously numbers the TSDUs in the sequence they are issued by the local transport user for transmission. The TSDU sequence number is conveyed in the TPDU header. Only if TSDU segmentation is selected, a TSDU block number is conveyed with each TPDU. The sequence number and block number uniquely identify a TPDU and are used to detect TPDU duplication, missequence and loss.

According to CCITT recommendation, ATM cells are not misordered or duplicated [3]. However, in reliable transport associations - especially multipoint ones - duplication and missequence may still occur due to retransmission at transport level.

Checksumming and forward error correction. Damaged TPDUs are caused by transmission interferences or hardware errors within the transport and network subsystem. Depending on the QoS requirements, Tempo ignores, indicates, discards or repairs TPDUs containing bit errors. A checksumming algorithm is used to detect bit errors in order to indicate the damage or simply discard the TPDU. A forward error correction method can be selected to repair bit or block errors.

Selective acknowledgment. An acknowledgment TPDU (*ack*) contains the complete TSDU window state of a stream sink. TSDUs in the window are either positively or negatively acked. A positively acked TSDU is completely received by the sink, negatively acked TSDUs are incomplete, i.e. TSDU blocks are still missing. Because of the detailed information contained in an ack, the stream source is able to retransmit only the missed TSDU blocks. An ack conveys the latest window state of the stream sinks as acks sent previously, i.e. if the latest ack arrives, all previous acks are superfluous. Thus, ack missequence, duplication and loss except of the last ack can be tolerated without any additional control expense.

The stream source requests an ack for each TSDU transmitted by tagging the TPDU with a ACK_REQ flag which contains the last block of a TSDU. A stream sink only sends an ack on the arrival of a ACK_REQ-tagged TPDU. The ack contains the sink's window state. Depending on the implementation, the sink simply sends its local state record as ack without any modification. The analysis for retransmission is done by the remote stream source.

Timer. For Tempo's retransmission scheme, a single ack timer is only needed at stream source site. The ack timer is reset, every time the stream source requests an ack. An expired ack timer indicates a lost ack or ack request. The last ack request is retransmitted. The ack timer only controls the ack of the last TSDU in the window. Previously requested acks which may or may not be lost will be subsumed by subsequent acks or, failing this, by the timer-controlled ack.

4.2.4 Flow Control

A stream sink keeps incoming TSDUs until they are consumed by the local transport user. A sink's buffer may run out of space when TSDUs arrive faster than the user consumes them. This is possible because the consuming rate is independent of the TSDU producing rate at the stream source. A window mechanism is used to prevent the sink buffer from overflowing. Window information is transferred with every response TPDU which is sent by the stream sink. The window mechanism is only used in conjunction with loss prevention. Otherwise rate control is sufficient to protect stream sink's buffer.

4.2.5   Rate Control

The stream source transmits TPDUs with a certain rate. A rate is used to determine the earliest transmission time for the next TPDU. For continuous streams, the TPDU rate is fixed.

Throughout an asynchronous stream's lifetime, the TPDU rate may be changed. The interval is multiplicatively increased once per window when TPDU losses are detected or a congestion is indicated by one of the remote sinks or by the network provider. It is additively decreased towards an initial interval after a complete window is sent without retransmission and no congestion indication has occurred meanwhile. This algorithm guarantees that all sources reduce their transmission rate fairly in case of a temporary congestion.

4.2.6   Inactivity Control

During stream-idle times or a closed window situation, the stream source periodically checks the liveliness of the stream sink by sending probe requests. The probe response of the stream sink contains window information which possibly indicate a newly opened window. The stream source controls an expected probe response with the normal retransmission mechanism.

4.2.7   Multicast Control

A stream may be directed to an anonymous or well-known group of stream sinks. For a well-known group, the source knows the individual transport addresses of each sink. If loss prevention is selected, a TSDU has to be received and consumed by all well-known sinks. Thus, the stream source has to process multiple acks which may arrive simultaneously. Stream sinks can be required to delay their ack generation randomly to avoid congestion by a large number of simultaneous acks. By an anonymous group, the stream source only knows the group address of the sinks. The first member which responds is selected as the stream sink. The ongoing reliable stream is only between stream source and the selected stream sink.

4.2.8   Traffic Control

In a continuous stream, TSDUs are issued by the producer and accepted by the consumer at a constant rate. In relaxing these constraints, independent workahead parameters have been introduced at both sites. The mechanisms to control the continuous stream of TSDUs are as follows.

Traffic shaping. At stream source site, the workahead defines the maximum number of TSDUs which the producer may issue at any time before their actual time slot in the continuous stream arrives. The stream source conveys TSDUs in one or more TPDUs which are transmitted at a constant rate. Workahead TSDUs are buffered until their actual transmission time. Thus, both TSDUs bursts and large TSDUs are smoothed for transmission.

Jitter compensation. At stream sink the workahead defines the maximum number of TS-DUs which the consumer may accept at any time before their actual time slot in the continuous stream arrives. TSDUs conveyed in TPDUs are transmitted with a constant rate but possible delay variations may occur during transmission. The stream sink may delay and buffer arrived TSDUs to compensate delay variations which are outside the defined workahead.

Jitter monitoring. The bounded delay variation which is negotiated between network provider, stream source and sink may be violated during the stream's lifetime. The network provider may introduce a larger delay jitter than agreed or the TPDU transmission rate of the stream source may drift from the defined rate. A timestamp mechanism allows stream sinks to detect jitter violations. The stream source conveys in each TPDU its local time. Comparing the timestamps of two subsequently arrived TPDUs enables the stream sink to detect delay jitter violations and pauses in the continuous stream. The TPDUs' sequence numbers avoid ambiguity by lost TPDUs.

Usage parameter control. The producer should be forced to use only the negotiated stream bandwidth. This is quite easily done in Tempo by allowing the producer to issue the next TSDU only if the previous one has been confirmed by the stream source. Thus, the stream source can block the producer if the defined workahead is exhausted.

5. State of Work

Channels, Tempo and Netglue have been specified and implemented. The specifications follow an object-oriented approach and the implementation is done in C++. A first prototype of a TIP system has been tested on a SUN Sparc station.

6. Conclusion

The high performance of ATM can easily be sacrificed through unsuitable protocols such as TCP/IP. The architecture and protocol proposed here allows applications to use the properties of ATM to a full extent. Tempo is not restricted to operate on ATM, but may use any type of subnet. In addition, TIP enables interconnection between ATM and different networks.

7. References

[1] CCITT Draft Recommendation I.364, „Support of connectionless data in a B-ISDN", Geneva June 1992.

[2] CCITT Draft Recommendation I.363, „AAL Type 5", Geneva January 1993.

[3] CCITT Recommendation I.150, „B-ISDN Asynchronous Transfer Mode Functional Characteristics", Geneva June 1992.

[4] CCITT Draft Recommendation Q.93B, February 1993.

[5] ATM Forum, „ATM User-Network Interface Specification Version 3.0".

[6] Contribution 94-0035R1 to the ATM Forum, „LAN Emulation Over ATM: Draft Specification", LAN Emulation Sub-working Group, March 1994.

[7] Contribution 93-784 to the ATM Forum, „TCP over ATM: Some Performance Results", Ottawa July 1993.

[8] ISO 8473, „Protocol for providing the connectionless-mode network service", 1988.

[9] ISO/IEC JTC1/SC6 N 7788, Second Draft, „Guidelines for enhanced Transport Mechanisms", December 1992

[10] Jean Cherbonnier, Jean-Yves Le Boudec, Hong. Linh Truong, „ATM Direct Connectionless Service", IEEE International Conference on Communications ICC 1993.

[11] Julio Escobar, „Future challenges for the adaptation layer", computer communications volume 16 number 2 february 1993.

[12] Jose C. Brustoloni, Brian N. Bershad, „Simple Protocol Processing for High-Bandwidth Low-Latency Networking", Carnegie Mellon University Technical Report CMU-CS-93-132, Pittsburgh March 1993.

[13] S. Böcking, „TEMPO: A Lightweight Transport Protocol", Proc. of the Third Workshop on Future Trends of Distributed Computing Systems, April 1992.

[14] S. Böcking, „The configurable transport service of Tempo", 11th Conf. on European Fibre Optic Communications and Networks, July 1993.

[15] S. Böcking, V. Seidel, „TIP's protocol run-time system", 12th Conf. on European Fibre Optic Communications and Networks, June 1994.

[16] User Datagram Protocol, Network Working Group, RFC 768; Transmission Control Protocol, Network Working Group, RFC 793; Internet Protocol, Network Working Group, RFC 791.

[17] XTP Protocol Definition, Protocol Engines, Rev. 3.6, January 1992.

Multimedia and Hypermedia Synchronization:
A Unified Framework*

Nikos B. Pronios, Theodoros Bozios
INTRACOM S.A.
Development Programmes Department
P.O. BOX 68, Peania 19002, GREECE
Tel.: (30 1) 6860000, Fax.: (30 1) 6860312, 6644379
e-mail: {npro, tmpo}@intranet.gr

Abstract. In this paper we present a unified layered framework for the study and assessment of multimedia/hypermedia synchronization problems and solutions. The basic objective of this framework is to include the various synchronization mechanisms as well as their inter-dependence. In the process we also identify the lack of universally accepted multimedia/hypermedia synchronization performance measures and the problems it creates for the performance evaluation of the various synchronization schemes and their comparison. Using this unified layered framework we present potential performance measures and parameter trade-offs. We position the various efforts within it, along with a proposed synchronization scheme for inter-stream synchronization at the communication system level.

1 Introduction

Synchronization, in general, is the temporal alignment of two or more events in stochastic environments. The term synchronization in the OSI RM refers to a single stream of discrete media with no real-time requirements. The introduction of multimedia computing and communications has added another dimension to the already complicated problem of synchronization in communications, that of multimedia/hypermedia synchronization. The accurate description of all multimedia/hypermedia synchronization aspects is a non-trivial task, because of the different levels (user-presentation, logical stream, packet) in which synchronization mechanisms are needed and can be implemented. We propose a unified layered framework for multimedia synchronization, we present potential performance measures and parameters trade-off, and we position the various efforts within it.

---

* The work described here is being carried out as part of the RACE II R2008 EuroBridge project

154

In this paper, we use the term *Synchronization* to describe the mechanisms that make presentation events in a distributed environment to occur, possibly at different locations, in a certain time order [1]. Thus, the term *Multimedia/Hypermedia Synchronization* will include the mechanisms that make the presentation of different media to one or more users, to follow the implied, or user-defined, spatio-temporal relations. New synchronization requirements are mainly caused by the introduction of *Structured Presentations* [2] containing *Continuous Media (CM)* like video, audio, and animation and the *multipoint* nature of many multimedia applications.

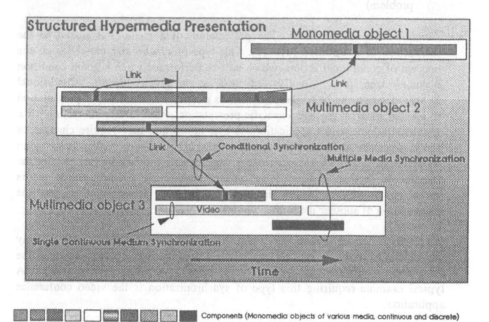

Components (Monomedia objects of various media, continuous and discrete)

Fig. 1: Composition of Structured Hypermedia Presentation

The *Structured Multimedia/Hypermedia Presentations*, see Fig. 1, imply new types of synchronization such as the *synchronization of a single continuous medium*, the *synchronization among related media*, the *Conditional synchronization* and the *Multipoint synchronization*.

- *Single Continuous Medium synchronization (intra-stream synchronization)*. Continuous media (CM)[1] require that their presentation at the receiver's side has the same temporal relation they had at the source (source can be a camera, a loudspeaker, a database, etc.) [3].

---

[1] The transmission of CM can be both real-time and non-real-time; the former imposes very strict performance requirements on the network, while the latter does not.

- *Multiple Media synchronization (inter-stream synchronization)*. Synchronization when two or more media (continuous or discrete) that must maintain a pre-defined temporal relationship with each other when they are presented to the user. The lip-synch between a video and the related audio is an example of this type of synchronization. We can distinguish two kinds of multiple media synchronization problems, depending on the locality of the sources:
  - The sources are on the same node ("one-to-one" synchronization problem).
  - The sources are on different nodes ("many-to-one" synchronization problem).

- *Conditional synchronization*. The presentation of a medium is linked to the satisfaction of a condition [4] - [7]. This type of synchronization relies on the concept of conditional action, which will be performed when a given condition becomes true (e.g., overlay of text at a given image). Conditional Synchronization can be triggered by events generated when the user-presentation of another medium reaches a specific presentation status, user's interaction, or a multimedia/hypermedia application. A user interaction may require changes of the presentation attributes, for example changing the audio quality from mono to stereo, or resizing a video window. Another user interaction may require the transfer of another medium. One type of conditional synchronization is the *serial synchronization* where the end of a medium presentation triggers the presentation of another medium.

- *Multipoint synchronization*. The *multipoint* nature of many multimedia/hypermedia applications requires presentation events to occur at the same time at different locations ("one-to-many" synchronization problem). A typical example requiring this type of synchronization is the video conference application.

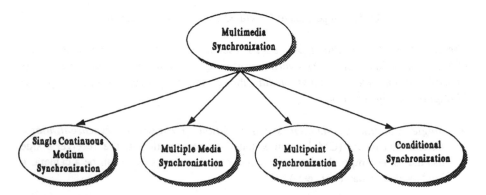

Fig. 2: The new types of synchronization introduced by distributed multimedia/hypermedia applications

The rest of this paper is organized as follows. In section 2 we describe the main multimedia synchronization issues. In section 3 we present a unified layered framework for the study and assessment of multimedia/hypermedia synchronization problems and solutions. We also identify the lack of universally accepted multimedia/hypermedia synchronization performance measures and the problems it creates for the performance evaluation of the various synchronization schemes and their comparison. In section 4, we describe a Multimedia Synchronization scheme and we position it within the proposed unified framework. Finally, in section 5, we provide certain conclusions and directions for future work.

2 Multimedia Synchronization Issues

As expected a variety of conceptual issues have to be addressed in multimedia synchronization, along with the interaction and integration of prototyping, analysis and design.

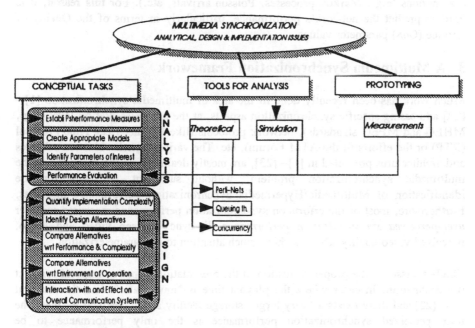

Fig. 3: Issues and techniques in multimedia/hypermedia synchronization

To arrive at the Performance Evaluation and analysis of multimedia or hypermedia synchronization, one can use three different paths as shown in Fig. 3. The first path is to use theoretical tools for the analysis, and, using the appropriate models, to evaluate the performance of the proposed scheme exactly, approximately, or within bounds. Naturally, the synchronization performance should be described via the identified parameters of interest.

An example of what we consider a desirable performance measure is the degree of synchronization (appropriately quantified) between two user-presentation events with a particular synchronization scheme, parameterized by the buffer sizes or the playout deadlines, for different network performances (e.g., delay statistics, accuracy of clocks, etc.). It is obvious that universally accepted and useful performance measures should be established prior to any attempt of evaluating the synchronization performance. The same naturally holds for the other two paths of Fig. 3, the performance analysis via simulation and the measurements of a prototype implementation.

It should be noted that the end-to-end network behavior and performance depend not only on the particular protocol stacks, but on the specific protocol implementations as well. This network performance cannot be accurately expressed beforehand in analytical forms. Furthermore, in order to obtain analytically tractable solutions, the various analytical techniques based on theoretical tools often require unrealistic assumptions (e.g., Markov processes, Poisson arrivals, etc.). For this reason, it is hard to predict the network's performance, particularly in terms of the Quality of Service (QoS) parameter values.

3 A Multimedia Synchronization Framework

Much work has been recently done in the area of multimedia synchronization [1] - [23] addressing specific synchronization aspects, at the user-presentation level (e.g., MHEG and MPEG standards), network protocols (like Internet Real-Time-Protocol (RTP) or the efforts in the ATM Forum), etc. The various synchronization schemes and architectures presented in [1] - [23], are mostly dealing with the modeling of the multimedia synchronization problem, without sufficient emphasis on the identification of Multimedia/Hypermedia Synchronization Performance measures. Furthermore, most of the efforts on synchronization performance measures consider *user-perceived synchronization performance* (lip-sync between video and audio, perceived video quality, etc.), without much attention to its parameterization.

The basic issue is the proper recreation of the presentation's spatio-temporal script at the destination. In cases where the playout time is "much later" than the retrieval time [22] and there exists a "very large" storage facility at the destination, then, the user perceived synchronization performance is the only performance to be considered. However, cases where either "small" storage facilities or tight palyout deadlines exist, the synchronization performance of the communication system should also be considered. There is therefore a variety of cases where the environment and scenario of operation require different performance measures.

The user-perceived degree of synchronization depends on the *combined synchronization performance* of the communication system and the multimedia application platform. Since the user-perceived synchronization performance depends heavily on the encoding methods used, as well as on the error concealment techniques and the multimedia application platform capabilities, *additional* performance measures should be established for the performance evaluation of multimedia/hypermedia synchronization schemes within the communication system.

The communication system's services should at least be capable of supporting the transfer of monomedia objects constituting a multimedia/hypermedia structured presentations from different logical channels. The transfer through different channels creates the need for maintaining the temporal relations between these objects across the network. With stochastic network behavior, something that might not be true with network resource reservation and admission control, the total network delay is non-deterministic and the data elements traveling through different logical channels may arrive in the destination with different end-to-end delays.

The capabilities (e.g., memory, OS, CPU) of the multimedia/hypermedia application platform is another issue of interest. The level of these capabilities could determine the degree to which the communication system would need to provide synchronization mechanisms. These mechanisms could be related to the user presentation of different media, and, depending on the application platform, they might not be used.

The lack of precisely defined multimedia performance measures and the associated parameters describing the environment of operation, creates a two-fold problem: the multimedia synchronization performance, while feasible, is not very meaningful; differences of opinions for various mechanisms and architectures cannot be resolved with objective criteria (meaningful performance measures). This problem is further exasperated by the lack of a sufficient amount of simulation or experimental results.

To address this problem, the Unified Layered Framework proposed herein and shown in Figure 4, separates the basic synchronization requirements and synchronization performance into two Levels: the user-presentation level (Multimedia Application Layer) and the Communication System level[2]. We note that the three layered abstraction of the Multimedia Synchronization proposed in [26], focusses on the Multimedia Application Layer of the model proposed herein. Thus, the two models can be considered as complementary. A mechanism that could be used for the mapping of the time-bases of theses two layers (so that the multimedia/hypermedia application requests to the communication system can relate to user-presentation events) is desirable. Note that in this model events triggering conditional synchronization take place at the Multimedia Application Layer.

Two approaches are being used for the re-establishment of the presentation's time-base by the receiver. The first approach attempts to keep all clocks absolutely

---

[2] The separation in Fig.4 is shown for reference purposes in relationship to OSI RM. However, the OSI Rms need not to be followed exactly in analysis or implementations (see Fig.5).

synchronized, using special synchronization protocols like the NTP [24]. Thus, *any* user-presentation event can be mapped to network events. This approach is being recommended by RTP and the Internet community, and could have difficulties for large networks. The second approach assumes unsychronized clocks and re-creates the Structured presentation's time-base at the receiver using a presentation-dependent time-base, taking into consideration differences among the presentation rates at source and destination, as well as the access and transmission (throughput) rates [22].

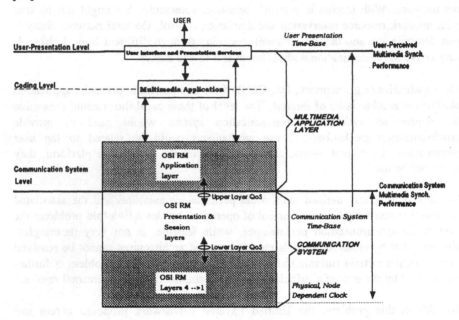

Fig. 4: A Unified Layered Framework for Multimedia/Hypermedia Synchronization

Synchronization Markers or Synchronization Channels, to transfer timing information pertaining to the handling and presentation of the various media, are strategies that can be used by both of these approaches. The concept of *synchronization markers* (SM) [7] can be used both for clock synchronization and for the control of the presentation rate at the user, while the Synchronization Channel (SC) periodically transmits timing and synchronization information.

Another solution that can be used for maintaining the temporal relations between the continuous media across the network could be the multiplexing of the different information streams onto a single logical channel in the appropriate ratio. However, the separate and parallel process and transmission of each medium help to achieve the real-time requirements and the simultaneous presentation to the user. Moreover, it potentially improves the network utilization, since monomedia data are transferred through logical channels with the appropriate QoS characteristics for their specific media type, and not with a combined QoS sufficient for the most demanding medium. In a distributed environment, it is very important to satisfy the need for continuous medium transmission through different channels, since the sources can be distributed across the network and multiplexing could not be possible [7] [8].

In general, the retrieval times (or the transmission times) have to be determined after the consideration of the relevant delays [21] which do not depend only on the communication system. The expected behavior of the network can be described through the QoS parameter values, one of which is the end-to-end delay. The handling of QoS is complicated by the fact that the QoS parameters *are not independent*. For example, the effect of delay jitter introduced by queuing in the intermediate network nodes may be reduced by the use of buffering at the receiver. However the use of buffers increases the overall end-to-end delay. Hence, the amount of delay jitter allowed must be traded off against the end-to-end delay requirements and buffer sizes. A similar trade-off exists between the reliability (error rate) that could be attained at the expense of reduced throughput and/or increased delay.

It is obvious that the type of QoS (Best Effort, Threshold, Compulsory) supported by the communication system creates different synchronization needs. For example, with Compulsory (or Guaranteed) QoS, the synchronization problem, can be reduced to a scheduling one, with associated resource management and admission control mechanisms [27].

For packet-switched networks, the inter-stream synchronization, particularly for the many-to-one synchronization problem, the estimation of delay within which "most" of the packets will arrive on-time for the playout deadline, affects not only the user-perceived quality of service, but the buffer size as well. In general, when the playout deadline allows, sufficient information should have arrived prior to the playout; this would allow for: a) the handling of "slow" packets, which otherwise would be lost, and b) the masking of long intervals with no arrivals, using the already delivered information. Note that the buffer size is again a parameter of interest.

Motivated by the fact that past results on multimedia synchronization were costrained on particular models and assumptions, and based on the earlier reasoning, we present a rather generic set of performance parameters which can be used for meaningful comparison of the various synchronization schemes and methods. At the user-presentation level the performance measures should be in real time units (ms, sec) as in [23] where the human perception is considered. However, the performance should be conditioned on both the availability of the information at the Multimedia Application layer and its capabilities. Thus, denoting by $S_{UPL}{}^{ij}$ the synchronization in ms at the user-presentation level (UPL) between Elementary Presentation Units (EPUs, [22]) of objects i and j of a multimedia/hypermedia presentation, we can write:

$$S_{UPL}{}^{ij}(T) = f(\beta_{MAL}, T-t_i, T-t_j)$$

where T is the playout time, ti, tj are the times at which the Elementary Data Units (EDUs, [22]) of objects i and j respectively are forward to the Multimedia Application Layer (MAL) which, in turn, has performance measure β_{MAL}; the later encompases the CPU, memory, OS, etc. combination.

We note that:
a) $S_{UPL}^{ij}(T)$ covers only Multimedia Application Layer issues;
b) the communication system's influence is introduced by $T-t_i$ and $T-t_j$;
c) $S_{UPL}^{ij}(T)$ addresses only a pair of EPUs, using the availability times $T-t_i$ and $T-t_j$ of the corresponding EDUs as provided by the communication system.

Thus, the communication system is expected to forward the EDUs to the Multimedia Application Layer at proper instances so that the user perceived multimedia synchronization $S_{UPL}^{ij}(T)$ is acceptable. Certain averaging should be expected in order to describe the long-term synchronization behavior, as opposed to specific pairs of elementary units.

4 An Example

This section describes how the multimedia/hypermedia synchronization scheme proposed in [23] can be positioned within the Unified Framework. This scheme is based on the idea that the communication system's role in multimedia synchronization is to provide a *coarse* synchronization (lower level synchronization) before the multimedia/hypermedia application layer can perform the *fine* synchronization (higher level synchronization). Thus, the communication system should forward the media streams, belonging to the same structured presentation, to the multimedia/hypermedia application layer with asynchronisity levels that the multimedia application layer can handle. The use of this, lower level, synchronization is *optional* and it depends on the QoS provided by the underlying networks. This two-stage synchronization concept exists, to our knowledge, in *all* synchronization schemes, since fine synchronization (or *tracking*) can be achieved at the cost of reduced dynamic range, that is the range of operation of the specific synchronization mechanism. At the same time, wide synchronization range results in reduced resolution-matching capabilities; this coarse synchronization (or *acquisition*) stage is used prior to the tracking.

The basic design objectives of this scheme, shown in Figure 5 in an ATM environment, are:
  1. Use of existing multimedia/hypermedia document architectures (such as MPEG and MHEG system level descriptions).
  2. Utilization of QoS formalism that is becoming prevalent in the high performance communication networks and services.
  3. Distinction of the synchronization problems into those caused by the communication system and those caused by other systems involved (e.g., DBMS, OS, playout devices).

4. The communication system synchronization mechanism will deal with the communication system's problems, and will perform a coarse inter-media synchronization, leaving the fine-synchronization issues to be dealt by the Multimedia/Hypermedia Application Layer Synchronization mechanism (e.g., MHEG-system).

5. Correlation of communication system and user-presentation events. This is achieved with the Incorporation of correlated SMs at two different levels (Presentation Time-Stamps or PTSs for the user-presentation level and Network Time-Stamps or NTSs for communication system), using the concept of *Synchronization Group*.

6. The optional use of the communication system's synchronization mechanism. The Multimedia Synchronization Service Element of a Dedicated Multimedia Session and Presentation Layer (DeMuPS), can and will be used only if the Multimedia/Hypermedia Application Layer considers it appropriate.

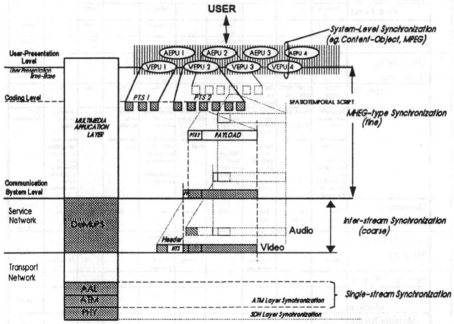

AEPU: Audio Elementary Presentation Unit
VEPU: Video Elementary Presentation Unit
PTS: Presentation Time-Stamp
NTS: Network Time-Stamp

Fig. 5: A Multimedia/Hypermedia Synchronization Architecture

A classification, based on the proposed Unified Layered Framework, of the various efforts on multimedia synchronization is shown on Table 1.

TABLE 1: MULTIMEDIA SYNCHRONIZATION CLASSIFICATION

MM SYNCH. WORK > / MM SYNCH. ITEMS \/	Tenet Group (U.C. Berkeley)	Internet RTP protocol	J. Escobar	ESPRIT OSI 95 project	IBM ENC	Little & Ghaffor	Qazi, Woo & Ghaffor	C. NIKOLAU	Ramanathan & Rangan	EUROBRIDGE
1 Synchronization Model	No	No	✓	No	MM object model	✓	✓			✓
2 Synchronization Architecture	Realtime "Tenet Protocol suite"	No (protocol)	Flow Synch. Protocol	✓	No	✓	✓	✓	No	✓
Single Level			✓							
Two Level	✓						✓			
Three Level				✓		✓	✓	✓		✓
3 Layer (OSI RM Mapping)		Above 3 or 4 layer	No specific layer			Not Appl.	No direct Correspondence		Not Appl.	
First Level	Layer 4			Layer 7		Layer 7		Layer 7		Layer 7
Second Level	Layer 3			Layers 5-6		Layer 5		Layers 5-6		Layers 5-6
Third Level				Between layers 4-5				Layers 1-4		Layers 1-4
4 Underlying Networks										
Specific	FDDI, ATM	UDP, TCP, TP1 ,TP4, ST-II	OSI TP,TCP UDP,AAL	ST-II, SRP		No	ATM	No	No	
Generic	Boundable Subnetwork Delay; Negligible Loss rate; Bounded Delay jitter or loose clock synch. at the nodes	Unreliable Multicasting is desirable	Synchronized network clocks. Use of other protocol for coordination and control	Rate based flow control transport protocol; Network level resource reservation protocol	Real-time scheduling; Resource management	Guaranteed level of service; Characterization of interarrival distributions of consecutive packets	Reliable data transfer; Guaranteed QoS	No specific assumptions	Bounded Delays using: Resource reservation; Admission Control; Real-time scheduling; Buffer reservation	QoS Formalism
5 Document Architecture	Not Appl.	Not Appl.	Not Appl.	No	No	No	No	No	No	Yes (MHEG -type)
6 Synchronization Strategy	Jitter-controlled scheme	SMs	SMs & SCs (control messages)	SCs	SMs & Multiplexing	Resource reservation & Playout Schedule	SMs	SMs (?)	SCs (Feedback Messages)	SMs
7 Synchr. Types										
Single Continuous Medium	✓	support				✓	✓	✓		support
Multiple Media	✓	support	✓	✓	✓	✓	✓	✓	✓	✓
Multipoint		support	✓						✓	
Conditional			✓					✓		✓
8 Synchronization Performance Measures	No	No	Yes (network level)	No	User-presentation level	No	No	No	User-presentation level	Comm. System via user presentation
9 Analytical tools	"Galileo Tool"	No	No	No	No	Petri-Net (OCPN)	Petri-Net (XOCPN)	No	No	No
10 Simulation	"Galileo Tool"	No	No	No	No	No	No	No	Preliminary	in progress
11 Implementation	Separate protocol prototypes	prototypes	Test Implement	No(?)	prototype	No	planned	in progress	in progress	planned

5 Conclusions

In this paper we presented a framework for the study and assessment of multimedia/hypermedia synchronization. The framework is generic enough to address various synchronization mechanisms and can be used for the classification of multimedia synchronization models and architectures; thus a classification has been provided in a tabular form. The basic feature of the framework is the separation of communication system multimedia/hypermedia performance issues from the one of the multimedia application layer; the effect of the communication system's performance on the user-perceived synchronization has also been addressed. We identified important parameters of interest in multimedia synchronization and we proceeded with the introduction of synchronization performance measures. More sophisticated performance measures will be required for a complete performance evaluation of multimedia synchronization, thus the alternative performance estimate will be based to the one introduced.

Acknowledgments

We would like to thank all the colleagues in the EuroBridge consortium. Our work depends very much on their efforts. Furthermore, we would like to thank F.Colaitis for the information regarding the status and the contents of the MHEG standard.

References

1. Synchronization Properties in Multimedia Systems," R. Steinmetz, IEEE JSAC, Vol.8. No 3. April 1990, pp 401-412.

2. "The Amsterdam Hypermedia Model: Adding Time and Context to the Dexter Model," L. Hardman, D. Bulterman and G. Rossum, Communication of the ACM, February 1994/Vol.37, No.2, pp. 49 -62.

3. "An Architecture for Real-Time Multimedia Communication Systems," C. Nicolaou, IEEE JSAC, Vol.8. No 3. April 1990, pp 391-400.

4. "Coded Representation of Multimedia and Hypermedia Information Objects," ISO/IEC JTC1/SC29/WG12, MHEG Committee Draft, June 1993.

5. "MHEG, the future international standard for multimedia and hypermedia objects," F. Colaitis.

6. "Coded representation of multimedia and hypermedia information objects: Towards the MHEG standard," F. Kretz, F. Colaitis, Image Communication 4, April 1992, pp 113-128.

7. "Standardizing Hypermedia Information Objects," F. Kretz and F. Colaitis, IEEE Communication Magazine, May 1992.

8. "Extending OSI to Support Synchronization Required by Multimedia Applications", Michel Salmony and Doug Shepherd, IBM European Networking Centre.

9. "A Continuous Media Transport and Orchestration Service," Campbell et al, OSI 95 project, August 92.

10. "Multimedia Synchronization Protocols for Broadband Integrated Services," T. Little and A. Ghafoor, IEEE JSAC, Vol.9. No 9. Dec. 1991, pp 1368-1382.

11. "A Synchronization and Communication Model for Distributed Multimedia Objects," N. Qazi, M. Woo and A. Ghafoor, Proccedings of the first ACM International Conference on Multimedia, 1-6 August 1993, Anaheim, California, pp 147-155.

12. "Adaptive Feedback Techniques for Synchronized Multimedia Retrieval over Integrated Networks," S. Ramanathan and P. Rangan, IEEE/ACM Transactions on Networking, Vol.1. No 2. April 1993, pp 391-400.

13. "Designing an On-Demand Multimedia Service," V. Rangan, H. Vin, S. Ramanathan, IEEE Communication Magazine, July 1992.

14. "Media Synchronization in Distributed Multimedia File Systems," V. Rangan et al, 4th IEEE ComSoc International Workshop on Multimedia Communications, California, April 1-4, 1992

15. Channel Groups: A Unifying Abstraction for Specifying Inter-stream Relationships," Amit Gupta and Mark Moran, Tenet Group, University of California, Berkeley.

16. "The Real-Time Channel Administration Protocol," A. Banerjea and B. Mah, Tenet Group, University of California, Berkeley.

17. "Flow Synchronization Protocol," J. Escobar, C. Partridge and D. Deutsch, Bolt Beranek and Newman Inc., Cambridge.

18. "RTP: A Transport Protocol for Real-Time Applications," S. Casner and H. Schulzinne, October 1993.

19. "Issues in Designing a Transport Protocol for Audio and Video Conference and other Multiparticipant Real-Time applications," H. Schulzinne, October 1993.

20. "Coding of Moving Pictures and Associated Audio, " Commitee Draft of Standard ISO11172: ISO/MPEG 90/176, December 1990.

21. "Synchronizing the Presentation of Multimedia Objects - ODA Extensions -," Petra Hoepner, Multimedia Workshop, Stockholm, April 1991.

22. "Multimedia Synchronization: The role of the Communication System," Th. Bozios, N.B. Pronios, "BROADBAND ISLANDS: Towards Integration," ELSEVIER 1993, Proceedings of the Second International Conference on Broadband Islands, pp 151-172.

23. "Multimedia Synchronization Techniques: Experiences Based on Different System Structures," R. Steinmetz, MULTIMEDIA' 92, California April 1992, pp 306-314.

24. Multimedia Synchronization Functionality in the EuroBridge Platform," RACE 2008 EuroBridge project, Th. Bozios, N. Pronios, March 1994.

25. "Internet Time Synchronization: The Network Time Protocol," D. Mills, IEEE Trans. on Com., Vol. 39, No. 10, October 1991.

26. "A Taxonomy on Multimedia Synchronization," T. Meyer, W. Effelsberg and R. Steinmetz, 4th IEEE International Workshop on Future Trends on Distributed Computing Systems, 22-24 Sep. 1993, pp. 97-103.

27. "A Quality of Service Architecture," A. Campbell, G. Coulson, D. Hutchison, ACM SIGCOMM, April 1994, pp. 6-27.

Multimedia Playout Synchronization Using Buffer Level Control

Daniel Köhler, Harald Müller

Lehrstuhl für Kommunikationsnetze, Technische Universität München,
D-80290 München, Germany

Abstract. This paper presents a new approach to the intrastream synchronization problem. The proposed technique is based on controlling the receiver buffer. The basic idea is to find realistic models for the control loop components and to apply automatic control methods to choose an appropriate control algorithm and to derive stability criteria and transient behavior of the whole system. The theoretical results are shown to have good accordance to real, measured system behavior. The synchronization method has been implemented and first experiences could be gathered with a video phone application.

1 Introduction

Many multimedia applications, like remote camera, video conferencing, and video on demand, require the delivery of continuous audio and video information in real-time. When using asynchronous networks for data transmission, timing information of the media units produced gets lost and a mechanism is necessary to ensure continuous and synchronous playback at the receiver side. The so-called intrastream or playout synchronization is one key concern in multimedia synchronization. In this paper, we present a playout synchronization technique that is based on receiver buffer control. By realistic modeling of the playout components, a suitable control algorithm can be determined and stability criteria and transient behavior of the system can be theoretically derived when using automatic control methods. This allows for the design and dimensioning of a well behaving system without having to rely on experiments.

The synchronization method was implemented using the PC-based multimedia system Action Media II (DVI). The first application to use the proposed technique is a video phone that allows bidirectional communication over local and high-speed metropolitan area networks (MAN) including audio and full motion video.

In Sect. 2, a classification of other intrastream synchronization solutions is given and the basic principles of the proposed technique are described. Section 3 gives a detailed description of the control loop components and how they can be modeled. Stability criteria and transient behavior of the system, including a well chosen regulator, are theoretically derived. In Sect. 4, some details concerning the implementation of the proposed synchronization technique using the Action Media II system are explained. Section 5 gives an impression of a sample video phone application that was realized based on the proposed synchronization method.

2 Playout Synchronization

The key concern of this paper is playout synchronization, which is also termed intrastream synchronization. A continuous media stream consisting of media units is produced synchronously at the sending station and transmitted over an asynchronous packet network, where jitter is introduced. On the other side, one or more receivers have to play out the continuous media units in a synchronous way again. When no synchronization measures are taken, the clocks of sender and receiver will be slightly different and buffer overflows or starvations will occur during the transfer process. Uncontrolled buffering compensates jitter effects to a certain extent, but cannot compensate clock asynchronism over longer periods of time.

2.1 Classification of Playout Synchronization Solutions

Solution space for playout synchronization consists of three almost orthogonal design criteria with two main choices in each dimension. The first decision is, whether the systems have an explicit common understanding of time or not. In the former case, some kind of clock synchronization takes place. The presentation time of a media unit can be calculated from an absolute or relative time stamp carried with every unit. If no clock synchronization takes place, playout synchronization can be achieved based on buffer control mechanisms.

The second criterion is the location of synchronization actions. They can be performed either at the sender or the receiver of continuous media information. Sender control always requires some kind of feedback.

The third dimension distinguishes the methods that are used to correct asynchronism. This can be done by speeding up or slowing down presentation or production speed of media units, or by stuffing. The second method is well known from bit or byte synchronization and means deleting or inserting media units in our context.

2.2 Related Work

Escobar et al. [5,6] suggest a clock synchronization method. The transmitted media units contain time stamps that allow the receiver to determine presentation time. A very similar technique using the notion of a common LTS (logical time system) for several media streams, is introduced by Anderson et al. [2]. Corrective actions are performed by skipping and pausing. The mechanism is mainly applicable to single-site workstations.

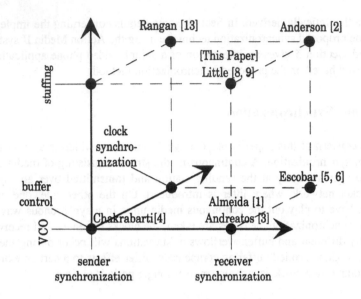

Fig. 1. Classification cube for intrastream synchronization and existing solutions

Another kind of clock synchronization used in conjunction with sender control, is proposed by Rangan et al. [13]. They assume less powerful receiver stations who feed back the number of the currently displayed media unit to the sender. There, corrective actions on each data stream can be taken, if necessary. To regain synchronization, media units are deleted or duplicated. Basic assumption here is a limited and known amount of delay jitter. This technique is mainly used as interstream synchronization for synchronous retrieval from networked file servers.

Chakrabarti et al. [4] suggest a solution, where the receiver clock drives the periodic, synchronous playout of media units. Depending on the amount of used buffer space in the receiver, the sender frame rate is controlled. As with other feedback methods, increasing network delay slows reactivity and stability cannot be proved in the case of unpredictable delay. The parameters of the control algorithm are derived experimentally.

PLL (phase locked loop) solutions are mainly known from bit synchronization. The basic principle is to compare a buffer level at the receiver to a nominal value. A loop filter forms the input voltage for a VCO (voltage controlled oscillator), which generates the buffer readout clock. Frequency usually drifts only by small amounts of some ppm. Almeida et al. [1] describe a PLL mechanism for bit synchronization of AAL services with synchronous timing relation. The byte level of the receiver FIFO is used to calculate the byte read clock and several loop filters are compared but not theoretically analyzed. A similar approach is taken by Andreatos et al. [3], who transmit scan lines of uncoded video signals over a 140 Mbit/s network with bounded delay jitter. A PLL is used to recover the line clock for the receiver. Control parameters of the PI controller are determined experimentally and neither stability nor dynamic behavior of the system is theoretically analyzed.

Little et al. [8,9] propose a receiver-based synchronization technique, which mainly realizes interstream synchronization, but some aspects are related to intra-stream synchronization. When the level of the receiver queue reaches a high or low threshold, frames are dropped or duplicated. Since the playout process is stated to be purely synchronous, stuffing is used instead of clock adjustment. Given that either sender or receiver clock speed is higher, the queue level will always tend to stay at one threshold, meaning either lower disturbance immunity or higher delay. Choosing the thresholds closer to each other leads to very frequent corrective actions and consequently errors. The correction function that controls frame drop and duplication is arbitrary, but only a constant rate-based function was investigated. The mechanism is based on the assumption of guaranteed network resources and has to react only on reservation violations or frame losses.

2.3 Proposed Synchronization Technique

The proposed synchronization mechanism is intended to act in an interconnected LAN and high-speed MAN environment, i.e. available bandwidth or delay cannot be guaranteed and transmission is based on best effort. Jitter effects are introduced by network transmission and by non-real-time operating systems in use. Figure 2 shows a typical distribution of frame arrival spacing at the receiver, that was measured on one LAN segment during normal traffic conditions. Every 33 ms, a frame containing multiplexed audio and video information is produced at the sending station, which explains the resulting peak. The slow reaction of the messaging system at the sender causes the process that reads out and transmits coded frames often to find two frames in the buffer. This explains peaks at very small frame distances around 5-10 ms and at 66 ms. The gaussian-like distortion of the peaks is caused by the operating system and mainly network jitter effects. Media units are transmitted without error correction and thus can be lost.

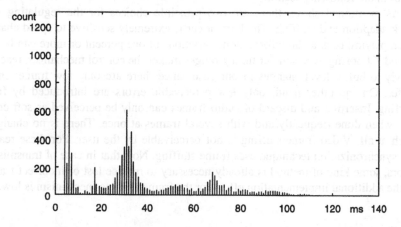

Fig. 2. Measured frequency distribution of the arrival spacing of received frames; frame rate 30 frames/s; transmitted over single ethernet segment.

We have chosen a receiver control that runs independently from the sender. This has the advantage that unpredictable delay cannot influence synchronization, and the method naturally extends to several receivers synchronizing independently to the same sender. Typical playback systems for continuous media and especially for video decompression with hardware support perform in a synchronous manner, i.e. after initialization frames are consumed at a fixed rate out of a buffer. The main purpose of the buffer is to compensate jitter effects introduced between frame production and delivery at the receiver. Because presentation time of the media units cannot easily be dictated and due to complexity of clock synchronization, our synchronization technique is based on controlling the level of the receiver frame buffer. The aim of the control is to keep the amount of media units in the buffer at a nominal value in the long term. The buffer control must not compensate short buffer fluctuations caused by jitter effects. In contrast to other solutions we count the amount of frames, not number of bytes, in the buffer. This is reasonable, since frames are produced synchronously but have variable frame sizes. Assuming constant delay, when the amount of frames in the buffer is held constant, sender and receiver are synchronized. Taking into account not only compressed frames but all media units in the receiver system up to the one currently being displayed, also compensates decompression jitter.

Since the mechanism should also be applicable to bidirectional communications, one of the main goals is to keep the delay caused by buffering and thus the buffer level as low as possible. On the other hand, network jitter dictates a certain amount of buffer for compensation. This contradiction suggests that deviations from an ideal buffer level to either side lead to disadvantages.

The main idea behind the proposed synchronization technique is the application of automatic control methods for buffer control. These allow the investigation of stability and transient analysis of the whole system. With these results the appropriate control algorithms and their parameters can be derived, as will be shown in detail in the following section.

As mentioned above, there are two principle choices for the regulating unit: clock adaption and stuffing. The human ear is extremely sensitive to speed changes when playing back audio information; variations of one percent or more can be perceived. Choosing even smaller tuning ranges makes the control mechanism react too slowly to buffer level changes in our case, since there are only few frames in the buffer. On the other hand, only few perceivable errors are introduced by frame stuffing. Insertion and discard of audio frames can only be perceived as soft crackling when done frequently and with several frames at once. There is no change in pitch at all. Video frame stuffing is not perceivable by the user. For these reasons our synchronization technique uses frame stuffing. Note that in case of transmission errors, some kind of method is already necessary to replace lost or incorrect frames, so the additional implementation effort for the frame stuffing mechanism is low.

3 Modeling the Control Loop

Due to the previously mentioned reasons, uncontrolled operation of the receiver leads to buffer overflow or starvation. The latter appears even more frequently in the case of uncompensated frame losses. These situations produce noticeable disturbances which have to be compensated by playout synchronization measures. As mentioned before, our solution is based on controlling the average level of the receiver frame buffer to a nominal value. This secures stable and fluent playout of the data and yields low delays.

In this section we describe a model for the control loop components that approximates the real playout system very close. A realistic model is essential to investigate system behavior and dimension the control components by application of automatic control engineering.

3.1 Components

Figure 3 shows the components of the control loop. The frame buffer is filled by incoming frames from the network. Due to jitter effects, the arrival spacing of the frames varies temporarily. However, the average is constant and is forced by the sender's clock. The buffer is emptied with a constant rate, driven by the receiver's playout clock.

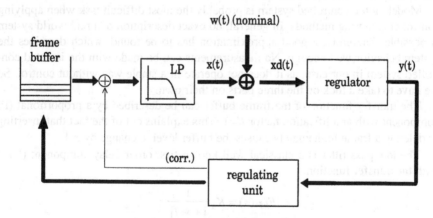

Fig. 3. Principle of the control loop

The level of the frame buffer is measured and filtered by a low pass component, before further processing is done. The buffer level includes compressed frames, frames currently being decompressed and already decompressed frames waiting to be displayed. This avoids introducing additional jitter due to variable decoding times.

Low pass filtering is necessary for two reasons. First it does an averaging of the measured buffer level since it passes only low frequent level changes. The sender's clock can only be derived from the average arrival spacing. High frequent changes caused by temporal delay and jitter must therefore be eliminated. The second reason for the filter is to keep the sampling theorem, which is important later when looking at the digital system.

A comparator subtracts the filtered buffer level $x(t)$ from the nominal value $w(t)$. Usually, $w(t)$ is a constant, ideal value. The calculated deviation $xd(t) = w(t) - x(t)$ becomes the input for the regulator which calculates a corresponding output value $y(t)$. The regulator is the only part which is not given by a system component but can be chosen in a more or less optimal manner, as we will see below. According to the chosen control function, the output value of the regulator gives the amount of deviation, which has to be corrected in order to adjust the buffer level to its nominal value. The actual corrective action is performed by the regulating unit by inserting or discarding frames. The shown correction output does not appear in the theoretical model but is an implementation detail described later.

The regulating unit, the frame buffer and the low pass filter can be grouped together to form the controlled system. The control loop therefore divides into two main sections: the controlled system and the regulator. To investigate the loop behavior, the transfer functions of both sections are needed. They are given in the complex variable domain, which has the advantage that the transfer function of the closed loop can easily be calculated using algebraic operations. Stability of the loop can be investigated in the complex domain and inverse Laplace transformation yields temporal behavior of the system.

Modeling the controlled system is probably the most difficult task when applying control engineering methods. In general, no exact description of a real world system is possible. Therefore, a good approximation has to be found, which describes the system behavior well enough. No measurements can be made with the isolated controlled system in our case, as it does not operate in a stable way without control. So we have to take a look on the three parts on their own.

The transfer function of the frame buffer can be described by a proportional (P) component with amplification factor $K=1$. This explains out of the fact that inserting or deleting a frame immediately causes the buffer level to change by ± 1.

The low pass filter is a classical well known first order delay component (PT_1) with the transfer function:

$$F_{LP}(s) = K_s \frac{1}{1 + s \cdot T_1} \; ;$$

K_s: low pass amplification, always set 1
T_1: low pass time constant

Finally, the regulating unit can also be modeled by a P component with $K=1$, if we assume that it can add or discard frames without delay. This assumption is true as long as the regulator dictates small changes of the buffer level in the area of a few frames (small-signal response), but becomes worse at high changes. As mentioned, the regulating unit acts by inserting or discarding frames to or from the data stream. Frames can only be discarded once when being received. To achieve symmetrical behavior and not to introduce visual or audible artifacts, doubling is done only once for each frame. Thus, corrective actions of more than one frame cannot take place immediately, but have to be taken in steps. This becomes even worse, if intercoded frames occur within the stream and not every frame can be doubled or discarded. Figure 4 shows the assumed response of the regulating unit to a step of height n_0 at time $t=0$. It is assumed that intraframes appear at fixed time intervals T_{SP}, which is a multiple of the frame time $T_F = 1/f_F$. Only one frame can be doubled every time interval T_{SP}, which results in the stepwise ascent until n_0 is reached. This behavior can be approximated in a worst case manner by a PT_1 component with time constant T_2. This leads to a PT_2 behavior of the whole controlled system.

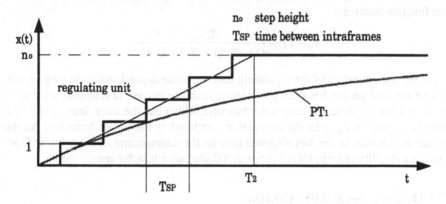

Fig. 4. Assumed step response of regulating unit and approximation with PT_1 component

3.2 Regulator Selection

With the knowledge about the controlled system, an appropriate regulator has to be found. We first assume that the contolled system has a first order delay (PT_1). This is a good approximation while the system is running and the buffer level changes within bounds of few frames (small-signal response at the operating point). To calculate stability in case of sudden high level changes as they can appear at connection setup or due to sudden network overload, the more exact model of a second order delay (PT_2) behavior is used for the controlled system (large-signal response).

After examination of several regulator types [7], an integrating (I) regulator turned out to be the best solution. It does not need a permanent deviation, but always regulates the actual value exactly to the nominal value. Further, the integrating behavior suppresses ripple of the input value and therefore supports the low pass filter in extracting the average filling level. Together with a PT_1 system, the I-regulator always yields stable system behavior. The transfer function of the I-regulator is:

$$F_r(s) = \frac{y(s)}{xd(s)} = \frac{K_i}{s} \quad ; \quad K_i: \text{ integrating constant [1/s].}$$

3.3 Reference Transfer Function

The reference transfer function describes the behavior of the input value $x(t)$ in response to changes of the nominal value $w(t)$. Note that all functions are given in the complex variable domain. For the given PT_1-I loop, the reference transfer function is:

$$F_w(s) = \frac{x(s)}{w(s)} = \frac{F_r(s) \cdot F_s(s)}{1 + F_r(s) \cdot F_s(s)} = \frac{K_i K_s}{K_i K_s + s + s^2 T_1} \; ;$$

In case of modeling the controlled system with an PT_2 behavior, the reference transfer function becomes:

$$F_{w2}(s) = \frac{K_i K_s}{s(1 + sT_1)(1 + sT_2) + K_i K_s} \; ;$$

T_2 is the time constant of the regulating unit. Its value depends on input step height, frame rate and period between intraframes. Stability in case of disturbances can be derived from the disturbance transfer function, which has the same denominator but consinsts only of $F_s(s)$ in the numerator compared to the above formulas. As the dynamic behavior of the loop depends only on the denominator of the transfer function, the stability criteria for reference and disturbance transfer are the same.

3.4 Damping and Stability Criteria

Small-Signal response. First we will analyze the system around the operating point with PT_1 behavior of the controlled system. In this case, the damping D of the loop can be derived from the coefficients in the denominator of the reference transfer function $F_w(s)$:

$$D = \frac{1}{2 \cdot \sqrt{K_i K_s T_1}} \; ;$$

The damping value describes the dynamic behavior of the system, the following cases can be distinguished:

1. $D > 1$: aperiodic behavior.
2. $D = 1$: aperiodic borderline case, quickest response without overshoot.
3. $0 < D < 1$: damped oscillation.
4. $D = 0$: constant oscillation.
5. $D < 0$: rising amplitude oscillation.

As K_i, K_s and T_1 are positive values, it is evident that D is always positive and the system is stable. K_s and T_1 are given by system components, whereas the integrating constant K_i can be chosen to dictate loop behavior.

Large-Signal Response. With the controlled system being modeled as a PT_2 component, the transfer function of the closed loop becomes more complex. One way to investigate stability in this case is using the Hurwitz criteria, which constrains:

1. All coefficients a_0 to a_n of the denominator of the reference transfer function have to be greater zero.
2. The determinant derived from a_0 to a_n and its subdeterminants must be greater zero.
3. If these constraints are met, the system is stable.

Application of these rules leads to the following stability criterion:

$$K_i K_s < \frac{1}{T_1} + \frac{1}{T_2}$$

In this relation, K_s is chosen to 1 (low pass amplification) and T_2 depends on the step height and the interval between intraframes. As $1/T_2$ is always positive, a secure stability criterion is at least $K_i < 1/T_1$.

4 Implementation

Finally, the modeled system is realized in a digital environment. The low pass filter, the regulator and the regulating unit are realized in software. The buffer level is an integer value which must be scanned periodically and can only vary in steps of whole frames. Besides the extraction of the buffer level average, the low pass filter has to act as an anti aliasing filter for the following control loop components. We chose a scan frequency of 4 Hz at which the buffer is scanned and all calculations are done. The sampling theorem therefore forces a filter cut off frequency of 2 Hz, which leads to the value of $T_1 = 0.5$ s. The low scan frequency is sufficient as we want an average regulation of the buffer level. Further, it does not afford much performance at the host system.

The low pass filter was formed by a 2nd degree FIR (finite impulse response) filter which is easy to implement and has stable behavior. Its output at instant k is:

$$y_{LP}(k) = a_0 [x(k) + x(k-2)] + a_1 x(k-1);$$

The constants a_0 and a_1 have been calculated by a filter software [10] and rounded to 0.25 respectively 0.5. The integrating behavior of the regulator is approximated by a trapezoid calculation. Its output at instant k is calculated by:

$$y(k) = x_d(k-1) + Ki \cdot T_a \cdot \frac{x_d(k) + x_d(k-1)}{2} \; ; \; T_a\text{: scan period (250 ms).}$$

The regulating unit finally decides, whether and how much frames need to be discarded or added. But some care must be taken to implement the control loop right. As can be seen in Fig. 3, the regulating unit offers a correction output whose value is added to the scanned buffer level value before it is fed into the low pass filter.

This is necessary because only whole frames can be added or discarded. In case the integrator 'wants' e.g. only a half frame correction, this is done by the correction output of the regulating unit, which simulates a level change as long as it is below a whole frame. Without this measure the regulator would always integrate up and down periodically with the amplitude of one frame. This process takes place within some seconds, which is not critical for the loop, but would introduce additional and obviously unnecessary disturbances.

Figure 5 shows the theoretically derived step response of the closed control loop for two different damping values compared to the measured response of the real system. The regulating unit was modeled as a PT_1 component. Note that jitter effects present in the measured curves produce statistical deviations.

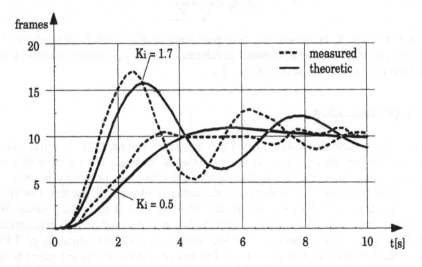

Fig. 5. Step response of the pT_2-I control loop and the real system to a step of 10 frames

The curves with integrating constant $K_i = 0.5$ are near the aperiodic borderline case and show almost no overshoot. The response of the real system is slightly quicker than theoretic behavior. This can be explained by the approximation of the stepwise ascent by a PT_1 component, which increases slower (see Fig. 4). With $K_i = 1.7$, the system comes closer to the stability bound. In this area, damped oscillation occurs. Again, the real system has a faster response and hence a higher oscillation frequency. The oscillation amplitude is reproduced almost exact. We also took measurements with higher K_i values and could verify the stability bound. For values with $K_i > 1/T_1 = 2$ [1/s], the system proceeds to oscillations increasing in amplitude. These results justify the choice of the PT_1 approximation for the regulating unit as basis for derivation of stability and dynamic behavior.

5 Application Example - Video Phone

The described control loop based playout synchronization is used in an application for bidirectional audio and video communication [11]. The video phone application was implemented under the operating system OS/2. Taking advantage of its multi-tasking capabilities, task priorities and graphical user interface, it is possible to run other applications in parallel. So, communication capabilities can easily be extended by e.g. data transfer.

For encoding, decoding and display of audio and video information, the IBM Action Media II system (DVI) is used. It offers the possibility to concurrently compress and decompress audio and full motion video with a frame rate up to 30 frames/s. For video coding, the real time video (RTV) algorithm is used, which offers a full color image resolution of 128x120 pixel, when doing coding and decoding in parallel. The algorithm uses both inter- and intraframe coding. The resulting data rate depends on the choice of several coding parameters and is between approximately 350 kbit/s and 1.2 Mbit/s in each direction. Audio coding is done either with a PCM or ADPCM algorithm and yields quality comparable to digital telephony.

For transport of the real-time audio and video information, the user datagram protocol (UDP) is used. As it is an unsecure datagram protocol, measures against frame corruption, loss and reordering have to be taken by the application program. Signaling for call management is separated from data transport and uses a special protocol. Besides call setup and release, parameters of the coding algorithms in both directions can be negotiated. This allows the user to change video or audio quality and respectively data rate even during an existing communication.

In the current implementation, the user can manually change the control loop parameters, i.e. the nominal buffer level and the regulator parameter K_i. Future investigations will include automatic adaption of these parameters to the current network situation, which can e.g. be estimated by online measurements.

At this point, we will sum up some experiences with the video phone in general and under different network conditions [7]. The control loop synchronizes the receiver within a few seconds at connection setup and keeps it stable during the communication. In usual LAN environments, corrective actions of the regulator are very seldom with intervals of some tens of seconds. In most cases the insertion and discard of frames cannot be perceived. Only in the case of low frame rates or high frame losses, tolerable crackle noises can be heard. Image artifacts can occur when losses or errors take place before intercoded frames. The round trip delay, which mainly depends on the nominal buffer level value, can be kept as low as 450ms with a frame rate of 30 frames/s. This still enables a comfortable communication.

We experienced with network loads on ethernet up to 60%, whereby only the jitter increased but the connection could be kept stable with approximately 1 frame loss per second. For testing the application over longer distances, a data mirror was installed at the remote end of a MAN, which just returned received frames to the sender. This is an even worse situation compared to having another video phone station at the remote end, because every media unit suffers twice the single transmission delay. Over a single distance of approximately 20 km inside the urban network, communication was possible without any noticeable disturbances. Only absolute delay and jitter increased, the latter could be compensated by choosing a slightly higher buffer level. Reasonable communication between two different MANs over about 250 km (Munich and Stuttgart) was only possible in non-busy hours. It should be noticed, that MAN interconnection is realized over a 2 Mbit/s link and that it is part of the german universities network, that extensively carries data traffic.

6 Conclusion

We described a control loop based model of an intrastream synchronization mechanism, which is not based on a specific system environment and may be used on multimedia playout devices working with frame oriented data. The synchronization mechanism is independent from the sender and based on regulating the level of the receiver frame buffer. Correction is done on frame level by inserting and discarding frames. Additional to the normal receiver synchronization, the control mechanism keeps the buffer level on a constant average level. This generates a small delay variance of the video and audio frames and enables a small amount of buffered frames, an important constraint within a bidirectional communication environment. In order to predict system behavior, realistic models of the control loop components were given. After choosing an appropriate control algorithm, stability and dynamic properties of the system have been derived. The results give good correspondance to practical measurements. Finally, the implementation of the synchronization method using a specific multimedia system has been described and experiences were reported.

References

[1]   N. Almeida, J. Cabral, A.Alves: End-to-End Synchronization in Packet Switched Networks. In: Proc. of the 2nd Int. Workshop on Network and Operating System Support for Digital Audio and Video, 1991. Lecture Notes in Computer Science, Springer Verlag, Vol.614, 1992, pp.84-93.

[2]   D.P. Anderson, G. Homsy: A Continuous Media I/O Server and Its Synchronization Mechanism. IEEE Computer Magazine, Vol.24, No.10, 1991, pp.51-57.

[3]   A.S. Andreatos E.N. Protonotarios: Receiver synchronization of a packet video communication system. Computer Communications, Vol.17, No.6, June 1994, pp.387-395.

[4]   S. Chakrabarti, R. Wang: Adaptive Control for Packet Video. Proc. of Int. Conference on Mulitmedia Computing, May 1994, pp.56-62.

[5]   J. Escobar, D. Deutsch, C. Partridge: A Multi-Service Flow Synchronization Protocol. BBN Systems and Technologies Division, 1991, pp.1-16.

[6]   J. Escobar, D. Deutsch, C. Partridge: Flow Synchronization Protocol. Proc. of the IEEE Globecom, Vol.3, 1992, pp.1381-1387.

[7]   D. Köhler. Optimierung und Weiterentwicklung einer Bildtelefonanwendung unter OS/2. Diplomarbeit am Lehrstuhl für Kommunikationsnetze, Technische Universität München, Munich, Germany, 1994.

[8]   T.D.C. Little, F. Kao. An Intermedia Skew Control System for Multimedia Data Presentation. In: Proc. of the 3rd Int. Workshop on Network and Operating System Support for Digital Audio and Video, 1992. Lecture Notes in Computer Science, Springer Verlag, Vol.712, 1993, pp.130-141.

[9]   T.D.C. Little: A framework for synchronous delivery of time-dependent multimedia data. Multimedia Systems, Vol.1, No.2, 1993, pp.87-94.

[10]   O. Mildenberger: Entwurf Analoger und Digitaler Filter. Vieweg Verlag, 1992.

[11]   H. Müller: Bewegtbildkommunikation über LAN und MAN. In: Fokus Praxis, Information und Kommunikation, Bd. 5, Verteilte Multimedia-Systeme. Hrsg.: W.Effelsberg u.a. München, Saur Verlag, 1993, pp.126-141.

[12]   S. Ramanathan, P.V. Rangan: Continuous Media Synchronization in Distributed Multimedia Systems. In: Proc. of the 3rd Int. Workshop on Network and Operating System Support for Digital Audio and Video, 1992. Lecture Notes in Computer Science, Springer Verlag, Vol.712, 1993, pp.328-335.

[13]   P.V. Rangan, S. Ramanathan, et al.: Techniques for Multimedia Synchronization in Network File Systems. Computer Communications, Vol.16, No.3, March 1993.

eXtended Color Cell Compression –
A Runtime-efficient Compression Scheme
for Software Video

Bernd Lamparter and Wolfgang Effelsberg

Praktische Informatik IV
University of Mannheim
68131 Mannheim, Germany
{lamparter, effelsberg}@pi4.informatik.uni-mannheim.de

Abstract. Multimedia applications require a compression and decompression scheme for digital video. The standardized and widely used techniques JPEG and MPEG provide very good compression ratios, but are computationally quite complex and demanding. We propose to use an extension to the much simpler Color Cell Compression scheme as an alternative. Our extension includes the use of variable block sizes, the reuse of color index values from previously encoded blocks, and Huffman encoding of the stream of blocks. We present experimental results showing that our scheme provides much better runtime performance than MPEG, at the cost of a slightly inferior compression ratio. It is thus especially suited for software videos in high-speed networks.
Keywords: multimedia, movie compression, block encoding, software video.

1 Introduction

The standardized compression techniques JPEG [12] for still images and MPEG [3] for motion pictures both include a Discrete Cosine Transform (DCT). This is a complex and computationally very demanding mathematical function. As a consequence, software motion pictures based on JPEG or MPEG are slow, even on the most powerful CPUs available today, and it is generally assumed that these compression schemes will only work well with special hardware. However, special hardware makes a movie system much less flexible and portable. It is thus desirable to develop alternative compression/decompression algorithms for movies which are optimized for computation in software on general purpose CPUs.

We propose an extension to the Color Cell Compression scheme for use in multimedia workstations. After a short introduction into Block Truncation Coding for monochrome images and Color Cell Compression for color images, we describe our eXtended Color Cell Compression (XCCC) scheme in detail. We have implemented XCCC and present experimental results on runtime performance and compression ratios.

2 Block Truncation Coding and simple Color Cell Compression

Our eXtended Color Cell Compression (XCCC) algorithm belongs to the family of block compression algorithms. Earlier examples from this family include Block Truncation Coding (BTC) and Color Cell Compression (CCC), brief descriptions of which preface the discussion of our algorithm.

2.1 Block Truncation Coding (BTC)

The Block Truncation Coding Algorithm [2] is used in the compression of monochrome images. When compressing color images, it can be applied separately to the three color channels.

The first step of the algorithm is the decomposition of the whole image into blocks of size $n \times m$ pixels. Usually these blocks are quadratic with $n = m = 4$. For each block P the mean value μ and the standard deviation σ is computed:

$$\mu = \frac{1}{nm} \sum_{i=1}^{n} \sum_{j=1}^{m} P_{i,j}$$

$$\sigma = \sqrt{\frac{1}{nm} \sum_{i=1}^{n} \sum_{j=1}^{m} P_{i,j}^2 - \mu^2}$$

where $P_{i,j}$ is the brightness of the pixel.

In addition a bit array of size $n \times m$ is calculated for each block. A one in this bit array indicates that the gray value of the corresponding pixel is greater than the mean value, a zero indicates that the value is smaller than the mean value:

$$B_{i,j} = \begin{cases} 1 & \text{if } P_{i,j} \geq \mu \\ 0 & \text{else} \end{cases}$$

The decompression algorithm knows out of the bit array whether the pixel is darker or brighter than the average. Last we need the two gray scale values for the darker and for the brighter pixels. These values (a und b) are calculated with the help of the mean value and the standard deviation, and are then stored together with the bit array:

$$a = \mu + \sigma \sqrt{p/q}$$
$$b = \mu - \sigma \sqrt{q/p}$$

Here p and q are the number of the pixels having a larger resp. smaller brightness than the mean value of the block.

During the decompression phase each block of pixels is calculated as follows:

$$P'_{i,j} = \begin{cases} a & \text{if } B_{i,j} = 1 \\ b & \text{else} \end{cases}$$

e. g. where the bit array shows a 1, the gray value a is used, where it shows a 0 the value b is used.

If the original image used one byte per pixel, we had a storage requirement of 128 bits for each 4×4 block. The compressed block can be stored with 16 bits for the bit array plus one byte for each of the values a und b. Hence we have a storage reduction from eight bits to two bits per pixel.

This basic version of BTC can be improved with a number of tricks [9]. Additionally, [10] describes a hierarchical version of BTC.

2.2 Color Cell Compression (CCC)

If BTC is to be used for color images rather than for gray scales, the components (red, green, and blue, resp. chrominance and luminance) may be compressed separately. However, the CCC method promises a much better compression rate [1].

Similar to BTC, the image is divided into blocks called "color cells". The two values a and b are now indices into a color lookup table (CLUT). The criterion for the bit array values is now the brightness of the corresponding pixel. The brightness of a pixel is computed in the following way, taking the human reception into account:

$$Y = 0.3 P_{red} + 0.59 P_{green} + 0.11 P_{blue}$$

The mean value of each block can now be computed out of these brightness values (analogous to the BTC method).

Let us define $P_{red,i,j}$ as the red component of $P_{i,j}$, $P_{green,i,j}$ the green component and $P_{blue,i,j}$ the blue component. The next step is then to compute the color values a_{red}, a_{green}, a_{blue} as well as b_{red}, b_{green}, b_{blue}:

$$a_c = \frac{1}{q} \sum_{Y_{i,j} \geq \mu} P_{c,i,j} \quad \text{bzw.} \quad b_c = \frac{1}{p} \sum_{Y_{i,j} < \mu} P_{c,i,j} \quad \text{with } c = \text{red, green, blue}$$

Again p and q are the number of pixels with a brightness larger resp. smaller than the mean value. The bit array is computed as for BTC.

The color values $a = (a_{red}, a_{green}, a_{blue})$ und $b = (b_{red}, b_{green}, b_{blue})$ are now quantized onto a color lookup table. In this way we get the values a' und b'. These values are stored together with the bit array (Fig 2.2).

The decompression algorithm works analogous to the BTC method:

$$P'_{i,j} = \begin{cases} \text{CLUT}[a'] & \text{if } B_{i,j} = 1 \\ \text{CLUT}[b'] & \text{else} \end{cases}$$

(CLUT is the color lookup table.)

The two values a' und b' can each be stored in one byte if the CLUT has 256 entries. Hence the storage needed for one block of size 4×4 is two bits per pixel as with the BTC (to be more exact we would have to add the storage needed by the CLUT (256×3 Bytes for the full image)).

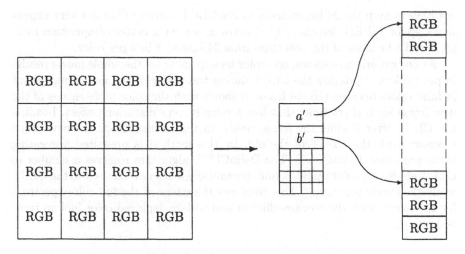

Fig. 1. Red-Green-Blue block and its CCC encoding

Color Cell Compression is not only one of the best compression algorithms, it is also one of the fastest [9]. All calculations can be done without floating point operations, and the asymptotic complexity is $O(N \cdot M \cdot (1 + \frac{\log k}{n \cdot m}))$ (image size $N \times M$, size of the block $n \times m$, size of the CLUT k). The decompression is also done without floating point operations with a complexity of $O(N \cdot M)$.

As in the case of the BTC algorithm, a number of possible improvements exist here as well [9]:

- If the two colors a and b are nearly equal, or one color dominates in frequency of occurrence, only one color is stored, and no bit array is needed.
- If an image contains large areas with only small differences in color, those areas may be encoded with larger blocks.
- For movies cuboids may be used, with time being the third dimension, if the changes from frame to frame are small enough.

Our algorithm "XCCC" is based on the second idea. It uses 4×4, 8×8, and 16×16 blocks. Larger blocks, tested in an earlier version, yielded no further improvement.

2.3 The DeltaCLUT-Technique

CCC (and XCCC) both use a color lookup table for storage-efficient color representation. The color lookup table typically has 256 entries, with the red, green, and blue components stored in one byte each. The color value of a pixel is then encoded as an index into the color lookup table, with one byte per pixel. Most of the color display adapters today use the color lookup table. The decompressed image already consists of index entries into the CLUT, hence the decoder does not have to map the 24 bit colors onto 256 CLUT entries (That is a very expensive step for MPEG decoders [8]). Moreover, we get a better compression ratio for a and b because of the reduction from 24 bits to 8 bits per color.

As our experiences shows, one color lookup table for the whole movie results in poor colors. Updating the CLUT during the replay of the movie may result in false colors because the old frame is shown with the color table entries of the new frame for a short time; that has a visually very disturbing effect. Loading the CLUT after loading the frame results in the reverse effect: the new frame is shown with the CLUT of the old. In [7] a method is presented preventing these problems: DeltaCLUT. The DeltaCLUT algorithm reserves a number of CLUT entries for color updates, and dynamically loads new colors into the CLUT while the movie is running. This combines the usage of the full color spectrum for the movie with the storage-efficient and widely deployed color lookup table technology.

3 XCCC: Extensions to CCC

Our XCCC scheme extends CCC in three steps in order to improve compression ratio and runtime performance.

3.1 First step: Adaptive block sizes

In many cases a image has large areas with small differences in colors (i. e. in the background). These areas can be coded with fewer bits.

We first investigate the optional use of large rectangles. If an image contains large areas with few colors, these areas can be compressed with larger rectangles [11].

In XCCC the images are first decomposed into large blocks (16×16) and, if necessary, these blocks are then subdivided. The algorithm for each block B is:

1. Calculate the CCC coding of the block
2. If the actual block has the minimal block size, then Done.
3. Calculate the mean difference Δe of the original pixel values and the values coded with CCC: $\Delta e = \sum_B |p - p'|_2$, where p is the pixel value, p' is the value of the same pixel after decompression.
4. If Δe is smaller than a given constant, then Done.
5. Divide the block into four subblocks and use the algorithm recursively for each of these blocks.

The data for simple CCC could be arranged in the data stream without any structuring information. But the output of the extended algorithm is a quadtree with color cells for each 16×16-block. Hence, we have to store a more complex data structure. This is done by adding a tag for each block. Figure 2 shows an example of the coding of an XCCC block. The tag is the logarithm of the length of the edge of the coded block. Blocks of minimal size need only one tag for four blocks because one minimal block is always followed by three more. After the tag we store the indices into the CLUT (a and b) and then the bit array. If the length or width of the image is not divisible by 16, we divide the residual rectangle into 4×4 blocks and encode some of these subblocks as rectangles.

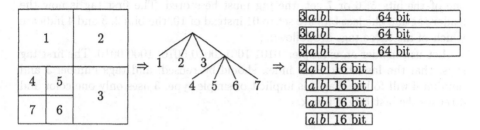

Fig. 2. Coding tree of the XCCC-algorithm

The use of adaptive block sizes introduces a small additional overhead for compression and results in much more efficient decompression for most images.

3.2 Second step: Single Color and Color Reuse

If an image has a large area with only one color, it is not necessary to store the bit array at all. XCCC does not store a bit array if the two color indices a and b are equal. There are two ways to let the decoder know that there is no bit array: First the encoder can store the two colors and the decoder will know from the equality that there is no bit array. Second, the encoder can use a special tag and store only one color. XCCC uses the second method.

If we implement bit array suppression for single color blocks, larger squares will not always improve the compression ratio. Instead of a 32×32 square, XCCC may use four 16×16 squares. But some of these squares will have no bit array and hence the total compression ratio may be better. Though this can also happen with smaller blocks, our experience shows that this is rarely the case.

Colors in the neighborhood are often equal in images. Because of this fact, we can sometimes reuse colors from the block coded previously. Color reuse is also stored in the tag.

For each tag we use one byte with the following bit encoding:

```
                    7 6 5 4 3 2 1 0
    4 × 4-block     - - - - - - 1 0
    8 × 8-block     - - - - - - 1 1
    16 × 16-block   - - - - - - 0 0
    single color    - - 1 - - - - -
    last dark       - 1 - - - - - -
    last bright     1 - - - - - - -
```

The bit for "last dark" indicates that the color value b for the dark pixels should be reused from the previous block, "last bright" indicates the usage of value a from the previous block.

In the first step the encoder used only one tag for four of the smallest blocks. But now the tags of these four blocks are possibly different. If one of them has one of the bits 5, 6 or 7 set, the tag must be stored. The first tag is now the leader tag and the last bits are set to 01 instead of 10. the bits 2, 3 and 4 indicate which of the three tags are following.

Let us consider an example: 0101 1001 0010 0010 1000 0010. The first tag says, that the last dark color index should be reused, and tags number 3 and number 4 will follow. Tag 2 is implicit of simple type, 3 uses only one color and 4 reuses the last bright index.

3.3 Third step: Further improvements

Some blocks are encoded in only one ore two bytes, namely the blocks without bit array. If one of the four subblocks of a 16×16 block is a single color block, then it is cheaper in memory to store the four subblocks instead of the 16×16 block. Some are even stored in one byte only, namely those where a or b are taken from the previous block.

The remaining redundancy in the bit stream could be further compressed with Huffman codes [4]:

- The bit stream consist of three parts: The tags, the colors and the bit arrays. All three parts still have redundancy in the stream.
- Some of the tags are used very often, others never or very seldom. The usage of a Huffman code will reduce the total size of the tags to 50%.
- The colors can be compressed only by about 10% with a Huffman code because of the usage of a color lookup table. This table is chosen in a way so that all colors are used about equally often. Hence only a small redundancy of about 10% remains.
- A large redundancy is in the bit arrays, especially in the bit arrays of the 4×4-blocks. The redundancy can be as large as 60%.

Hence compression with three different Huffman codes would divide the size into half, but slow down the decompression.

From frame to frame colors are usually changed infrequently. So we could use a second "same color" flag to signal the reuse of the corresponding color in the last image.

A third improvement has been implemented: All blocks are test-wise subdivided into 4×4 blocks. Step 4 of the XCCC algorithms is then changed to the following:

4. If Δe is smaller than a given constant, and encoding of the actual block is better than encoding of the subdivided block, then Done.

188

4 Experiences: Compression ratio and decompression speed

The main goal of the XCCC scheme is to allow rapid decompression with software decoders. Table 1 shows the decompression speed in images per second. The measurements of MPEG [3] were done with the MPEG player of the University of California at Berkeley [8]. We used three movies: The first and second movie (butterfly) is a raytraced sequence of 350 frames sized 320 × 256 and 780 × 576 resp. The third movie was digitized from an analog video showing the University of Mannheim. It consists of scenes of buildings of the university (a palace) and other scenes depicting university life. Due to the analog origin, this movie consists of many different colors and color shadings. It is 320 × 256 in size and has 2000 frames.

The decompression was done on a DEC/alpha workstation with a 133Mhz CPU. The speeds are the real speeds viewed on the screen. The display adapter uses a color lookup table with 8 bits per pixel. XCCC uses the same technique internally thus requiring no conversion. In contrast, MPEG uses full color internally. Hence, the MPEG player has to dither in real-time. The player has several built in dithering methods. For the tests we used the fastest color dithering available ("ordered dithering").

Movie	Size (pixels^2)	MPEG (frames/s)	XCCC (frames/s)
Butterfly	320 × 240	10.5	42
Butterfly	780 × 576	2.1	6.8
Castle	320 × 240	7.8	24

Table 1. Decompression speed of software decoders (in frames/s)

Movie	Size	MPEG	JPEG	XCCC
Butterfly	320 × 240	0.80%≙0.19bpp	2.49%≙0.60bpp	3.0%≙0.72bpp
Butterfly	780 × 576	0.53%≙0.13bpp	1.54%≙0.37bpp	1.83%≙0.44bpp
Castle	320 × 240	1.5%≙0.36bpp	5.9%≙1.42bpp	6.3%≙1.51bpp

Table 2. Compression ratios (compressed size/full color size and bits per pixel)

The compression speed was measured on a DEC5000/133. For the small butterfly movie we got about 6.5 seconds per image with MPEG. Before XCCC can be started, the color lookup tables has to be computed with DeltaCLUT. This step needs about 7 seconds per image. XCCC then needs about 2 seconds per image. Together our compressor uses about 10 seconds per image.

Our experiments show that XCCC can decompress images very fast. The quality of the XCCC compressed images is comparable to the quality of MPEG compressed images. The great advantage of MPEG is the bit rate of the compressed movie. The size of XCCC compressed movies is about three to four times larger than the size of MPEG movies. Hence the domain of XCCC are local area networks with color workstations using the color lookup table technique. In this environment XCCC performs significantly better than MPEG.

5 Conclusions and Outlook

We have presented XCCC, an algorithm to decompress and play digital movies on standard color workstations at a reasonable speed without special hardware for the decompression. We have shown, that our algorithm is much faster in decompression than MPEG when implemented in software. On the other hand, MPEG gives a better compression ratio.

The next step will be experiments with Huffman tables for the three parts of the compressed data streams (tags, colors and bit arrays). We expect better compression, but we will have to pay the price of slower decompression.

The XCCC decompressor has been integrated into the XMovie system [6, 5], a test bed for the transmission and display of digital movies developed at the University of Mannheim. It is based entirely on standard hardware, and uses standard network technology and standard graphics adapters with color lookup tables.

References

1. G. Campbell, T. A. DeFanti, J. Frederikson, S. A. Joyce, A. L. Lawrence, J. A. Lindberg, and D. J. Sandin. Two Bit/Pixel Full Color Encoding. *Computer Graphics*, 1986.
2. E. J. Delp and O. R. Mitchell. Image Compression using Block Truncation Coding. *IEEE Transactions on Communications*, 1979.
3. D. Le Gall. MPEG: A Video Compression Standard for Multimedia Applications. *Communications of the ACM*, 34(4):46–58, 1991.
4. D.A. Huffman. A method for the construction of minimum reduncancy codes. *Proceedings IRE*, 40:1098–1101, 1962.
5. R. Keller, W. Effelsberg, and B. Lamparter. Performance Bottlenecks in Digital Movie Systems. In D. Shepherd, editor, *4th International Workshop on Network and Operating System Support for Digital Audio and Video, Lancaster, November 1993*, pages 163–174, 1993.
6. B. Lamparter and W. Effelsberg. X-MOVIE: Transmission and Presentation of Digital Movies under X. In R. G. Herrtwich, editor, *2nd International Workshop on Network and Operating System Support for Digital Audio and Video, Heidelberg, November 1991*, volume 614 of *Lecture Notes in Computer Science*, pages 328–339. Springer-Verlag Berlin Heidelberg, 1992.

7. B. Lamparter, W. Effelsberg, and N. Michl. MTP: A Movie Transmission Protocol for Multimedia Applications. In *Multimedia92, 4th IEEE ComSoc International Workshop on Multimedia Communications, Monterey, California*, pages 260–270, April 1992.

8. K. Patel, B. C. Smith, and L. A. Rowe. Performance of a Software MPEG Video Decoder. In P. Venkat Rangan, editor, *Proceedings of ACM Multimedia 93*, pages 75–82. Addison-Wesley, Aug 1993.

9. M. Pins. *Analysis and choice of algorithms for data compression with special remark on images and movies (In German)*. PhD thesis, University of Karlsruhe, Germany, 1990.

10. J. U. Roy and N. M. Nasrabadi. Hierarchical Block Truncation Coding. *Optical Engineering*, 30(5):551–556, May 1991.

11. A. Urban. *ECCC - Implementation of extensions to the Color-Cell-Compression (In German)*, 1993.

12. G. K. Wallace. The JPEG Still Picture Compression Standard. *Communications of the ACM*, 34(4):31–44, April 1991.

QoS Adaptation and Flow Filtering in ATM Networks

Nicholas Yeadon , Francisco García, Andrew Campbell and David Hutchison

Department of Computing,
Lancaster University.
Lancaster LA1 4YR, U.K.
E.mail: mpg@comp.lancs.ac.uk

1. Introduction

The integration of multimedia systems which handle text, graphics, digital audio and video together with other media types is being encouraged by the development of international compression standards [JPEG 92, H.261 90, MPEG-I 93, MPEG-II 93]. Graphics, still images, high quality digital audio, and moving pictures in their uncompressed forms consume digital storage and network resources which far exceed likely availability in the near future. Compression can considerably support developing multimedia applications and will be a prerequisite in future multimedia systems engineering. Together with technological developments in high speed networks and multimedia workstations, new classes of distributed applications such as distance learning, desktop video-conferencing and video on demand are now possible. These applications place new and diverse Quality of Service (QoS) requirements on operating systems support, communication protocols and networks. The task of meeting these diverse QoS requirements is being assisted by the use of compression.

Our previous work in the area of QoS support for distributed multimedia applications has concentrated on resource management strategies for an extended Chorus micro-kernel [Robin, 94], enhanced transport services and protocols [Campbell, 93], and a Quality of Service Architecture (QoS-A) [Campbell, 94] which proposes a framework to specify and implement the required performance properties of multimedia applications over ATM networks.

The research reported in this paper is motivated by the need to provide flexible *flow management* support for a wide variety of continuous media applications in heterogeneous environments. The Lancaster environment consists of PCs, workstations and specialised multimedia enhanced devices connected by ATM, Ethernet, mobile, and proprietary high-speed networks [Lunn, 94]. The work outlined in this paper describes how flow management in a QoS-A may be engineered over an ATM network.

Two general aspects of flow management which will be addressed are *QoS adaptation* and *flow filtering*. QoS adaptation relates to the monitoring and adjustment of flows at the edge of the network to ensure that the user and provider QoS agreed on is maintained. Filtering operations are applied to codec generated data where structural composition of media traffic can be exploited and adapted in order to meet application, end-system or network capabilities and characteristics. Again, these filtering operations are instantiated and performed at the edge of the network.

2. Motivation

Heterogeneity issues are present in both end-systems and networks. The flow management design will attempt to meet diverse QoS requirements through the utilisation of services and mechanisms which will help bridge the heterogeneity gap. The following QoS related issues motivate our research:

- The range of multimedia applications and user requirements is likely to be quite diverse. For example, multimedia conferencing may require only low resolution video but high quality sound. Industrial applications, involving for example the output from highly specialised monitoring devices such as microscopes, will require very high image resolution. Generally, the perception of video and audio quality is user-dependent and hence users may express different requirements in playout qualities. This will be encompassed in the specification of distinct QoS requirements by disparate users.

- Considering end-system hardware, heterogeneity is present in: CPUs, I/O devices, storage capabilities, compression support (dedicated boards / software), internal inter-connect architecture, communication protocol support, network interfaces, etc. These issues place limits on the end-system's capabilities to process, consume and generate multimedia data.

- End-systems are likely to be connected to different networks which not only have different bandwidth capabilities but also varying access delay characteristics. For example: medium access control mechanism, maximum and minimum data unit size, service types, packet loss rates, propagation delays, congestion, etc.

While in general our research will address all three issues outlined above, in this paper we focus on how the flow management model will attempt to meet application and end-system heterogeneity over ATM networks.

3. Flow Management Model

The flow management model consists of services and mechanisms for QoS management and control of continuous media flows in end-systems and multiservice networks. The most fundamental architectural concept we use is the notion of a *flow* [Partridge, 92]. A flow characterises the production, transmission and eventual consumption of a single media stream as an integrated activity governed by a single statement of QoS. Flows are always simplex but can be either unicast or multicast.

The flow management model incorporates three QoS policies: (i) a resource selection policy for resource allocation and sharing; (ii) a media filtering policy to perform encoding conversion and scaling; and (iii) QoS adaptation policy to alter encoding characteristics at source encoders. These policies are expanded upon in the following sections.

3.1 Resource Selection Policy

In [Zhang, 93] a number of reservations styles describe how individual reservation requests are aggregated inside the network and shared by multicast groups. The flow management protocol interacts with group management and the routing protocol to provide receiver specified resource allocation. The flow management protocol commits end-system and network resources using a *flow-spec* [Campbell, 94]. Essentially the flow-spec provides a way of conveying stream characteristics and user requirements to network and end-system resource management protocols. The QoS-A flow-spec contains the following quantitative QoS parameters: user data unit size, user data unit rate, end-to-end delay, end-to-end jitter. These parameters are also given qualitative meaning by the use of a commitment clause which identifies the required level of service commitment for each of the above parameters. The levels of commitment considered in the QoS-A include: best effort, statistical and guaranteed. Other qualitative parameters include the identification of encoding type (i.e. MPEG-I, JPEG, H.261 etc.), media type of sub-stream, overall flow-id, sub-stream-id etc.

3.2 Media Filtering Policy

Filtering policies will be applied to translate one encoding scheme to another. In [Pasquale, 93] filtering is introduced as a general method for supporting heterogeneous receivers for multicast communications. Pasquale filters can be used to aggregate resources inside the network to either discard or enhance media, and propagate along the data path to an optimum point where the committed network resources are minimised.

Our approach is to perform all filtering at the network edges such that the network has no knowledge that filtering is taking place. Translation filtering services will be provided by *filter servers* ideally located within the local environment. These servers may perform conversions on behalf of the various receivers. Another option is to download the filter translation code to receivers capable of performing this functionality on the local host and hence eliminating the path through the filter server.

Other filtering policies will be applied to hierarchical encoding schemes. Such schemes generally provide a *base layer* and one or more *enhancement layers*. The base layer provides a subset of the full bit stream which can be employed for the reconstruction of the lowest resolution or quality signal. The enhancement layers will provide value added quality which will ultimately depend on the receiver and network capabilities. Scalability tools offered through certain compression standards (e.g. MPEG-II) allow data partitioning, Signal to Noise Ratio (SNR), spatial and temporal scalability.

3.3 QoS Adaptation Policy

Many continuous media applications can tolerate small variations in the QoS delivered by the network without any major disruption to the user's perceived service. In some cases quite severe service fluctuations can be accommodated. In such cases, however, it is often appropriate to inform the application of the service degradation so that it can take a suitable action. For example, codecs may allow for changes in sampling rates, quantization matrices, quantization steps, etc. These operations will allow oscillation between qualities in media playout which can be employed to reflect the dynamic nature of the underlying communications environment. QoS adaptation will involve the monitoring of a flow to detect QoS degradation and in response to this, trigger appropriate actions including the renegotiation of a new user supplied flow-spec.

4. Flow Management Architecture

Central to the flow management architecture is the notion of a *flow manager* which comprises of three functional components responsible for *group management, call management and control* and *filter management.* Sources and receivers interact with the flow manager over well-known meta-signalling channels using the flow management protocol defined in [Campbell, 94].

Group management is responsible for the setting up and maintenance of communication group membership lists and resource sharing lists based on user supplied share specs. Here we couple group management with network level resource and routing mechanisms. Some early work on group management[1] in the QoS-A project uses out of band signalling channels to broadcast group address notifications to potential group members. While the group manager validates group membership, a group becomes *active* (i.e. data being transferred) only after the connection procedure is completed (as described below).

Call management and control is responsible for setting up flows, resource reservation and routing. A single call can potentially consist of a number of connections, akin to a virtual path consisting of a collection of related virtual connections. Call management and control, which is implemented using ATM signalling in the local ATM environment, interacts with filter management. Call control also interacts with filter management if filters are required upon reception of joinRequest or later during a changeRequest or leaveRequest. Filter management keeps track of the location of filters in an active communication group session.

Filter management is responsible for the identification of filtering operations such as scaling and translation mechanisms. This is achieved by the interpretation of the source flow-spec (the source flow-spec also identifies sub-stream flow-specs) and the clients' flow-specs (this identifies the clients' capabilities, including media decoding capabilities). Filter management not only acts as an agent between source and respective clients but can also provide routing information to call management so that a particular flow can be routed via a *filter server* to perform any required translations or mixing operations.

4.1 Network Programming Interface

The network programming interface primitives includes two important QoS metrics: *(i)* a *flow-spec*, which characterises the nature of the source traffic or receiver requirements; and *(ii) QoS commitment*, which specifies the level of service commitment required by both source and receiver as explained in section 3.1.

The network service user (which is the transport protocol in the QoS-A) interacts with the flow management protocol for the establishment and management of flows with an agreed network level service contract. This is achieved via the following set of primitives which conform to the traditional request, indication, response, confirm interactions between a user and provider:

- **flowRequest**(nsap_t *source, nsap_t *sink, flow_spec_t *flowSpec, commitment_t *cmt).

  The flowRequest primitive informs flow management that a source is ready to establish a call with one or more receivers.

- **joinRequest**(nsap_t *source, nsap_t *sink, flow_spec_t *flowSpec, commitment_t *cmt).

  The joinRequest primitive requests flow management to add the receiver to the active session. The receiver's capability is captured by the flow-spec and commitment

---

1 Group communication management and control is currently being investigated by the GCommS project at Lancaster [Mauthe, 94].

parameters in the call. The receiver moves into the data transfer state when it receives a joinConfirm from flow management.

- **changeRequest**(nsap_t *source, nsap_t *sink, flow_spec_t *flowSpec, commitment_t *cmt).

The changeRequest is issued by the QoS adaptation mechanism at the receiver in response to violations specified in the adaptation policy. The new flow-spec and required commitment are included in the primitive. The adaptation phase is complete when the end-system receives a changeConfirm from flow management.

- **leaveRequest**(nsap_t *source, nsap_t *sink).

The leaveRequest primitive can be issued by any active receiver when it requires to unilaterally leave an active communications session. This is complete upon reception of a leaveConfirm message from flow management.

- **terminateRequest**(nsap_t *source, nsap_t *sink).

The terminateRequest can only be issued by a sender in the active session. Each receiver is gracefully released from the session before the session is terminated. Note that in the case of multiple senders in a group session the session can only be terminated when the last active sender issues the terminateRequest. In all other cases the sender is released and the remaining communication group is left intact.

In section 5 we present a simple flow management scenario to illustrate the network programming interface and the operation of the flow management in setting up *QoS filters*.

4.2 QoS Filters

QoS filters provide us with the potential to bridge the heterogeneity gaps outlined in the introduction of this paper. Currently we have identified a number of baseline-filters as the initial building blocks for simple filters [Yeadon, 93] or more complex encoding transformation filters [Pasquale, 93]. Note that these building blocks are not mutually exclusive.

- *hierarchical-filter.* Operates on a sub-stream basis [Delgrossi, 93] where a flow is split into a number of related sub-streams (or scalable extensions). Each sub-stream can then be associated with a particular QoS and handled independently at the end-system and network. At the end-system we can associate distinct communication protocol funtionality (i.e. reliable data transfer, flow control, multicast etc.) with each sub-stream, while in the network the sub-streams may take independent paths and again distinct QoS characteristics may be associated with these paths.

- *frame-dropping-filter.* This is a media discarding filter used to reduce frame rates. Generally this filter is only applied to compressed video but not audio. The filter has knowledge of the frame types (i.e. Intra- or Inter- frame coding type) and drops frames according to importance. For example, the pecking order for an MPEG-I stream would be B-, P-, and finally I-frames[2].

---

[2] I-pictures are intra- coded pictures. P-pictures are predicted using motion compensation from past pictures. B-pictures are bidirectionally predicted using motion compensation from past and future I- and P-pictures.

- *codec-filter*: This filter performs compression, decompression, encoding translation, colour to monochrome translation etc. Some of these operations may be more demanding than others, thus specialised hardware support based on Multimedia Enhancement Network Device (MEND) technology is being developed for this purpose [Yeadon, 93]. Note that some compression standards are backward compatible (i.e. Baseline JPEG is becoming the 'de facto' intraframe compression standard [Wallace, 91]) and it may be possible to perform simple encoding translations. For example, an MPEG-I video stream can be converted to a motion JPEG video stream by simply discarding B- and P- frames at the cost of a reduced frame rate.

- *mixing-filter*. This filter is used to mix streams together [Schulzrinne, 93]; for example mixing stereo audio into mono or mixing a 6 person audio conference into one flow.

4.3 QoS Adaptation

In the end-system a QoS adaptation policy is included to implement dynamic adaptation of the data flow. QoS adaptation is based around an event/action model. On defining a flow-spec a potential event and subsequent action to be taken are specified. Events are violations of agreed QoS values such as loss, delay, throughput and jitter. Actions taken in the event of QoS violation include renegotiate or adapt to a new QoS, generate a QoS indication to the upper layers, disconnect flow, no action or some combination of all the actions. Renegotiation may mean new end-to-end QoS values or the introduction of a new filter by *filter management* .

5. Simple Flow Management Scenario

In order to clarify our discussion of flow management we will consider three scenarios which require distinct filtering operations: scaling, translation and mixing. While all three operations may be taking place concurrently and acting on a single flow these operations are highlighted individually. All three scenarios are depicted pictorially in Figure 1. Individually they illustrate the flow of data through an ATM network after respective module interaction between group management, call management and control, and filter management has taken place and the flow has been established.

5.1 Flow Establishment

An active group communications session is initiated by a source S sending a flowRequest.connect to the flow manager over the meta-signalling channel specifying a flow-spec, service commitment, and a source and *target group* address (if any). This flow-spec captures the highest quality of the generated stream from the source. It also outlines individual qualities of any encapsulated sub-streams which are uniquely identifiable.

The flow manager relays this flow-spec to the receivers in the target group in the form of a flowRequest.indication. If a receiver wishes to participate in the data transfer phase it responds to the flow manager with a flowRequest.response specifying its QoS capability in terms of its own individual flow-spec and service commitment. If the source and receiver flow specs are compatible (i.e. the receiver can consume the sources stream in its entirety), the flow manager generates a flowRequest.confirm.

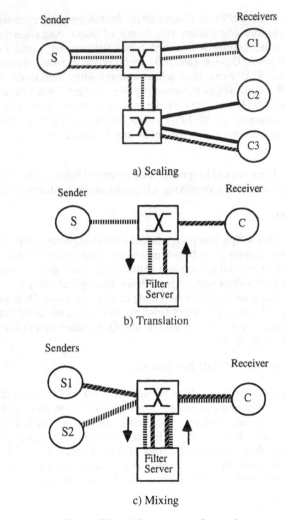

Fig. 1. Flow Management Scenarios

If the flow-specs are not compatible the flow manager attempts to negotiate a compatible flow for the source and client via the filter manger. The outcome of this negotiation could be a rejection of flow establishment or one of the possible scenarios illustrated in Figure 1.

• a) *Scaling scenario*. Under this scenario the filter manager has interpreted that the flow-specs of the source and receiver are compatible provided scaling can be performed. That is, the receiver can consume a subset of the full stream without the need for any kind of translation. This is applicable to hierarchically encoded data streams. The filter manager informs the call manager which sub-streams should be dropped. Using this knowledge the call manager ensures that downstream switches perform this dropping before the sub-stream reaches the receiver.

- b) *Translation scenario*. This is a case where the flows-specs are not compatible. Typically this is when the source's encoding scheme is different from a receiver's decoding capabilities (i.e. sources' and receivers' codecs are distinct). The filter manager provides the location of a filter server which may act as an intermediate agent to perform the required translation. This location is passed to the call manager which has the option to route the flow via the filter server. The call manager may of course reject the establishment of this flow via the filter server if the overall end-to-end QoS required can not be satisfied (i.e. characteristics such as end-to-end delay, throughput, jitter, etc. can not be met).

- c) *Mixing scenario*. Again flow-specs are found not to be compatible but the filter manager realises that by bringing in another source stream and mixing it with the original source stream compatibility is attained . This mixing may also be performed by an identified filter server as in the case of translation.

It must be emphasised that for both cases b) and c) the filter server may also down load the translation or mixing agent to the appropriate receiver if this has sufficient processing capabilities. This will eliminate the need for an intermediate filter agent in the network which inevitably add to the overall experienced end-to-end delay because of queuing and propagation delays through the various routes.

5.2 New Receiver Join

If a new receiver wishes to join an ongoing flow, it issues a joinRequest to the flow manager with its own flow-spec outlining its QoS capabilities and the service commitment required. The flow manager checks the receiver's flow-spec by communicating with filter management and tries to find a compatible flow traversing one of the upstream switches (preferably the nearest upstream switch). At this stage the three scenarios described in flow establishment section once again come into play. Paths will need to be established or torn down in order to scale up or scale down respectively. Alternatively, filter servers will have to be introduced into the flow path or else filter agents will have to be downloaded to the new receiver to perform translation or mixing operations.

5.3 Flow Renegotiation

Flow renegotiation is initiated at the receivers if the QoS adaptation mechanism tries to adjust the QoS because of either the current network conditions have degraded or as instructed by the receiver. The QoS adaptation mechanism automatically issues a changeRequest to the flow manager should the QoS degrade below a set of receiver specified criteria. For example, if the throughput drops below a predefined level or the maximum end-to-end delay is repeatedly violated, etc. The flow manager treats a changeRequest exactly the same as a joinRequest except that the receiver's current flow is replaced by a new flow without the initial connection being torn down.

5.4 Receiver Leaves and Shut-down

To complete the scenario if a receiver wishes to leave an active group it issues a leaveRequest to the flow manager. The flow manager then issues instructions to tear down connections and free resources on any filter servers currently allocated for use by the leaving receiver only. Again, this will include the three scenarios outlined in flow establishment, that is, scaling paths down, removal of translating angents and removal of mixing agents.

A source can terminate a complete active session by issuing a terminateRequest to the flow manager. Here all receivers are gracefully terminated - being informed of the pending close via leaveIndications - and resources deallocated.

6. ATM Based Implementation

Lancaster's experimental infrastructure is illustrated in figure 2. This infrastructure consists of several 486 PCs which are running the Chorus or Linux operating systems. These PCs are connected to the ATM environment either directly (using Fore or ORL ATM cards) or via our proprietary MEND network. The MEND network is a high-speed mini-cell (6 octets) switching network with a potential bit rate of 100Mbps.

The multimedia storage server is a high capacity device composed of two large disk bricks using RAID technology [Lougher, 92]. This storage server is central to our experimentation as it will store many compressed images and sound for the support of distributed multimedia applications. Compression of these images and sound is achieved through the use of real-time MPEG and JPEG encoders. These real-time encoders, because of expense, are only used to provide inputs for the storage server and not for individual PCs.

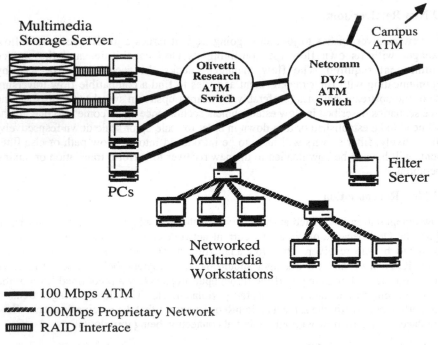

Fig. 2. The Lancaster Experimental Infrastructure.

Decompression at the various PCs is achieved through a mix of software and inexpensive hardware decoders. Compression software (which does not function in real-time) is also available for H.261, MPEG-I, MPEG-II and JPEG. The experimental filter server also contains a mix of these modules.

The ATM interface to be employed is an AAL5 interface. Encoded video/audio frames will be individually delineated by single AAL5 PDUs, in other words, the bits belonging to one frame will not be encapsulated with the bits of another frame. These AAL5 PDUs will be uniquely tagged using the user-to-user byte to identify the type of encoded frame encapsulated by the AAL5 protocol. As AAL5 PDUs can encapsulate up to 64KBytes of data, we feel that this data unit size will be sufficiently large to contain all the bits of a single compressed frame. Such a scheme will allow us to implement smart frame discarding filtering services at the ATM network edges.

The individual bit streams of a hierarchical encoding will be assigned to unique VCs. This approach will allow us to associate distinct QoS characteristics with the identified sub-streams and we can exploit this break up for the implementation of scaling services. At each stage of the flow, the highest quality stream available (represented by a unique bundle of VCs) can be encapsulated within a single VP.

Stored encoded data will be augmented by information describing the stream contents and structure as in [Rowe, 94]. This information will allow us to demultiplex a stream into its component types (e.g. I-, P-, B- frames, etc.) and assign these to distinct VCs if required. Also, it will allow us to instantiate filtering services as part of the media retrieval process. For example, from an MPEG encoded stream, we can request that only I frames be transmitted, only the audio data, etc. In general, while demultiplexing a stream may add complexity at the end-system in terms of synchronisation, the information held on the stream structure and characteristics can be better utilised in resource reservation and allocation.

7. Summary and Status

In this paper we have described work in progress in the area of QoS adaptation and flow filtering, and placed this work in the context of Lancaster's efforts towards a Quality of Service Architecture (QoS-A) for multimedia communications over ATM networks.

The next stage of research is the implementation and development of the filter server outlined in this paper. Work will concentrate on the scalable properties of the server. In particular the processing required for each of the filter operations described in section 4.2 will be evaluated, so that the number and types of stream the server will be capable of handling can be assessed.

We feel that in our QoS-A, filtering mechanisms will provide essential modules to bridge heterogeneity issues found in networks, end-systems and applications. This is particularly important when considering dissemination and conferencing applications. All the proposed filtering operations will be performed at the network edges.

In a forthcoming paper we will be presenting initial results and an analytical evaluation of the flow management architecture.

8. Acknowledgements

Nicholas Yeadon is a CASE PhD student supported by the EPSRC and BT Labs. The work reported in this paper is carried out in the context of the QoS-A project funded by the EPSRC (grant number GR/H77194) and in collaboration with GDC Advanced Research (formerly Netcomm Ltd).

9. References

[Campbell, 93] Campbell, A., Coulson G., and D., Hutchison, "A Multimedia Enhanced Transport Service in a Quality of Service Architecture", Proc. Fourth International Workshop on Network and Operating System Support for Digital and Audio and Video, Lancaster, UK, October 1993, and ISO/IEC JTC1/SC6/WG4 N832, International Standards Organisation, UK, November, 1993.

[Campbell, 94] Campbell, A., Coulson G., and D., Hutchison, "A Quality of Service Architecture", ACM Computer Communications Review, April 1994, and Internal Report No. MPG-94-08 Department of Computing, Lancaster University, Lancaster LA1 4YR, March 1994.

[Delgrossi, 93] Delgrossi, L., Halstrinck, C., Hehmann, D.B., Herrtwich R.G., Krone, J., Sandvoss, C., and C. Vogt, "Media Scaling for Audiovisual Communication with the Heidelberg Transport System", Proc. ACM Multimedia '93, Anaheim, August 1993.

[H.261, 90] H.261, "ITU-T: Video codec for audovisual services at p c 64 Kbit/s", Internatinal Telecommunications Union- Telecommunications Standardisation Sector, ITU-T Recommendation H.261 1990.

[JPEG, 92], JPEG, "Information technology - Digital compression and coding of continuous-tone still images", British Standards Institute, Draft International Standard ISO/IEC 10918, 19-10-92, 1992.

[Lougher, 92] Lougher, P. and D. Shepherd, "The Design and Implementation of a Continuous Media Storage Server", Proc. Third International Workshop on Network and Operating Systems Support for Digital Audio and Video, San Diego, California, November 1992, pp. 63-74.

[Lunn, 94] Lunn, A.S., Scott, A.C., Shepherd, W.D. and N. J. Yeadon, "A Mini-cell Architecture for Networked multimedia Workstations", Proc. International Conference on Multimedia Computing and Systems, Boston, Massachusetts, May 14-19, 1994.

[Mauthe, 94] Mauthe, A. et al., "From Requirements to Services: Group Communication Support for Distributed Multimedia Systems", to be presented at IWACA'94, IBM European Networking Center, Heidelberg, Germany.

[MPEG-I, 93] MPEG-I, "Information technology - Coding of moving pictures and associated audio for digital storage media at up to about 1,5 Mbit/s.", British Standards Institute, International Standard ISO/IEC 11172, 1-8-1993, 1993.

[MPEG-II,93] MPEG-II, "Information technology - Generic coding of moving pictures and associated audio information.", British Standards Institute, Draft International Standard ISO/IEC 13818, 1-12-93, 1993.

[Pasquale,92] Pasquale, G., Polyzos, E., Anderson, E. and V. Kompella, "The Multimedia Multicast Channel", Proc. Third International Workshop on Network and Operating System Support for Digital Audio and Video, San Diego, USA, 1992.

[Robin, 94] Robin, P., Campbell, A., Coulson, G., Blair, G., and M. Papathomas, "Implementing a QoS Controlled ATM Based Communication System in Chorus", To be presented at: 4th IFIP International Workshop on Protocols for High Speed Networks, and Internal Report No. MPG-94-05 Department of Computing, Lancaster University, Lancaster LA1 4YR, March 1994.

[Rowe, 94] Rowe, L., Patel, K., Smith, B. and K. Liu, "MPEG Video in Software: Representation, Transmission, and Playback", Proceedings of High-Speed Networking and Multimedia Computing, San Jose, California, 8-10 February 1994, pp 134-144.

[Schulzrinne, 93] Schulzrinne, H. and S. Casner, "RTP: A Transport Protocol for Real-Time Applications", Work in Progress, Internet Draft, <draft-ietf-avt-rtp-04.ps>, October 1993.

[Wallace, 91] Wallace, G., "The JPEG Still Picture Compression Standard", Communications of the ACM, April 1991.

[Yeadon, 93] Yeadon, N.J., "Supporting Quality of Service in Multimedia Communications via the Use of Filters", *Internal Report* No. MPG-94-10 Department of Computing, Lancaster University, Lancaster LA1 4YR. March 1993.

[Zhang, 93] Zhang, L., Deering, S., Estin, D, Shenker S. and D. Zappala, "A New Resource ReSerVation Protocol" Draft available via anonymous ftp from parcftp.xerox.com:/transient/ rsvp.ps.Z, August 1993.

Optimal Resource Management in ATM Transport Networks Supporting Multimedia Traffic

Jochen Frings   Thomas Bauschert

Lehrstuhl für Kommunikationsnetze
Technische Universität München
D-80290 München
Email: frings@lkn.e-technik.tu-muenchen.de

Abstract. The existing or soon to be built multimegabit multimedia islands have to be tied together with a public broadband ISDN transport network, which will be based on the asynchronous transfer mode (ATM). Although this transfer mode is generally well suited to support multimedia requirements, there is a strong need for an optimal resource management to make the best possible use of the installed system and network components.

In this paper we present a method for resource management in ATM wide area backbone networks entirely based on virtual paths. The principle is to adapt the load sharing call routing strategy and the virtual paths bandwidths to changing traffic demands. The task will be defined as the mathematical programming problem of minimizing the maximum call blocking probability. Numerical results for a 23 node network and possible solutions to the problems of network availability and to statistical multiplexing will be shown.

1 Introduction

The transition from industry towards communication society is accompanied by the growing demand for multimedia communication on the developing global marketplace. Today local multimegabit multimedia islands are planned or already installed in many places, but to obtain worldwide connectivity, public wide area networks are needed.

Multimedia traffic is known to make highly diverse demands on the transmission network, like variable or constant bitrates of several magnitudes and with real time constraints or very bursty data traffic. Because of its flexibility the asynchronous transfer mode is generally considered to be a good compromise to satisfy all these demands in a future broadband integrated services digital network (B-ISDN).

One important feature of multimedia traffic will be the large variation of the load distribution over time. For example during the day a lot of people will use videophones for long distance business calls between metropolitan areas. In the evening many people like to see some video on demand programmes from a near by server or call old friends in more rural sites. ATM wide area networks will manage to cope with a wide range of load distribution variations with a minimum of installed resources, if these resources are properly managed. The complexity of resource management will be reduced significantly, if only virtual paths (VP) with dedicated bandwidth are used in the wide area network [1].

As a contribution to this resource management problem we suggest a method for global optimization of these networks. Load sharing will be used to route arriving calls to one of at least two VPs for each pair of access nodes. The decision to use the load sharing call routing algrithm was made, because load sharing is the only nontrivial routing strategy for which the call blocking probability can be calculated exactly on an analytical basis.

The concept of solving a global resource management optimization problem was used for example in [2], [3] and [4]. In [2] however only one VP per demand pair exists, thus reducing the achieveable optimization gain, because breaking up the bottlenecks is not possible via load rearranging. In addition to this the network availability is reduced significantly. Girard [3] does not use the concept of VPs. In [4] a global resource management optimization problem is solved for an eight node network using a heuristic approach.

The optimization is aimed to minimize the maximum call blocking probability and to guarantee fair service for all calls. This means that all users are provided with the same best possible grade of service.

In the next section the wide area ATM network and the assumed traffic characteristics are modeled. In addition the formulas for the multiservice call blocking probabilities will be given. The management algorithm will be introduced in Sect. 3 followed by the definition of the optimization problem in Sect. 4. As this is a high order noncontinuously differentiable problem we use an extended least square optimization programme for numerical solution, as given in Sect. 5. As todays public life heavily depends on telecommunication services, network availabilty is a serious topic. For single failures this availability can be guaranteed by the suggested resource management algorithm as shown in Sect. 6. Some numerical results for a 23 node network show the optimization gain in Sect. 8.

Finally the extension to statistical cell multiplexing is sketched.

2   Modelling

Throughout this paper we consider wide area ATM networks with high transport and switching capacity. At least one ATM switch is connected to each wide area ATM crossconnect node to manage the access of virtual channels (VC) to the wide area network (Fig. 1). Only virtual paths are known in this network and VCs are multiplexed into VPs. Multiple parallel VPs that use distinct links connect every pair of access nodes (demand pair d). Virtual paths are allowed to carry any kind of traffic, which permits transport of all VCs belonging to one multimedia session in one VP. This, for example, eases the synchronization of lip movement and voice.

ATM Switch, here access node

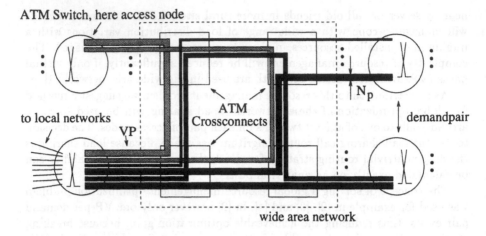

Fig. 1. Main elements of a wide area network with access nodes

Because of the high capacity of the transmission facilities and their ability to carry lots of calls, bandwidth will be treated as a continuous value. Bandwidth N_p is assigned to VP p exclusively. Therefore the VPs build a virtually full meshed layer upon the presumably just partly meshed physical network and thus allow virtualization of network resources. In addition, VPs with dedicated bandwidth enable fast call setups, because all routing information is located at the VPs access nodes.

The following discussion is done under the assumption of peak bitrate reservation. The transmission of variable bitrate connections with statistical cell multiplexing will be discussed briefly in section 8.

To get a handle to the calculation of call blocking probabilities for calls needing bandwidths between some kbit/s and several Mbit/s, the notion of service classes is used. Calls with peak bitrate r with $r_{s-1} < r \leq r_s$ are members of service class s. Bandwidth r_s will be used to determine call blocking. Obviously this is a worst case approach, if call acceptance algorithms use the real peak bitrate r.

Call blocking probability $B_{p,s}$ for a call of service class s on VP p with bandwidth N_p will be estimated with Labourdette's [5] approximation. It has been derived from the well known product form formula for multiservice call blocking in circuit switched networks

$$B_{p,s} \simeq \begin{cases} \frac{1-\alpha^{r_s}}{1-\frac{N_p}{M}} \cdot B\left(\frac{M}{d}, \frac{N_p}{d}\right), & \text{if } \alpha \neq 1 \\ \\ \frac{M \cdot r_s}{V} \cdot B\left(\frac{M^2}{V}, \frac{M^2}{V}\right), & \text{if } \alpha = 1 \end{cases} \qquad (1)$$

206

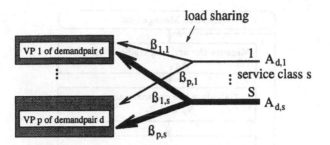

load sharing

Fig. 2. Load sharing of several service classes

with α being the definite solution to

$$\sum_{s=1}^{S} \frac{r_s a_{p,s}}{N_p} \cdot \alpha^{r_s} = 1 \tag{2}$$

and

$$M = \sum_{s=1}^{S} r_s \cdot a_{p,s} \tag{3}$$

$$V = \sum_{s=1}^{S} r_s^2 \cdot a_{p,s} \tag{4}$$

$$d = \frac{ln\frac{N_p}{M}}{ln\alpha} \tag{5}$$

with $a_{p,s}$ being the path traffic intensity for service class s. B is the continuous Erlang blocking function which can be calculated following [6]. The approximation holds very well for large VP bandwidths in the presence of Poisson call arrival processes for each service class. Both prerequisites are commonly assumed to be fullfilled in the case of wide area networks for public use.

Because of the multiple paths for each demand pair, a routing decision is required for every single call. This is achieved by using load sharing routing with one load sharing coefficient $\beta_{p,s}$ for each path p and each service class s of the demand pair. In this way the demand pairs traffic intensity $A_{d,s}$ for service class s is divided into several path traffic intensities $a_{p,s}$ (Fig. 2).

$$a_{p,s} = \beta_{p,s} \cdot A_{d,s} \tag{6}$$

The end to end call blocking probability $B_{d,s}$ for the traffic of class s to the demand pair d will be calculated as

$$B_{d,s} = \sum_{\forall p \in d} \beta_{p,s} \cdot B_{p,s}. \tag{7}$$

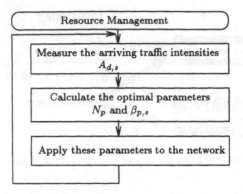

Fig. 3. Flowchart of the resource management algorithm

3 Management Algorithm

As shown in Fig. 3 the management algorithm is an infinite loop. It starts with measuring the traffic intensities $A_{d,s}$ for each demand pair d and each service class s. Of course there is the problem of measuring traffic intensities which depend on the frequency of arriving calls. Due to the permanently high loads of wide area networks these measurements can be done to a satisfying accuracy.

Based on this input an optimization problem aimed to minimize the maximum call blocking probability is solved. The result is call blocking fairness and most demand pairs will suffer the same call blocking probability in numerical evaluations.

Adapting VPs to the new bandwidths may be another problem, if in the moment of adaptation more bandwidth of the VP is used than will be assigned to it. The methods of achieving an optimal adaptation are subject to further research.

4 Definition of the Optimization Problem

We define the problem of finding the minimal maximum call blocking probability as a problem of mathematical programming.

The objective is

$$f(x) = \max_{\forall d,s} \left[B_{d,s}(a_{p,1}, \ldots, a_{p,s_{max}}, N_p) \right] \to \min \qquad \forall p \in d \qquad (8)$$

with f beeing the objective function and $B_{d,s}$ representing the end to end call blocking probability for demand pair d and service class s, which is calculated according to (7). The optimization variables are the path traffic intensities $a_{p,s}$ and the path bandwidths N_p with the boundaries

$$0 \le N_p \qquad \forall p \qquad (9)$$

$$0 \le a_{p,s} \le A_{d,s} \qquad \forall p \in d \tag{10}$$

Of course the total amount of offered traffic to the VPs of one demand pair cannot exceed the total traffic offered to the demand pair.

$$\sum_{\forall p \in d} a_{p,s} \le A_{d,s} \qquad \forall d, s \tag{11}$$

The bandwidth of all virtual paths using the same link l cannot be larger than the link capacity C_l.

$$\sum_{\forall p \in l} N_p \le C_l \qquad \forall l \tag{12}$$

There is no optimization algorithm, which can handle the numerous noncontinuously differentiable points that are introduced by the maximum function in (8). In [8] a method of transforming the objective function into a linear one is given, but this transformation introduces numerous nonlinear boundaries. Furthermore no optimization algorithms capable of dealing with this transformed problem effectivly, if more then a few optimization variables and constraints are needed, could be found.

5 Extended Least Square Optimization Algorithm

Because of the arguments given above any transformation should produce an optimization problem with continously differentiable objective and constraint functions and without nonlinear constraints.

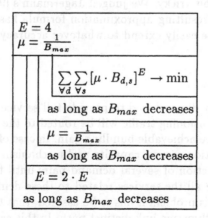

Fig. 4. Algorithm for the scaled dynamic least square function

An easy programmable algorithm was developed, which solves the given problem (Fig. 4). In the case of call blocking optimization the desired maximum blocking probability can be easily given: Zero blocking. Therefore the mimimum least square objective function is

$$f(x) = \sum_{\forall d,s} (B_{d,s})^2 . \tag{13}$$

The aim is to diminish the maximum values and therefore the exponents may be greater than two. But soon, as all blocking values are to be expected much smaller than one, f(x) will be set to zero for numerical reasons. This can be avoided by scaling $B_{d,s}$, which results in

$$f(x) = \sum_{\forall d,s} (\mu \cdot B_{d,s})^E \tag{14}$$

with

$$\mu = \frac{1}{B_{max}} \tag{15}$$

and E being the exponent.

Due to the large number of optimization variables and constraints and the usually not fully meshed physical networks, the central minimization problem has to be solved by a sparce programmed algorithm. The combined quasi Newton and reduced gradient algorithm of Stanford University's optimization package MINOS was used to compute the minimum.

To achieve high efficiency each optimization programme should be provided with the derivatives of the objective and constraints functions. While differentiating Labourdette's formulas is just lengthy, obtaining derivative of the continuous Erlang function is a bit tricky. We judged Jagermann's [9] method to be most practical one, as the resulting approximation formula has only low computing complexity and can be easily extend to whatever accuracy needed.

6 Availability

If a network has been dimensioned to provide full service for minimum investment, the whole dimensioning traffic will be routed to the shortest path of each demand pair to use the achievable bundling gain. In case of link or node failures, whithout link restoration reserve capacity and mechanism supplied, this will result in total disconnection of several demand pairs until the next optimization run is finished. Hence all the services related to these demand pairs will not be available. To avoid this problem it has to be assured, that the dimensioning traffic is routed to more than one link distinct path. In this case a single failure will only reduce the remaining transmission capacity, which will of course increase the blocking probability. Nevertheless the services will remain available.

Taking into account these features in the optimization problem, an additional boundary has to be introduced

$$\beta_{p,s} \le V_{p,s} < 1 \qquad \forall p,s \tag{16}$$

with

$$\sum_{\forall p \in d} V_{p,s} \geq 1 \qquad (17)$$

and $V_{p,s}$ being the share of the total traffic to the demand pair.

By defining suitable $V_{p,s}$ parameters, several availability classes concerning minimum remaining bandwidth after a failure event, could be defined. For example mission critical telephone or videophone services would be divided to even parts, whereas video on demand services could be for instance divided only 90 to 10 and hence reduce the overall bandwidth need.

Of course additional bandwidth and switching capacity has to be installed to get the same performance as in the nonrestricted case. But this investment pays off with expanded availability and with better network adaptability, as can be seen in the following section.

7 Numerical Studies

We investigated a 23 node ATM network, as shown in Fig. 5, which could be a wide area backbone network for Germany. This was done in two scenarios: First for a minimum investment network, where the whole dimensioning traffic is routed to the shortest path of each demand pair and second an advanced availability network, where the traffic to each demand pair will allways be divided into several link distinct streams.

There are up to 5 virtual paths for each of the 253 demand pairs. The paths were routed to the physical network with a shortest path algorithm. The virtual paths of each demand pair do not use a single link together. For each case the network is dimensioned to bear exactly the design load with a call blocking probability of one percent.

Traffic patterns were derived very roughly from population density and todays telephone line usage matrices of Germany. Hence, traffic intensities are not symmetric. Three service classes with 1Mbit/s, 3Mbit/s and 10Mbit/s are considered with a traffic intensity relation of 100:50:30. We assumed the total traffic intensity being larger than today, because a single multimedia session may contain more than one call.

7.1 Minimum Investment Network

To demonstrate how the network can cope with traffic variations the offered traffic for each demand pair is either increased or decreased randomly by a certain percentage t. Therefore the traffic volume will be approximately constant. Figures 6 and 7 show that the call blocking probability can be reduced by up to 86 percent compared to the nonoptimized case. For a traffic variation of more than 15% the optimization gain decreases, because the bandwidth needed to resolve the congestions cannot be found by traffic load rearranging.

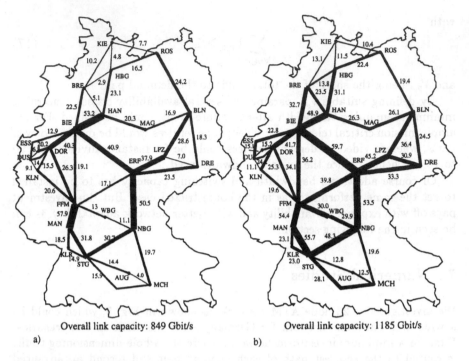

Overall link capacity: 849 Gbit/s   Overall link capacity: 1185 Gbit/s

a)   b)

Fig. 5. German wide area network: a) minimum investment and b) advanced availability scenario (Link capacity in Gbit/s)

7.2 Advanced Availability Network

Now the network be dimensioned for advanced availability. As an example at most half of the traffic to each demand pair and of each service class be carried on a single path. Due to the lost bundling gain, about 40 percent additional bandwidth has to be installed. Even in the nonoptimized case it can be shown that this must result in decreased sensitivity to changing traffic (Fig. 6). In the optimized case, the call blocking probability increases very slowly with increasing traffic variation (up to about 40 % traffic variation), because traffic has to be divided in any case for nondesign load and additional bandwidth exists compared to the minimum investment network. This can also be seen in terms of the relative optimization gain (Fig. 7), where the relative gain in the advanced availability network is allways higher than in the minimum investment network.

The reason for this behaviour can be seen in Fig. 5a and 5b . The minimum investment network on the one side has of course mainly a tree structure with numerous bottlenecks. The bandwidth of the advanced availability network on the other side is much more distributed on the links.

Advanced availability networks thus facilitate network dimensioning, because traffic proposals do not need to meet the real traffic exactly.

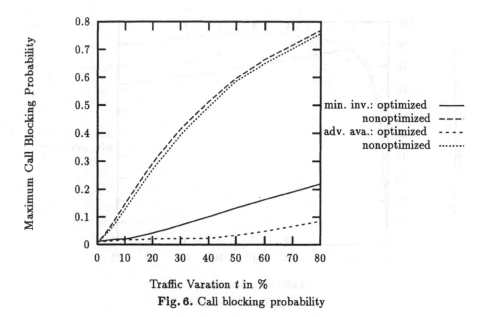

Traffic Varation t in %

Fig. 6. Call blocking probability

7.3 Calculation Time

Optimizing the network to changing traffic takes about 4 to 10 hours on a SUN sparcstation 10. This is obviously too long compared to the update interval of 2 to 10 minutes recommended in [10].

The first attempt to reduce this long calculation times would be to use more specialized algorithms, as our programme was primarily based on general libraries. In the following step, as the optimization problem is separable, one could think about hierarchical optimization.

If some minor approximations are allowed, another promising attempt would be to simplify the optimization problem to a linear one, as shown in [11].

Obviously there are cases, where some demand pairs do not use any of the bottlenecks that evolve from changing traffic distributions. These demand pairs could be optimized to carry the maximum flow, so that they do not exceed the maximum call blocking probability of the whole network.

8 Statistical Cell Multiplexing

It is being expected that the majority of multimedia datastreams is coded for bitrate reduction. Therefore multimedia traffic will be variable bitrate traffic. Via statistical cell multiplexing this feature can be exploited to reduce the needed average transmission capacity. This can be included into the suggested resource management method as follows: The call acceptance algorithms have to use the effective instead of the peak bitrate. The effective bitrate r_e is the lowest bitrate

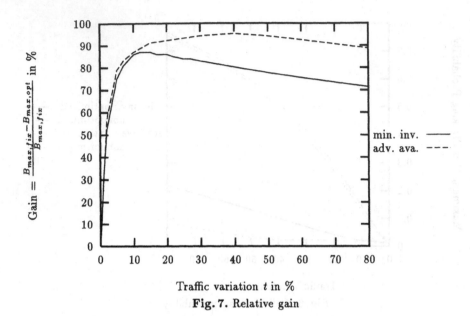

Traffic variation t in %

Fig. 7. Relative gain

that guarantees a maximum cell loss rate of e.g. 10^{-9}, if used for resource allocation. Therefore, it has to rely on the cellstream statistics. Several suggestions of approximations exist, to simplify this calculation. Lindberger for example presents in [12] a linear equation (18) for the calculation of the effective bitrate r_e based on path bandwidth N_p, peak bitrate r and average bitrate r_a.

$$r_e = 1.2 \cdot r_a (1 + 50 \frac{r - r_a}{N_p})$$ (18)

If

$$\frac{r}{N_p} \leq 0.1$$ (19)

which represents the requirement to multiplex a lot of relatively small data streams to a single VP and

$$\frac{r}{r_a} \leq 20$$ (20)

or

$$\frac{r}{r_a} \geq 2$$ (21)

are not fulfilled then one should use the peak bitrate r for the effective bitrate, as either the maximum cell loss probability cannot be guaranteed or the calculated effective bitrate is larger than the peak bitrate.

Hence, r_e has to be calculated for each virtual path separately and statistical multiplexing service classes $\sigma(r, r_a)$ should be introduced in order to calculate call blocking for demand pairs.

Concerning the optimization problem, the multiplexing service classes σ will be used as service classes as before. Each multiplexing service class will have a different bandwidth need on every path of the demand pair and this bandwidth need will change during the optimization. Another problem are the discontinuities (19),(20),(21) that make optimization a lot harder. But we expect, that this problem can be overcome by continuous interpolation in a small intervall around the discontinuities. Call blocking can be calculated as before with Labourdette's formula.

Variable and constant bitrate traffic should be routed to different paths, because there is no multiplexing gain for constant bitrate traffic.

9 Conclusion

In this article we could show that global optimization of networks with realistic complexity is possible and highly improves the overall network performance. Also service availability can be increased because it can be guaranteed, that a demand pair will never be disconnected in case of single failures.

The presented method should be extendable to statistical multiplexing with the notion of effective bitrate.

This optimization problem was done to satisfy the users demand for bandwidth. The results could be combined with a following traffic volume optimization under the constraint of not increasing the already minimized maximum blocking probability. Hence, a multimedia communication network for public use with call blocking fairness between demand pairs and maximized throughput could be set up, which guarantees service availability for single failure events.

Acknowledgements

We would like to thank R. Siebenhaar for valuable discussions.

References

[1]   S. Ohta, K. Sato, I. Tokizawa: A Dynamically Controllable ATM Transport Network Based on the Virtual Path Concept. Proceedings of the *Globecom 1988* 1272–1276.

[2]   S.P. Evans: Optimal bandwidth management and capacity provision in a broadband network using virtual path. Performance Evaluation, North Holland No. 13 (1991) 27–43.

[3]   A. Girard: Revenue Optimization of Telecommunication Networks. IEEE Transactions on Communications Vol.41 No. 4 (1993) 583–591.

[4]   M. Logothetis, S. Shioda: Centralized Virtual Path Bandwidth Allocation Scheme for ATM Networks. IEICE Transactions on Communication Vol. E75-B No.10 (1992) 1071–1080.

[5]   J-F. P. Labourdette, G. W. Hart: Blocking Probabilities in Multitraffic Loss Systems: Insensitivity, Asymtotic Behavior, and Approximations. IEEE Transactions on Communications Vol. 40 No. 8 (1992) 1355–1366.

[6]   R.F. Farmer, I. Kaufmann: On the Numerical Evaluation of Some Basic Traffic Formulae. Networks, John Wiley & Sons (1978) 153–186.

[7]   A. Girard: Routing and Dimensioning in Circuit-Switched Networks. Addison-Wesley (1990).

[8]   P.E. Gill, M.H. Wright: Practical Optimization. Academic Press, Hartcourt Brace & Company Publishers (1993).

[9]   D.L. Jagermann: Some Properties of the Erlang Loss Function. The Bell System Technical Journal, AT&T Vol.53 No.3 (1974) 525–551.

[10]   J. Burgin, D. Dormann: Broadband ISDN Resource Management: The Role of Virtual Paths. IEEE Communications Magazin No. 9 (1991) 44–48.

[11]   R. Siebenhaar: Optimized ATM Virtual Path Bandwidth Management under Fairness Constraints. to be presented at *IEEE Globecom '94*, San Francisco.

[12]   K. Lindberger: Analytical Methods for the Traffical Problems with Statistical Multiplexing in ATM-Networks. Teletraffic and Datatraffic, ITC-13, Elsevier Publishers (1991) 807–813.

Incorporating Security Functions in Multimedia Conferencing Applications in the Context of the MICE Project

Knut Bahr, Elfriede Hinsch, Guenter Schulze

Institut für TeleKooperationsTechnik, GMD
Darmstadt

Abstract. This report briefly describes the multimedia conferencing infrastructure provided by project MICE (Multimedia Integrated Conferencing for European Researchers), an Esprit Project of the European Union. This infrastructure has been used for applications such as project meetings and seminars. Based on the experience gained with the MICE tools and the applications, the paper discusses the need for and the planned incorporation of security functions in the conferencing tools in order to provide a platform for the development of secure multimedia conferencing.

1 The MICE Project

MICE ('Multimedia Integrated Conferencing for European Researchers') [1] is an Esprit Project of the European Union with the objective of providing for European scientists means and ways for multimedia conferencing.

The project started as a one year project in December 1992 and has subsequently been extended for about another year. The project also receives national supports; in Germany it is supported by the DFN, Berlin.

Partners of the project are: University College of London UCL (UK), project leader, GMD Darmstadt (Germany), INRIA (France), Norwegian Telecom Research NTR (Norway), Nottingham University (UK), ONERA (France), Oslo University (Norway), Rechenzentrum der Universität Stuttgart (Germany), Swedish Institute for Computer Science (Sweden), and Universite Libre de Bruxelles (Belgium).

2 Technology

The basic philosophy for MICE multimedia conferencing is to make good use of available equipment and tools and to require as little as possible in addition to general purpose workstations and data networks. Participating in MICE conferences is possible from either a single workstation or a conference room. A workstation is usually found on an individual's desktop. For the purpose of MICE, it has to be equipped with video camera, microphone, loudspeaker, and appropriate conferencing software. (Figure 1)

Fig. 1. Multimedia Workstation performing video coding and decoding in software

A conference room on the other hand, allows a group of people to take part in a conference and mostly offers more and higher quality audio–visual equipment. A conference room workstation is used to connect it to MICE conferences. Multimedia workstations and conference rooms can interoperate.

The infrastructure for the data transport is mostly based on packet switched networks with Internet protocols. In addition there is support of some ISDN lines and their interworking. Project MICE offers two alternatives for conferencing:
(1) a centralized solution with the CMMC in the center (Conferencing Multiplexing and Management Center, developed and operated by UCL) [2] with unicast or point to point connections from each user to the conference hub and
(2) a decentralized solution with the MBone as a virtual network [3] on top of Internet, which allows users to work in a multicast mode (one–to–many connections) without the need for any central service.
The two solutions may be mixed.

Among the software tools used in MICE are the following: VAT is the conference tool used for audio [4], WB is a whiteboard which substitutes for a regular whiteboard as well as a slide projector [5], SD is a tool for creating, announcing and joining a conference [6]. IVS is used for the video part in workstations. It does software compression and decompression of video according to H.261 [7,8]. For higher performance video coding a H.261 hardware codec may be connected to a workstation [9]. UCL has developed a codec controller software for this case.

An example of a part of a workstation screen in a conference is shown in fig. 2.

VAT Control Window   images from remote site

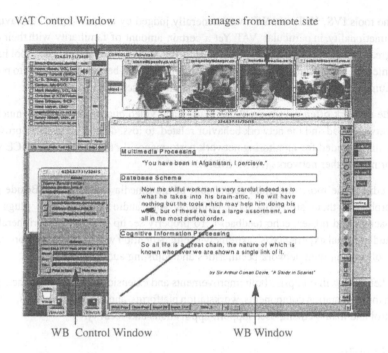

WB Control Window   WB Window

Fig. 2. Part of a workstation screen in a conference

3 Applications and Experience

The MICE technology has been successfully utilized for about a year in a large number of demonstrations, project meetings and seminars [10].

Demonstrations took place for example at the JENC conference in Trondheim in May 93, the IETF conference in Amsterdam in July 93, the Interop in Paris in October 93. There was a project demonstration for the European Commission, as well as a number of demonstrations at the site of project partners, each time with connections to remote sites.

The project held weekly project tele–meetings using MICE tools among its partners from 6 European countries. These provided a major field of trials and experiences.

During the last quarter of 1993, a MICE seminar series was started. About a dozen lectures were given with speakers from USA, UK, Sweden, Norway, Germany and France and audiences of 10 to over 50 attendants from virtually all countries reachable via Internet. Both the tele–meetings and the tele–seminars were carried out over Internet and MBone. While the tele–meeting is a typical example of a closed user conference, the seminars are typically widely open for attendance.

Experience has been very positive overall, but has also shown areas which need further research and development:

- the tools IVS, VAT, and WB were generally judged by the participants as having good functionality, in particular VAT. Yet a certain amount of familiarity with their use does greatly improve both comprehension and the flow of conferences. Each tool has a user interface of its own. An integrated user interface would be welcome at least to reduce the number of tool windows crowding on the monitor screen.

- the network often causes problems. Audio and video quality depend very much on the network load and the network behavior related to loss, delay, and jitter. Improved tools will be needed for monitoring network performance. In the next phase, MICE will also bring in other networks (ATM, ISDN).

- audio is the most important component of multimedia conferencing. Beside network problems, there still exist local problems with audio (microphones, plugs, cables, background noises, echo feedback). To properly set up and use the peripherals, some audio–visual experience is extremely useful. Attention will be given to improved audio tools coping with network peculiarities and adjusting audio levels.

Further work will comprise both improvements and extensions. Major work packages deal with the workstation components, workstation platforms, and conference rooms, conference control and management, user support, applications and security.

4 Security

Secure multimedia conferencing is needed in open telecooperation when confidentiality and privacy are an issue of multi–party communication and when it is desired to authentically identify parties.

For the continuation of MICE in 1994, GMD has taken charge of the security work package; contributing partners are INRIA, UCL and NTR. The following description gives the requirements of MICE technology with respect to authentication and confidentiality, presents an overview of the used security functions and defines how these functions could be applied to enhance the MICE services.

4.1 Requirements

The OSI Reference Model has been extended by a security architecture [11,12], which defines the following functions:
- Authentication
- Access Control
- Data Confidentiality
- Data Integrity
- Non–Repudiation

The MICE tools should be enhanced by these security functions as far as the technology at hand allows extensions at a reasonable effort. Though the transport technology has some problems (e.g. traffic flow confidentiality...), we feel that it will be possible to offer secure conferencing in an acceptable manner.

One characteristic of the present MICE technology is its openness. Once someone has started a conference anyone who gets knowledge of the used addresses and port numbers can participate. Since the conference tools make use of the Internet MBone broadcast facilities, there is no support by the system to prevent other (unauthorized) users from taking part in a conference.

Some applications may require that a conference be restricted to a determinable closed user–group. Users outside such a user–group must not be able to take part. Examples are (tele–)seminars where only those who have paid the fee for it can participate or (tele–)meetings which are restricted to the members of a project. Such applications need some kind of access control to prevent unauthorized users from taking part in the conference. Other applications may require in addition, that information be kept confidential.

One possibility to fulfill these requirements is to add security functions to the MICE technology, to encrypt the information broadcast by MBone and to keep the encryption key confidential among authorized participants.

Encryption allows both access control and confidentiality. In case of centralized conference services with point–to–point connections to the clients, it would be sufficient to restrict the conference access by an authenticity control function. Encrypting the transmitted data is only required for confidential conferences.

If in case of decentralized conference services and multicasting over MBone, a conference shall be restricted to a specified set of participants, the broadcast information will have to be encrypted, since MBone does not offer any access control facilities. Audio is most sensitive and thus a prime candidate for encryption. The Information distributed by Shared Workspace tools such as WB is often nearly as sensitive as audio information and thus should also be encrypted. Similar methods may apply in either case.

Encrypting realtime video information needs significantly more processing power, so that probably different encryption methods will be required.

Where MICE tools have already provisions supporting encryption –such as VAT and WB– these shall be utilized as far as possible.

4.2 Used Security Functions

Symmetric cryptography is used for encrypting large data volumes. In order to achieve authenticity and integrity and to guarantee confidentiality of datastreams, they will be encrypted by means of a so called session key. This key may be valid for one or more sessions, depending on the particular situation , and it may differ depending on the type of data.

The session key used to encrypt and decrypt the audio and Shared Workspace datastreams, must be distributed to all conference participants. The key has to be distributed confidentially so that unauthorized users cannot get it.

Several methods may be used to distribute small amounts of confidential information. Common to most of them is the application of a Personal Security Environment (PSE). The

PSE comprises the management of an asymmetric key pair which represents the kernel of all security functions. This key pair consists of a public key and a private key. The public key is normally certified by a trusted third party – the certification authority – in order to confirm the correct identity of the key owner. The private key must be kept secret by the owner. Any user who wants to take part in confidential conferences needs a PSE.

Everyone needs a key pair of his/her own and a certificate for the public key. For this purpose there have to be certification authorities and software in every workstation to generate an asymmetric key pair and to communicate with the corresponding certification authority. This procedure can be performed application independent and has to be done before starting the conference. It is suggested to use the formats and procedures of the European PASSWORD project [13,14,15]. Software for this purpose exists at INRIA, UCL and GMD (e.g. the SECUDE [16,17] software package of GMD).

Available security packages, like SECUDE or others, allow to generate an asymmetric key pair for the user to do encryption and decryption, to support the certification of the public key, and to generate and to read PEM (Privacy Enhanced Mail [18]) letters. The PASSWORD project has set up a European infrastructure for certification. It suggests to use PEM for key distribution.

4.3 Realization in MICE

As a first step in implementation, PSEs will have to be created for each user, providing everyone with an asymmetric key pair and a corresponding certificate. MICE conferences shall be run with session keys.

As a next step, a conferencing user agent (CUA) will be implemented, to support users in maintaining conferences, PEM letters, and session keys. For this purpose, PEM letters have to be standardized so that functions can be performed by the CUA automatically.

Usually when a conference is announced, the organizer informs participants of the details of the scheduled conference, like e.g. the date and time, the agenda, the tools to be used and the addresses and port numbers for each tool. This information is often distributed by plain e–mail.

One possibility to realize a secure conference is to follow the procedure as described above. So mail may also be used to distribute session keys for a conference. But in order to keep the key confidential, the mail content must be encrypted such that only the intended recipients can decrypt it. It is proposed to use the standardized PEM (Privacy Enhanced Mail) for this purpose. The originator of a PEM would encrypt the mail content with a key encryption key and attach it to the list of recipients. The attached key encryption key is encrypted with the public key of the corresponding recipient. Each recipient needs its own key attachment. It is decrypted with the recipient's private key. This mechanism ensures that the key attachments can only be decrypted by the intended PEM recipients who are thereby enabled to decrypt the mail content. A PEM is of no value to users outside the list of recipients.

Another possibility is to offer a Conferencing Access Manager CAM. This is a service like a simple directory service. It manages information needed for confidential conferences and for protected communication with authorized users. CAM will only be offered as a centralized service. More than one CAM may be used for managing information of several conferences. If so, one CAM does not know of the existence of the others and there will be no communication between CAMs. It is completely up to the user to address the proper CAM. Since a CAM is a peer communication entity to a user exchanging confidential information with him, it needs a security environment similar to that of the user. Thus any CAM needs a key pair of its own with an appropriate certification of its public key.

The use of a CAM will free the user from the responsibility to maintain conference announcements in his local environment and to keep the session key confidential. At the time when a user wants some information he can request it from the CAM. In addition, the session key must not be kept in the local environment rather the CUA can request it just before starting a conference.

5 Conclusion

It has been presented that MICE multimedia conferencing has been successfully utilized for several applications. All of these have been carried out in a completely open environment. Although the openness is desirable and valuable for many applications, there are others where a more restricted and secure environment is preferred. For these cases section 4 of the paper discussed ways and strategies for augmenting the MICE technology with the necessary security functions.

With these security enhancements plus the additional technical enhancements realized by different project partners, the project aims to achieve an improved quality of conferencing services. We feel that the MICE environment is particularly well suited as a test field for security functions in multimedia conferencing.

References

1. P. T. Kirstein, M. J. Handley, M. A. Sasse: Piloting of Multimedia Integrated Communications for European Researchers (MICE), Proc.INET 1993.

2. M. J. Handley, P. T. Kirstein, M. A. Sasse: Multimedia Integrated Conferencing for European Researchers (MICE): piloting activities and the Conference Management and Multiplexing Centre, Computer Networks and ISDN Systems, 26, 275–290,1993

3. S. Casner: Frequently Asked Questions (FAQ) on the Multicast Backbone (MBONE), available by anonymous ftp from venera.isi.edu in the mbone/faq.txt, May 6th93.

4. V. Jacobson: 'VAT' manual pages, Lawrence Berkeley Laboratory (LBL), February 17th 93, available by anonymous ftp from ee.lbl.gov.

5. V. Jacobson: 'WB' README file, Lawrence Berkeley Laboratory (LBL), August 12th 93, available by anonymous ftp from ee.lbl.gov.

6. V. Jacobson: 'SD' README file, Lawrence Berkeley Laboratory (LBL), March 30th 93, available by anonymous ftp from ee.lbl.gov.

7. C. Huitema, T. Turletti: Packetization of H.261 video streams, INTERNET–DRAFT, December 5, 1993.

8. T. Turletti: H.261 Software Codec for Videoconferencing Over the Internet, Research report No 1834, INRIA, January 1993.

9. Video codec for audiovisual services at p x 64 kbit/s, CCITT Recommendation H.261, 1990

10. C.-D. Schulz, R. Nilsen, M. A. Sasse, T. Turletti: Multimedia Conferencing for Remote Events: A User's Guide, submitted to ACM Multimedia '94, October 15–20, 1994, San Francisco, California

11. ISO/IEC DIS 10745: Information technology – Open Systems Interconnection – Upper layers security model.

12. ISO/IEC DIS 10181–4: Information technology – Open Systems Interconnection – Security frameworks in Open Systems.

13. P. Kirstein, P. Williams: Piloting Authentication and Security Services within OSI Applications for RTD Information, Computer Networks and ISDN Systems 25, 1992 pp. 483 – 489.

14. W. Schneider, PASSWORD: Ein EG–Projekt zur pilotmäßigen Erprobung von Authentisierungsdiensten, Kommunikation und Sicherheit, Teletrust Deutschland e.V.,1992.

15. PASSWORD Reports, available by anonymous ftp from cs.ucl.ac.uk.

16. W. Schneider (Hrsg.): SecuDE Overview, Version 4.1, Arbeitspapiere der GMD 775, Sept. 1993

17. SecuDE Documentation, available by anonymous ftp from darmstadt.gmd.de: W. Schneider (Hrsg.): SecuDE, Vol.1 Principles of Security Operations, Vol.2 Security Commands, Functions and Interfaces, Vol.3 Security Application's Guide

18. US Internet RFC 1113 – 1115

Secure Multimedia Applications and Teleservices - Security Requirements and Prototype for Health Care

H. Bunz, A. Bertsch, M. Jurecic, B. Baum-Waidner

IBM European Networking Center
Vangerowstraße 18
D-69115 Heidelberg
BUNZ at VNET.IBM.COM

C. Capellaro

Siemens AG
Otto-Hahn Ring 6
D-81730 München

Abstract: Due to technological advances, distributed multimedia applications and teleservices, such as multimedia mail, video-conferencing, joint-viewing, or access to multimedia databases are becoming a reality. However to get accepted by the users and to meet business requirements, these teleservices have to include appropriate security mechanisms and services to limit the risk and the extent of damage caused by attempted fraud.

This paper describes security requirements of teleservices and their mapping to strong security services. An integrated application scenario demonstrates on the one hand the use of managable security services by multimedia applications, and on the other hand a solution for security management based on a security architecture developed by the RACE II project SAMSON (Security and Management Services in Open Networks). As example the electronic exchange of patient records in health care has been chosen, requiring distributed multimedia applications as well as strong securitymechanisms.

Multimedia applications and teleservices are very useful to provide just-in-time patient information in time-critical situations and enable better treatment of the patients at lower cost. Due to the nature of the collected data (medical results, diagnoses etc.) guaranteeing privacy and integrity of the data is a must. In addition the processing of personal and medical data is the subject of many restrictions and requirements imposed by laws, the users of the system, and the patients. As indispensable security requirements, the authentication of medical staff, the control of access to patients records, the generation, distribution and maintenance of underlying cryptographic keys and a strong audit facility providing a quick but convincing overview of all security relevant events and alarms have to be provided.

Furthermore, a couple of privileged persons acting as network administrators must be supported by a set of strong management tools, enabling the control of data processing and information exchange as well as the management of the security services. Given the strong security requirements, the health care scenario is well suited as a testbed for secure multimedia applications and teleservices. The results and concepts are also applicable to less security sensitive multimedia applications like group work in consortia, aircraft maintenance and joint design eg. in the automotive industry. All these issues are addressed by the RACE II project SAMSON, and the concepts are verified in a prototype implementation which demonstrates how multimedia applications - namely electronic information storage and retrieval system - can provide appropriate security levels for commercial use in an IBC environment. Furthermore it is possible to build such a system using existing teleservices, security and management services conforming to open architectures. In order to demonstrate a smooth migration path towards networked multimedia applications, existing distributed applications as well as multimedia applications are combined in the prototype. A key point in this context is that security aspects have to be covered early in the design. In addition, it is crucial to integrate an overall security management concept.

1 Health Care and their Security Requirements

The aim of the RACE II project SAMSON is to design and implement a security management architecture applicable to the emerging European broadband networks and validate the development by integrating the managed security services in secure applications. The approach chosen consists in

- Analyzing requirements for security features and their management, for various application scenarios

- Prototyping of integrated solutions (applications, security services, security management)

As application area to be considered for the prototype,the health care area was chosen for the following reasons:

- Due to the sensitive nature of patient records, with respect to privacy as well as to patient's life and health, the health care area seems to imply a representative set of security requirements (caused by system, users, and law) which may be a superset of those implied by several other areas. Therefore the experiences made in this area are be useful for other areas, too.

- The health care area has practical relevance also for involving multimedia applications for processing text, graphics, forms, images (e. g. X-rays), audio. The latter is currently more important than video to support quick annotations.

It is obvious that appropriate clinical computer systems may improve the treatment of patients and reduce costs: Patient data can be transferred electronically from one doctor to another (e. g. in a different location), and thus multiple medical examinations may be avoided. On the other hand, without sufficient protection mechanisms, patient information could easily be abused or processed in a patient damaging way. Hence, a secure solution is needed combining the advantages of open computer networks and dealing with their disadvantages by appropriate means.The scenario chosen enables the exchange of patient records (e. g.multimedia documents) between general practitioners and hospitals. In the following we shortly scetch the scenario chosen for the demonstration application (see figure hostarc).It supports the exchange of patient records (e. g. multimedia documents) between general practitioners and hospitals, and within the hospital.

Here we focus on

- the admittance of a patient,

- consulting doctors of remote hospitals,

- processing, exchange, and storing of documents within one hospital

as interesting actions with respect to the security requirements.

The scenario consists of a secure hospital LAN and an external public WAN, separated by a security firewall. To enforce obedience to data protection laws in all processing steps, the LAN is physically protected, due to local policies, whereas security for the exchange of data over the WAN has to be achieved aided by open standards and protocols. Other hospitals and locations of general practitioners may be interconnected over the WAN with the hospital, e. g. to send patients to the hospital via electronic mail, or for remote access to the archive containing all patient relevant data. All security relevant events in all components cause logging information which can be used for separate auditing. Secure desktops are used to enforce obedience to the security policy for information processed locally in a workstation. The organizational directory is used for access control to the archive. All components are described below in more detail.

In collaboration with a university hospital, end-user requirements of a clinical information system have been elaborated [1,2]. Resulting recommendations for an integrated solution interconnecting several environments are derived:

- to use a uniquely structured network with high band-width e.g. for exchange of multimedia documents

- to use communication protocols according to open standards (DCE, OSI)

- to use standardized (distributed) applications

- to provide user-friendly user interfaces enforcing obedience to the security policy

- to enforce obedience to data protection laws in all processing steps

- to manage the system efficiently

- to provide a system security management covering all subsystems

Thereby important security aspects have to be taken into account, e. g.

- The creator of a medical document has the responsibility for the obedience to data protection laws

- It should be possible to delegate access rights and responsibilities

- Recovery of data concerning delegation of rights and responsibilities should be possible
- Access rights to a document may change, depending on its status
- Access rights should depend on the accessor's role and function (e. g.doctor, nurse, secretary) and on the document type
- All steps of patient treatment have to be documented
- Security checks should be able to be bypassed in case of emergency (with special audit activity only).

On designing and implementing a prototype for an application scenario as illustrated in (figure 1), all requirements given above have been considered in order to obtain a secure solution.

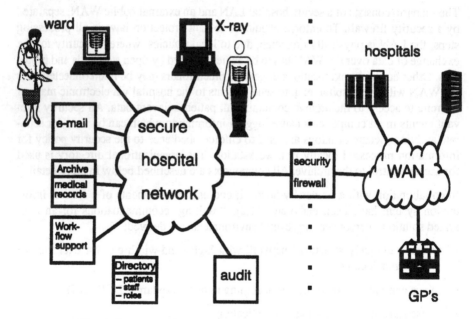

Fig. 1. Communication of medical records within and between hospitals

2 Secure Multimedia Applications and Teleservices

As shown in figure 2 the prototype is composed of three parts
- Multimedia applications and teleservices
- Security Services
- Security Management

In order to have minimal dependencies between these parts of the system, the platforms communicate via suitable protocols and programming interfaces. Security services and security management generate only normal data traffic not requiring

multimedia transport systems. In SAMSON, existing open protocols and applications
are used to the extent possible (e. g. CCITT X.500, CMIP, OSF DCE). Interfaces are
used and extended (e. g. XMP, XDS, GSS API) for appropriate layers.

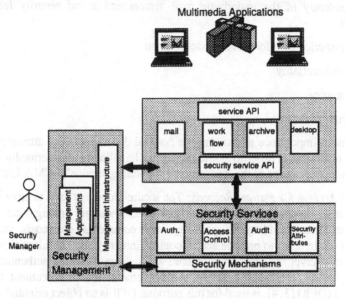

Fig. 2. Secure teleservices and security management

Because of legal requirements in health care it is crucial to enforce the security policy
in all processing steps and to document all steps in the treatment of a patient. Therefore strong security services have to be included in different parts of the applications,
and all security relevant events have to be logged. This security audit information can
be processed by other applications, e. g. to collect all diagnosis signed by a certain
doctor in a certain time period, or to collect all (signed) modifications in the archive
for one day and to sign and store them in a trusted audit trail for the purpose of wit-
nessing. Moreover, logging information may be used to trigger hospital internal workflow applications.

A representative set of services has been implemented:

- *Secure Desktop:* The secure desktop enforces the security policy for all documents
  stored and processed locally on a workstation. Users have to identify themselves to
  get access to the desktop - protected by the secret key of the user. After the login
  users are able to do all actions according to their current role and the security policy. Attempts to do other actions are rejected and audited. Confidentiality is provided through encryption of selectable document classes. Documents are signed to
  provide integrity. Access control policies are enforced between different services,
  e.g., by restriction of the distribution of an archived patient record via electronic
  mail. Different integrity and confidentiality mechanisms can be configured for the
  various document classes. The implementation is using labelled objects as defined
  by ECMA 138.

- *Electronic Mail*: Both for hospital internal and for external communication, mail messages have to be protected against modification and other fraud related attacks. Moreover, privacy requirements have to be met especially for medical findings. Independently of the underlying mail system end-to-end security features are needed:

  - •Restriction of information dissemination

  - •Confidentiality

  - •Integrity

  - •Proof of origin

It is of crucial importance to transmit the hospital defined security attributes together with each document in order to assure the correct handling in all intermediate systems. For this purpose features of the standardized "Labelling service"(ECMA 138) are used.

- *Secure Archive for Patient Records:* The secure archive is used to store all patient data caused in connection with their treatment. These are among others: personal data, anamnesis, findings (e. g. X-rays with comments), diagnosis, and medicamentation. In order to provide access control, integrity and confidentiality, a secure document server has been implemented which supports strong authentication, rule based authorization, and auditing. An implementation of Document Filing and Retrieval (DFR) [3, 4]. is used for this purpose. DFR is an object oriented service for the storage of a large collection of documents in a distributed client server environment. The DFR base standard allows to include external security services like strong authentication (X.509) and access control (X.741). Therefore this OSI standard is well suited to form a testbed of the emerging OSI security architecture and services. In addition multimedia extensions for audio are included within the Berkom GAMMA project.

- *Workflow Support:* Workflow support is needed in a hospital to control the course of events relating to the treatment of patients. For example, an incoming message from a doctor hospitalizing a patient may have to cause examinations and operations to be prepared (depending on the diagnosis involved in the message), a bed to be reserved etc. Thereafter, a message confirming the admittance may be sent to the external doctor. Since external events have to be integrated securely, audit information created by the security gateway is used. This audit information is well suited to be used for workflow support and start the required hospital internal actions. For the purpose of demonstration, the prototype includes a notification service: Each incoming admission of a patient causes an electronic mail to be sent to the responsible doctor.

- *Security Gateway:* The security gateway is the only point of connection between the hospital network and the external network. In order to avoid break-in attempts, the following features are supported:

•protocol mapping

•verifying user identity and mapping of privileges

•confidentiality of the patient records exchanged over the public network (see "electronic mail"),

•separate audit of all external communication

The gateway currently allows external access to the patient archive (accordingly to the security policy), to selected information contained in the organizational directory, and workflow support by providing logging information (see "workflow support").

• *Organizational Directory.* As an essential part of the application, the Organizational Directory stores all information on all organizational units and persons. A secure directory implementation - as implemented within SAMSON (X.500-92) [5-8].- is needed to prevent unauthorized access. All organizational units and persons of the organizational hospital structure are represented by corresponding entries in the Organizational Directory. These entries contain information on the person and his/her current role (e. g. of a doctor in a certain department, being on duty this day), and store certificates for the person's public key. If any user likes to get access to applications, e. g. to the archive, he/she makes a strong authentication bind to the Organizational Directory (X.509). The access is done successfully if the certificate stored in the user entry is valid and the user is not disabled to access the directory (e. g. by a revocation list). Then, to get access to several applications, the user requests a so-called PAC (Privilege attribute certificate) from the PAC-Authority which controls the PAC-Server as part of the Organizational Directory. This PAC will prove the user's right to access the required information and is necessary to a successful access to this information. It contains detailled information, signed by the PAC-Authority, on the user's privileges. The PAC-Authority signs and returns the PAC representing information on the user and the user's current role described by several security privilege attribute values. On the other hand security relevant objects, to be accessed using the PAC, carry information on which privileges are needed to access which part of the object in which way. Depending on the role information of this PAC, and on the privileges required by the object considered, the application allows the user to access this object or not. The main advantage of this concept is that

•each change of a role or each modification of the security policy is included in configuration data of the Organizational Directory and therefore immediately available in the provided PACs

•in this way, each change of a role or each modification of the security policy will immediately be reflected in the access control decisions.

So granting of access rights is actually based on the current state of roles and their connections to the organizational structure as well as to the security policy.

For integration of security services into an application, an interface is crucial which hides details of the security mechanisms from the applications, but still offers all security services they need. Additional configuration information is required to select the appropriate security mechanisms and may be part of the system security management and not under the control of the enduser. These interfaces allow changing selected security services easily (e. g. cryptographic algorithms) in order to comply with new requirements that may arise from changes in data-protection laws, or from the use of new application programs.

3 Security Architecture

In general, security policies depend on

- legal regulations
- security awareness of involved organizations
- value of processed information
- costs of security versus risk and extent of damage.

All security services selected have to be used consistently by each multimedia application and teleservice involved. This requires an organization specific modelling of a security policy covering

- Organizational Model
- Classification of processed information (security labels)
- Access control policies
- Audit, accounting policies
- Business processes modelling

Currently teleservices handle mostly public information. However, they have to include appropriate strong security mechanisms required for sensitive information. In order to be backwards compatible, the security service interface used by the applications as well as the underlying security architecture must be extendable. Moreover, a choice of several standards and algorithms have to be given to allow the adaptation of security levels to graded user requirements. The security architecture and services selected in the SAMSON application demonstrator are based on available implementations of open standards (OSI, ECMA):

- *Authentication:* (X.509) - This standard defines simple authentication mechanisms (name, password) as well as strong ones. The latter are basing on asymmetric digital signature schemes where each user holds a key pair: one key is kept secret and is used for signing exclusively by the user, the other is made public and may be used by any party to verify the user's signature. The public key certificate is accessable in the X.500 directory using the DAP protocol.

- *Authorisation*: (X.741) -Access control policies in the hospital environment are quite complex and depend on the identity of the person, its current role, the document type as well as its current status. In addition the environment is very dynamic - e.g. during the night a different doctor may be responsible, patients may be transferred to another department. To address such a complex behavior, access control lists are not sufficient. Therefore a functional subset of X.741 - originally specified for network management - was implemented which covers the complexity mentioned above.

- *Labelling service:* - Labelled objects, as defined in ECMA 138, rely on encryption and digital signatures for protection against modifications and misuse. Any objects may be included in the certificate, and security labels are attached, used for invocation appropriate security services. Labelled objects provide a flexible way to secure migrating objects, e.g., security over unsecure mail network as well as locally for a file system. For this purpose, an object oriented interface has been implemented. In an efficient way, it can be configured which classes of documents have to be secured by which confidentiality and integrity algorithms.

- *Security Audit:* -Auditing is essential within any secure environment. It allows-monitoring proper usage of the system and detecting security failures and potential breaches of the system security. However,auditing is a pervasive service - it has to be implemented and consistently used by all applications, security services, and network resources. In the prototype, logging information arises from security relevant events. Therefore standardized data structures (X.721, X.740) are used. This information is managed e. g. by configuration of event forwarding descriptors (EFDs) as used by the SAMSON management. All information arising this way can be used for analysis and evaluation of all events (e. g. detecting attempted logins), and for recovery. In addition, management extentions allow for application specific information exchange.

- *Cryptographic Support:.* - Cryptographic mechanisms are the basis for all applied security services. However, the algorithms and mechanisms applied are not fixed but can be chosen accordingly to practicability and security policy requirements.In our application, this choice can easily configured for each relevant function of each application. Generally,

    •Data integrity can be achieved by the use of cryptographically sealed checksums, or by signatures (using the secret key, to be checked by the public key)

    •Data confidentiality is based on encryption mechanisms using either symmetric or asymmetric techniques

    •Digital Signatures are realized by means of asymmetric cryptographic techniques

    •Certificates are protected datastructures which prove the authenticity of public keys and privilege attributes (i.e. authorizations)

4 Security Management

The profit of standardized networking, as it is proposed by the International Organization for Standardization (ISO), is the opportunity to have a maximum of flexibility for the use of possible communication infrastructures or the employment of different end systems. The variety of different components and the evolutional development of a network during its life-time enforces the introduction of administrative tools that provide a quick overview about the networks current state and that offer a way to configure or change network entities where it is necessary. By this reason network management can be seen as the set of activities which are outside the normal instances of communication, but which are required to support and control this communications, and the information exchanges necessary for performing this control. The online administration of a network is one aspect of network management, the other one deals with offline activities such as hardware and software installation or other manual work.

Security management covers all administrative tasks that are related to security aspects within a network. In the terms of ISO 7498, Open Systems Interconnection (OSI), the security aspects of a network environment can be divided in those of the network itself, those of security services which are provided within this environment, and those of security mechanisms which are employed to realize the security services. This distinction has been expressed in the introduction of three categories of security management, called system security management, security service management, and security

mechanism management. Another aspect, that has to be taken into account, is the security of network management itself. This task has a fundamental meaning, since remotely configurable security components in a network enable all ways of unauthorized manipulation.

Offline management activities are strongly dependent on the different components that are employed in the network. By this way it is hard to find common aspects or general instructions for this kind of management. This is the reason why this subject is hardly elaborated in standardization work. On the other hand side the remote administration of components in a multi-vendor environment requires the use of a common communication protocol and a way to describe the different managed resources. These tasks have been addressed with the Common Management Information Protocol (CMIS, ISO 9596) and with the Guidelines for the Definition of Managed Objects (GDMO, ISO 10165). Furthermore ISO has done some investigations to describe general management activities as they are required in any networked environment. The result of this work can be found in the ISO 10164 series which is entitled as System Management Functions.

Before we go to a more detailed description of the different categories of security management mentioned above, we should introduce the concepts of security policy and domains.

4.1 Security Policy

The ISO security management gives network administrators the means to change the networks properties in a certain way. A policy serves as a certain kind of coordination and target-orientation of the network administrators activities. A security policy in special can be defined as a set of rules that are applicable to a set of network entities. On one hand side these rules express authorized and prohibited management activities on the security components within the network, and on the other hand side a description of how the security related attributes and properties of network entities should be given. A security policy should cover both legal and company-specific predicaments, furthermore it should take into account technical aspects and user requirements. All these different aspects that have influence on the definition of a security policy show the difficulties that have to be overcome until a complete and consistent description has been found. It gets necessary to bring specialists of various areas to one table and to find an agreement between their different view-points and interests. The resulting security policy should be represented in the managed network by taking use of the same means that are used to describe the network entities it is related with. This makes it possible to compare the current system state with policy rules. As well it is possible to apply management principles to the policy itself, this enables a maintenance of a security policy by taking use of a security policy management which is part of the system security management as it is introduced in the next section.

The set of network entities that are object of the regulations of one security policy is called a security domain. Interactions of network entities that belong to different security domains get critical if the desired kind of interaction is governed in different ways by the two involved policies. In this case the installation of an interdomain service gets necessary. A service of this kind provides mappings between the different terms of security that are used within the involved domains. Furthermore it can be used to find a least sets of common understanding and abilities in the area of security.

4.2 System Security Management

The system security management concludes the administration of the aspects of security within a network as a whole. This means, that it is responsible for the maintenance of a security policy and the control of the consistency between security policy rules and system properties. Furthermore the system security management has to take care that security management and other network management decisions and activities are coordinated. It is often the case that a non-security related management activity has its influence in security management. As an example the introduction of a new user, which is in the area of responsibility of system configuration management, implies security management activities like fixing access rights and creating authentication information for this user.

Another aspect of system security management is the handling of events, alarms and audit records. Especially a security administrator is interested in ongoing activities in his area of responsibility. ISO introduced different tools for managers in order to support monitoring. One way to get informed is the concept of event forwarding. Here the administrator has a possibility to install discriminators within end systems which are in his area of responsibility. These discriminators contain filters that can be configured

in a way that events occuring in this system which belong to the context specified in the filter are immediately forwarded to the manager. This concept has its legitimacy especially in the case when critical events are observed. Here ISO introduced a certain kind of events called alarms. Another way to collect information about things occuring in the network is audit trailing. Here events are collected in certain logs within the end systems. The records in these logs can be examined via a set of management functions which are concluded as audit trail management. The decision whether a certain event is logged or not is made by the use of discriminators as they have been introduced above.

4.3 Security Service Management

In terms of ISO, a service is an abstract feature that is provided to users in the network with the aim to give them a certain level of comfort, so that the users accept the communication infrastructure. Security services are data confidentiality and integrity, user authentication and the provision and proving of privileges. Beneath general management activities on services, such as switching on and off, a typical management operation on security services is the selection of security mechanisms which have to be used

to realize the service. In the case that there is more than one possible mechanism available, there must be a way to find a negotiation between the parties involved. This is as well a matter of security service management.

4.4 Security Mechanism Management

The security services introduced in the last section can be realized by security mechanisms such as encryption, digital signatures, access control decision functions, authentication mechanisms or notarization functions. These mechanisms themselves are instantiated via implementations of certain algorithms, access control lists, password files, etc. These applications have their own management requirements such as the provision of parameters or the notification about error states, furthermore products of different vendors may show different properties. All in all the management of security mechanisms depends very much on the kind of mechanism and on the way it is realized. Nevertheless there can be found common ways. Investigations in this area have been done e. g. in SAMSON.

4.5 Secure Network Management

Since the management has a remarkable impact on security assets in the network, it has to be made secure itself, to prevent unauthorized use. Taking a closer look on the network management architecture you can recognize three vulnerable points. These are the management applications, the stored management information and the exchanged management data. A set of well known security measures can be applied to protect these components:

- The access to network management applications should be restricted to authorized persons.

- The management information being distributed in the network must be secured. There is less the requirement of confidentiality of these data than of integrity and availability. Since every security service has two interfaces, one to users and one to the management, there must be a strong distinction between these two entrances.

- There must be a certain level of reliance for the administrator that the management actions he initiates are performed in the way he expects. This can be ensured by mutual authentication between the management application and the respective application residing on the managed system which is responsible for the component forming the target of the management request. Furthermore data integrity services must be involved during information transmission.

The security requirements of network management stated here can be fulfilled by the use of security services which are installed in the network. Since these security services are object to security management, as it has been pointed out in "Security Service Management", we run into a bootstrapping problem: The security management itself is dependent of having performed certain security management activities previously. This problem can be solved only by doing the first installation and configuration of security services and mechanisms offline before. This is another sort of offline management activity as it has been mentioned at the beginning of this chapter.

Acknowledgments

The RACE II project has been partly sponsored by the CEC under the contract R2058. In addition we would like to thank all SAMSON partners for valuable discussions and contributions.

References

[1]   Marjan Jurecic, Herbert Bunz: *Exchange of Patient Records - Prototype Implementation of a Security Attributes Service in X.500*, to be published

[2]   Marjan Jurecic, Ulrich Kohl, Ernst Pelikan: *Datenschutz und Datensicherheit fuer verteilte Klinikanwendungen*. Arbeitstreffen Entwicklung und Management verteilter Anwendungssysteme, Oktober 1993. GI/ITG Fachgruppe Kommunikation und Verteilte Systeme.

[3]   Information technology - *Document Filing and Retrieval (DFR) - Part 1: Abstract Service Definition and Procedures* ISO/IEC 10166-1

[4]   Information technology - *Document Filing and Retrieval (DFR) -Part 2: Protocol specification*, ISO/IEC DIS 10166-2

[5]   *The Directory - Overview of Concepts, Models and Services*, CCITT 1992 Recommendation X.500

[6]   *The Directory - Models*, CCITT 1992 Recommendation X.501

[7]   *The Directory - Authentication Framework*, CCITT 1992 Recommendation X.509

[8]   *The Directory - Abstract Service Definition*, CCITT 1992 Recommendation X.511

The BERKOM Multimedia Teleservices

Michael Altenhofen, Joachim Schaper, Susan Thomas

Digital Equipment GmbH,
CEC Karlsruhe Vincenz-Priessnitz-Str.1,
D-76131 Karlsruhe,
E-MAIL: {altenhof, schaper, sthomas}@kampus.enet.dec.com

Abstract. This paper reports the BERKOM Multimedia Teleservice project, which is a joint effort between the German PTT and various companies and research organizations to build prototypes for future broadband public services. The goal of this paper is to give an introduction to the overall design of the services, their architecture and to describe the current implementation state. After a problem description of multimedia applications according to open heterogenous network environements, the specific project context of BERKOM is introduced. The services Multimedia Collaboration (MMC), Multimedia Mail (MMM), and the Multimedia Transport system (MMT) will be described. A summary on our current experience in providing an implementation on top of Alpha AXP and OSF/1 of the 3 subsystems provides the Digital expierence within the consortium. An outline of the future work gives directions for the focus for the next few years.

1.0   Introduction

1.1 Background

The motivation to start the BERKOM programme was the fact that larger organizations such as corporations or administrations are becoming more and more geographically distributed. As a result, increasing communication needs -- both intra- and inter-organization -- make the use of high bandwidth communication facilities (e.g. high speed network links) necessary as well as useful. At the same time, the increasing power of computer systems expands their scope from the traditional datatypes, such as text and graphics, to multi-media datatypes such as image, video, and audio. In combination with the advent of high performance networks, the ongoing evolution of multi-media workstations with a reasonable price/performance ratio builds the technical foundation to introduce new public multi-media teleservices as a standard telecommunication infrastructure.

BERKOM [5] (the BERliner KOMmunikationssysteme) is one of the most prominent Broadband ISDN trial projects world wide.The first phase of the project ended in 1991 after 5 years. Unlike the first phase, which explorred numerous independent broadband network solutions, the second phase, which lasts until the end of 1994, concentrates on providing a single uniform communication infrastructure. This infrastructure should be the base for future broadband multimedia applications. The goal is to provide three prototypes of teleservices which are implemented in a standards based open, and heterogenous environment of networked multimedia workstations and PCs. The 3 subprojects, which are outlined below, comprise the areas of network transport, realtime conferencing, and multimedia extensions to mail.

1. The Multimedia Transport Service (MMT) provides the communication platform for audiovisual communication. MMT is based on the Internet Stream Protocol ST-2. It allows the creation of multi-endpoint connections with guaranteed throughput and delay. The current transport media which the consortium are focusing on are Ethernet, FDDI and ATM.

2. The Multimedia Mail Service (MMM) allows the exchange of multimedia messages consisting of text, image, video, audio and structured documents. It provides extensions to the X.400 functionality as well as gateways to other multimedia mail systems such as Internet MIME.

3. The Multimedia Collaboration Service (MMC) supports concurrent cooperative work of persons in a distributed network environment. The services enable the user to share applications among the participants and to conduct an audiovisual conference using desktop computers.

1.2 Partners

The consortium lead by the DeTeBerkom consists of

Partner	Location	Platform
DeTeBerkom	Berlin	
Danet	Darmstadt	HP 7000
Digital/CEC and TU Karlsruhe	Karlsruhe	Alpha OSF/1
FhG	Darmstadt	SUN
GMD Fokus	Berlin	SUN, NeXT
HP	(TU) Berlin	HP 7000
IBM/ENC	Heidelberg	RS/6000

Table 1. Project Partners

Partner	Location	Platform
Liebing & Ullfors	Berlin	PC
MacConsult	Berlin	Apple
Siemens	München	SGI, UNIX-PC's
SIEtec	Berlin	SUN, SGI

Table 1. Project Partners

The consortium has agreed to design and implement the teleservices under contract with the consulting group of the German Telecom (Deutsche Bundespost Telekom), DeTeBerkom which supervises the overall project. The objective of the group is to provide teleservices targeted to the working environments of large, geographically distributed organizations (e.g. office administration, medicine , or manufacturing). Within these scenarios, the use of broadband WANs and the interconnection of heterogenous LANs allows communication to take place from the desktops of the individual users. No special video conferencing room full of specialized expensive equipment is needed.

The final goal is to create an open, homogenous framework for collaborative computing running on a variety of hardware platforms. So, already during the design phase the team had a strong focus on using international standards (ISO/CCITT) wherever feasible.

The following chapters will describe the services to a first level of detail. Current publications have been completed for MMC [6], MMM [7] and MMT [8]. For more details please refer to the specification documents [1], [2], and [3].

2.0 Multi-media Mail

2.1 Overview

The Berkom MMM Teleservice uses the X.400 e-mail standard to define an open multi-media mail system which supports text, structured office documents (ODA) [10], image, audio (G.711 [11], G.721 [12]) and video (SMP [9], MPEG [13].

Beyond the X.400 extensions, the teleservice has two distinguishing features which will be described at length in later sections.

1. support for the creation and resolution of external references, including real-time viewing of the external data.

2. support for structuring single mail messages; this structure can range from the relatively simple, e.g. cross referencing or annotating a text, to the more complex, e.g. hyper-media networks.

X.400 and New Body-Part Types

For those interested, this section gives more information about X.400 and how it is used by the Berkom Teleservice. For those less interested, all but the next paragraph can safely be skipped.

X.400 structures a message into a header and a sequence of typed body parts. It also specifies how to encode some types such as text and ODA. In addition, it specifies how to define new types by using a type called externally-defined. The Berkom Teleservice uses this mechanism to define the types image, audio, video, link and external reference. The last two are used to implement message structuring and external references, respectively.

Supporting a new type requires the addition of viewers/editors or converters for the new type to the User Agent (UA), the program which enables the user to send and receive mail. But, it requires no changes to the X.400 MTS (Message Transport System) which transports the mail from UA to UA. See Figure 1 for a diagram of the X.400 mail system components.

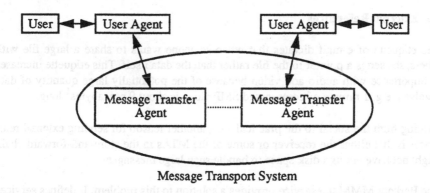

Message Transport System

Fig. 1. The X.400 Mail System, is a Store and Forward System, where each message is copied by an MTA before forwarding starts

Figure Note: X.400 '88 is needed because '84 does not have the body type 'externally defined' which is used to define new types.

Information about the encoding of body parts is given in the following table:

Component	Representation
text	IA5 (8 bit)
document	ODA/FOD26 + Corrigenda
audio	Industry Implementation of G.711, G.721
image	Image Interchange Format
audio/video	Phase 1: SMP[a]
	Phase 2: MPEG, MPEG, or M-JPEG
external reference	Distinguished Object Reference
link	textual

Table 2. Supported Types of Body Parts (include encoding, etc.)

a. The SMP [9] Software Motion Picture software codec is a privat format from DEC which has been ported to all partner platforms.

Table Note: (about ODA) The ODA-encoded documents are converted from/to the
Digital Document Interchange Format (DDIF) by the DIGITAL UA.

2.2 External References

The etiquette of e-mail dictates that when someone wants to share a large file with
others, she sends a pointer to the file rather than the data itself. This etiquette increases
in importance with audio and video because of the potentially large quantity of data
involved, e.g. 1 minute of compressed SMP video is about 9 megabytes[1] long.

Turning from the social to the practical side, another reason for sending external references is that either the receiver or some of the MTAs in the store-and-forward chain
might not have enough disk space to handle such large messages.

The Berkom MMM teleservice provides a solution to this problem. It defines services
which enable a mail sender to move data to a server, create an external reference to it
and mail the external reference encoded in an X.400 body part.

The receiving user can resolve the external reference and copy the data from the server
to a local file. Alternatively, the receiver can take advantage of a real-time viewing ser-

---

1. This figure is based in an assumed data rate of 150
KBytes/second, a typical CD-ROM data rate.

vice to view the data as it arrives from the server. This is an extremely useful feature for receiving systems with limited disk space. (Note: The transport service defined by MMT can be used to implement a real-time viewing service with guaranteed quality, e.g. throughput and delay.)

Myriad uses for external references can be envisaged, e.g. an e-mail video-of-the-week club whose subscribers receive a weekly mailing of video previews (trailers) and external references to the previewed videos. The next figure shows the sequence of actions required to provide and use such a service using external references.

(Note: one interesting problem is how to prevent users from stealing the videos and passing them on for free)

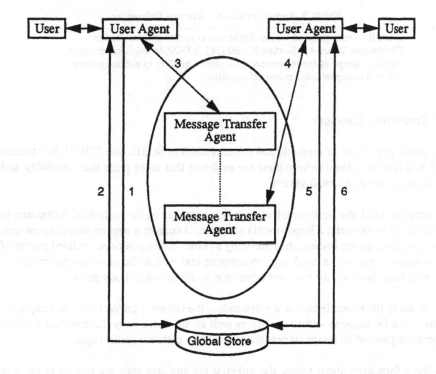

Fig. 2. USING EXTERNAL REFERENCES: If a mail contains an external reference body part, the following steps occur to pass the external data to the user:

Step	Action
1.	The User Agent gives data to the Global Store which stores it.
2.	The Global store returns a DOR[a] containing a reference to the stored data.
3.	The User Agent mails the DOR as a bodypart in an X.400 message.
4.	The receiving User Agent receives the X.400 message.
5.	The receiving User Agent uses the DOR to request the data from the store.
6.	The store returns the requested data.

Table 3. Steps for using the External References

a. DOR - Distinguished Object Reference is part of the standard on Distributed Office Application Model [14]. A DOR contains information on the datatype of the referenced data object, qualitiy of service parameters and transport mechanism information.

2.3 Structured Messages

The rising popularity of methods and systems, such as SGML and WWW, for structuring and linking related information are evidence that users prize this capability and would welcome it in a mail system.

With this in mind, the Berkom teleservice superimposes on the basic X.400 structure (a sequence of body parts) a hyper-media structure. It enables a user to superimpose networks of links on the bodies, thus allowing a reader to jump between related pieces of information. This can be used, e.g., to annotate text with audio or video comments by defining links between the text and other (e.g. audio or video) body parts.

As shown in the example below the linkends -- the information the user can jump to or from -- can be sections within a body as well as the entire body. In this case the link ancors are parts of an image, where the target body parts are audio clips.

All the information about where the linkends are and how they are related is put into the link body-part. This decision has at least two desirable consequences.

1. Multiple ways of relating the same information can be sent in one mail message by sending a link body-part for each.

2. Mail programs which do not support linking can ignore the link body-part, but still process all the other parts since they are sent exactly as they would be if the message had no links.

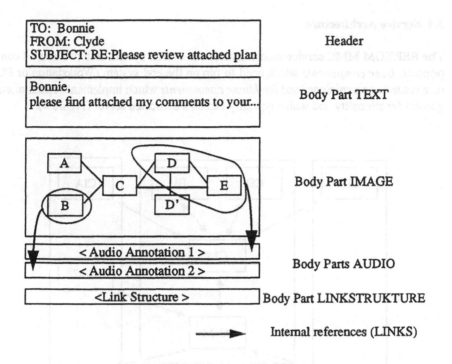

TO: Bonnie FROM: Clyde SUBJECT: RE:Please review attached plan	Header
Bonnie, please find attached my comments to your...	Body Part TEXT
A C D E B D'	Body Part IMAGE
< Audio Annotation 1 > < Audio Annotation 2 >	Body Parts AUDIO
<Link Structure >	Body Part LINKSTRUKTURE
——————▶	Internal references (LINKS)

Fig. 3. enumerates the different body parts of an example mail.

Additionally to the internal link structure an external reference mechanism will allow to store parts of a mail on a global server (e.g. hugh image or video data) and give the user the option to decide if he wants to view the data or not. If he decides to view it, the data would be transmitted at presentation time. This concept allows to extend the current store and forward mail systems to a more flexible system which is designed to easily cope with large amounts of data (giga up to tera byte).

3.0 Multi-Media Collaboration

As opposed to Multimedia Mail, which supports *asynchronous* message-based communication, Multimedia Collaboration provides *synchronous*, real-time interactions between members of a geographically dispersed work group. To accomplish this, the BERKOM MMC Service implements a shared computer-based workspace augmented by real-time audio/video conferencing.

The shared computer-based workspace is based on the *WYSIWIS (What You See Is What I See)* metaphor, i.e. users cooperate by sharing the same *view* of an application, e.g. a document processor.

Audio/video communication is completely integrated to the end system and is used to supplement the conferencing context in the shared workspace.

3.1 Service Architecture

The BERKOM MMC service model divides the overall system into two sets of components: *User components* which need to run on the end system (Workstation or PC) of a conference participant and *Backbone components* which implement central management functionality and which typically run on a dedicated server (Figure 4).

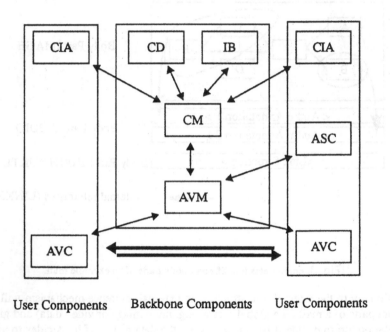

| User Components | Backbone Components | User Components |

Fig. 4. MMC Service Architecture

Users interact with the system via the *Conference Interface Agent (CIA)*. The CIA exports all necessary service operations and conveys conference status information to the user. Audio/video communication between conference participants is handled by *Audiovisual Components (AVC)* which implement an abstraction of the audio/video capabilities of an end system. The functionality that is needed to support shared workspaces are provided by *Application Sharing Components (ASC)*. CIA and AVC comprise the minimum set of user components which need to run on an end system. Only those users who want to share applications within a conference should have an ASC running.

A MMC backbone consists of four components: The core module of the service that runs and administrates conference is the *Conference Manager (CM)*. The *Conference Directory (CD)* implements a "white pages" service by storing address information of (potential) MMC users (and their default user components). All management functionality concerning audio/video communication is handled by the third central component, the *Audiovisual Manager (AVM)*. *Invitation Brokers (IB)* are used to store and forward invitations to a particular (set of) user(s).

3.2 Conference Management

In order to be able to support a large variety of conferencing situations and scenarios conference management in the BERKOM MMC Service is divided into several subtasks, namely

- participant administration,
- conference administration, and
- token administration.

Different group structures can be modeled in the system by assigning different roles to participants in a conference. For conferences which happen on a regular basis, the service also supports the concept of "conference groups" which make the setup of such "group meetings" very easy. Furthermore, conferences may be be run in different modes (open, joinable, or closed), thus giving users the freedom to specify their conference type within a broad spectrum.

Within running conferences, *token* or *floor* policies need to be applied to control access to the shared workspace. Again, the service has been specified to support a large variety of different strategies [2].

Last but not least, the system provides flexible mechanisms to deal with the problem of how people can find out what conferences are running and in which conferences they could/should participate. On one hand, the service has a model of location independent invitations. Rather than sending an invitation directly to an user the Conference Manager delivers the invitation to a designated Invitation Broker. On the other hand, the Conference Directory will also store information about ongoing conferences, so that users can use this "blackboard" to find conferences in which they are interested.

3.3 Application Sharing

Computer-based shared workspaces come in two different flavors: Systems either provide them by turning off-the-shelf single-user applications (e.g. word processors, spreadsheets) into multi-user applications or they come with collaboration-aware, true multi-user applications (e.g. multi-user whiteboards). The BERKOM MMC Service supports the former approach, usually termed *application sharing*, for the following reasons: People can work with the applications they are used to, i.e. they don't need to learn a new interface. And, with this approach, a huge number of applications is readily available! The fundamental approach to accomplish this is well-known and has been described in detail in [15]: The I/O path between the application and the window system is intercepted, output is multiplexed to all participants and input is demultiplexed to the application, thus keeping the application collaboration-transparent.

The notion of tele-presence can be further enhanced by introducing so-called *telepointers*: With a telepointer, the movements of one participant's cursor (pointer) are visible to other conference participants.

3.4 Audio/Video Communication

In an open interoperable conferencing environment provision have to be made for coping with different end systems. This holds especially for audio/video capabilities of different platforms. To ease conference administration in the Conference Manager and to add flexibility in terms of audio/video quality negotiations, the Audiovisual Manager introduces a new abstraction called *AV groups*. All the Conference Manager has to do is adding/removing participants to a AV group. Based on the quality parameters set (mode and policy) for that group, the Audiovisual Manager does the actual quality negotiation (in terms of encoding, frame size, etc.) and the wiring between the Audiovisual Components which, in turn, deal with the actual data exchange.

4.0   Multi-Media Transport

4.1 High Speed Networks

As one of the goals of the BERKOM II project is to explore the use of high speed network infrastructure for the common use in companies and organizations, the bandwidth that is needed in order to use the teleservices defined above ranges from S-ISDN (small band ISDN starting at 2 x 64kbit channels) up to B-ISDN providing 155 Mbit channel or even more using ATM technology. The current focus of BERKOM II is to use 155 Mbit fiber optic links between the major development and test sites, but the migration to ATM is part of the BERKOM strategy.

4.2 MMT Overview

The Multi-Media Transport (MMT [3]) subproject will provide the layers 1 to 4 that are necessary to efficiently support the fast transport of huge amounts of multi-media data. Referring to the ISO/OSI reference model layer 3 is provided by the ST/2 protocol, where as layer 4 are provided by a modification of XTP called XTP-Lite. The data link layer for the LAN will be based on FDDI.

Note:
In Phase 1 of the project Ethernet has been used as infrastructure based on TCP/IP will be used to implement the MMM and MMC prototypes. This phase was completed in mid 1993.

The following picture shows the network topology provided for the BERKOM II project.

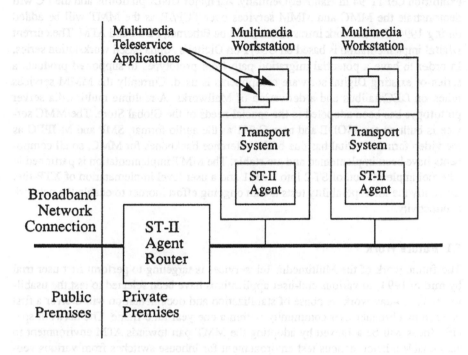

Fig. 5. Network Topology

4.3 Requirements for MMT

The major requirements identified for MMT are to

- allow the user to specify the desired **quality of services** for the communication and provide guarantees on it, to
- support **isochronous traffic**, which reflects the behavior of audio and video devices, as well as
- **high-speed data communications** as well as audio and video communication, and
- support for **real-time communication** where the reaction time to given events in the system is critical.

5.0   Status

The current implementation of all partners will be presented at the spring computer exhibition CeBIT'94 in Hanover/Germany. All major UNIX platforms and the PC will demonstrate the MMC and MMM services over TCP/IP, as the MMT will be added during 1994. The network infrastructure will be Ethernet, FDDI and ATM. The current Digital implementation is based on OSF/1 on Digital's Alpha based workstation series. In order to have a potential migration path from prototypes to supported products a series of existing Digital software components is used. Currently the MMM services relies on DECmailbus and a derivation of Mailworks. A realtime multimedia server prototopye has been adopoted to the special needs of the Global Store. The MMC service is built on top ISODE and uses G.711 as the audio format, SMP and M-JPEG as the video format. Digital has one of the reference backbones for MMC, so all components have been implemented and are stable. The MMT implementation is partioned in a kernel implemtation of ST-2 into OSF/1 and a user level implemetation of XTP-lite. Currently the interoperability tests are an ongoing effort inorder to enable the network connectivity.

5.1 Future Work

The future work of the Multimedia Teleservices is targeting to perform first user trial by mid of 1994, so various end-user applications have been selected to test the usability in day-to-day work. A phase of stabilization and documentation will deliver a first version to a broader user community within a one year time frame. The network specific focus will be achieved by adopting the MMT part towards ATM environment to also enable a heterogenous test environement for inhouse switches from various vendors (DEC, IBM, HP, Siemens). The goal is to use the MMTS as well within the German B-ISDN pilot that also starts during spring 1994. The next exhibition is planned for the International Switching Symposium in Berlin 1995.

References

1. DeTeBerkom[1], The BERKOM-II MultiMedia Mail System (MMM) Version 2.4

2. DeTeBerkom, The BERKOM-II MultiMedia Collaboration System (MMC) Version 3.1

3. DeTeBerkom, The BERKOM-II MultiMedia Transport System (MMT) Version 3.0

4. DeTeBerkom, The BERKOM Multimedia -Mail teleservice API description of External Reference Manager, Distinguished Object Reference and Referenced Data Transfer Guide, Release 1.0

5. H. Ricke, J. Kanzow: BERKOM - Breitbandkommunikation im Glasfasernetz, R. v. Deckers Verlag, Heidelberg, 1991

6. Michael Altenhofen et.al.: The BERKOM Multimedia Collaboration Service, Proc. ACM Multimedia, 1993 Anaheim CA

7. Eckard Moeller, Angela Scheller, Gerd Schuermann: Der BERKOM-Teledienst 'Multimedia Mail', PIK, Praxis der Informationsverarbeitung und Kommunikation 3/93

8. S. Böcking, a. al.: The BERKOM Multimedia Transport System, IS&T/SPIE 94

9. Burkhard Neidecker-Lutz, Software Motion Pictures, Digital Technical Journal Volume 5 No. 2 1993, ISSN 0898-901X

10. ISO/IEC 8613: 1989, Information processing - Text and office systems - Office Document Architecture (ODA) and Interchange Format

11. CCITT Recommendation G.711: 1988, Pulse Code Modulation (PCM) of Voice Frequencies

12. CCITT Recommendation G.721: 1988, 32kbit/s Adaptive Differential Pulse Code Modulation (ADPCM)

13. ISO/IEC DIS 11172: 1992, Information technology - Coding of moving pictures and associated audio for digital storage media up to about 1.5 Mbit/s - MPEG

14. ISO/IEC 10031-2; 1991 Information technology - Text and office systems - Distributed-office-application-model - Part 2 Distinguished-object-reference and associated procedures

15. Michael Altenhofen, et.al: Upgrading a Window System for Tutoring Functions, ARGOSI Workshop on Distributed Window Systems, Abingdon, 1991.

---

1. For all references to the consortium documents please contact:
DeTeBerkom, Peter Egloff
Voltastr.5
D-13355 Berlin

The CIO Multimedia Communication Platform

Andreas Rozek, Paul Christ

Stuttgart University Computer Center
Allmandring 30, D-70550 Stuttgart, Germany

Abstract. Within the context of an european networking project (RACE 2060 CIO) the University of Stuttgart is developing a common "Communication Platform" which can be used for (real-time multimedia) data transmission and runs on top of several operating and transport systems.

Primary intention is to allow for development of multimedia applications that make use of advanced networking features which will soon become available. While early implementations of the Communication Platform will have to simulate any features missing in current transport protocols, future advances in network research may be included without a need for changing the service, its programming interface, or any applications relying on the CIO Platform.

Beginning with a short overview of the whole project this paper briefly describes the Communication Service and its interface focusing on special characteristics of the CIO Communication Platform.

1 RACE 2060 CIO

CIO ("Coordination, Implementation and Operation of Multimedia Services") is a network research project in the context of RACE (Research and technology development in Advanced Communications technologies in Europe), situated in project line 8 ("Test, Infrastructure and Interworking") with a main emphasis on "interworking".

1.1 Project's Objectives

Main technical goal of CIO is the realization of *a common communication and service platform* based on *standard interfaces* - and with implementations for a number of *different workstations* - in order to distribute (input and output of) existing applications which originally had been designed to be executed and controlled on a single computer.

1.2 Computer Supported Collaborative Work (CSCW) by means of Teleservices

Following CIO's approach "computer supported collaborative work" becomes possible without the need for new or extended applications. Instead, many programs people are already familiar with may still be used - all that needs to be done is to replace the module implementing one of the supported standard interfaces with the appropriate CIO *Teleservice* (see figure 2).

As a consequence, a client may run *any* application that uses such an interface (e.g. XWindows) and - although the program itself is completely unaware of being collaboratively used - share any data passed down through that interface (e.g. an X-based GUI) among a number of other users which are also connected to the same teleservice.

Any detail of the distribution process itself is hidden from the application. This includes any necessary communication control (like initiating and terminating a conference) - the client gets a separate user interface for that purpose.

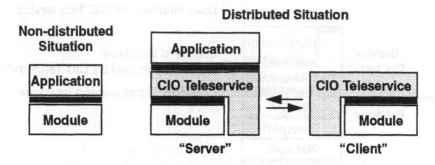

Standard Interface (to be distributed by a CIO Teleservice)

Fig. 1. Role of CIO Teleservices

1.3 CIO Service Platform

For a widespread use of teleservices it has been tried to keep them independent of particular computer or operating systems and network technologies. A small set of *basic* teleservices which already allow to distribute a significant number of applications leads to a *service platform* that can be used in a great variety of environments.

As a starting point, two such teleservices have been specified and implemented within CIO:

* a *Joint-Viewing and Tele-Operation Service* (*JVTOS*) and
* a *Multimedia Mail Messaging Service* (*MMMS*)

JVTOS has been developed in order to share input and output of X-Window applications. This allows for distributed display of graphical data as well as for *remote control* of a program (e.g. for the purpose of *joint editing*). An additional telepointer tool provides a means for unambiguously pointing to objects within shared X windows. Furthermore, JVTOS offers a picturephone for direct audio and video communication - this service may be used independently, i.e. without having to share an X application.

The Multimedia Mail Messaging Service is built around X.400 and X.500 which have been extended with respect to "Multimedia Data" (such as Audio and Video). User agents are available for a number of workstations and try to keep any platform-specific "look-and-feel" - only adding those features which are necessary in order to achieve the whole MMMS functionality.

For the purpose of data transmission CIO teleservices rely on a common *Communication Platform* (CPf, see figure 3).

1.4 The CIO Communication Platform

In view of the real-time, bandwidth and connectivity requirements of modern multimedia applications new transport systems become necessary. Many research groups are currently working in that area developing appropriate concepts and protocols. However, the variety of different approaches makes it hard for an application programmer to use these new protocols and to exploit their benefits as this reduces the circulation area of his/her product. In addition, the individual solutions often focus on certain aspects (e.g. Qualities-of-Service, multicasting, etc.) postponing other important issues (such as address management, routing, group integrity, a.s.o.) - requiring continuous updates in order to keep track with any advances in networking.

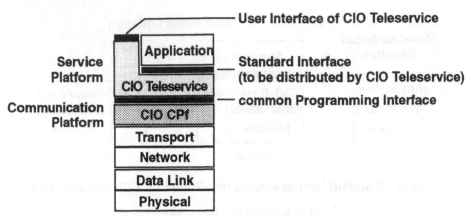

Fig. 2. CIO Communication and Service Architecture

In order to overcome this problem CIO has tried to create a stable Communication Platform which combines and integrates several ideas and concepts of current network research (e.g., arising from projects like OSI 95, MICE (MBone) and the Tenet group or transport and routing protocols such as RTP, XTP(X), IP/Multicast, MTP, ST-II, or MOSPF, DVMRP and PIM) within a common Communication Service.

The CPf - as seen from an Application (or Teleservice)

Applications will use the Communication Service like an *enhanced transport service* (in the sense of OSI terminology) which is well suited to the needs of multimedia conferences and real-time data transmission. Its programming interface has been designed to be implementable on every computer and operating system foreseen in the project.

The CPf - as been supported by the Transport System

As the Communication System itself has to rely on an appropriate transport system, within CIO, a new *Transport Service* has been defined [8] that keeps track of comparable developments in OSI 95 [10]. As underlying protocol XTPX has been developed [9] which is an extension of XTP and covers transport and network layers of an OSI protocol stack.

CPf - the "Glue" between Application and Transport System

The first implementation of the Communication Platform runs on top of TCP/IP (for the sake of a widespread use) and XTPX (in order to exploit its Quality-of-Service capabilities). We will also try to integrate IP/Multicasting in order to connect to the MBone.

Every element of the Communication Service that is not already part of a particular transport system has to be simulated by the Communication Platform itself. However, this versatility offers the opportunity of introducing experimental protocols and of testing them under practical conditions without having to change any application.

1.5 Expected Results of CIO

In the end, CIO will have established (prototypes of) a complete protocol stack for transparent distribution of numerous distribution-unaware standard applications for a few workstations and network technologies. These prototypes will have been installed (tested and operated!) in a number of local environments with (at least) some of these isolated islands being interconnected through public networks.

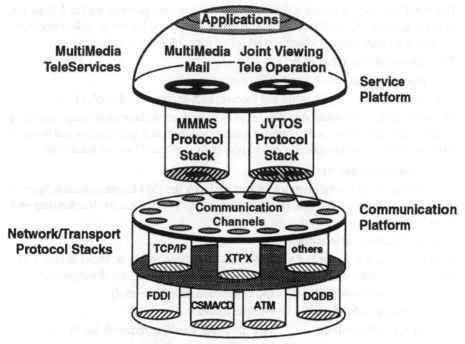

Fig. 3. CIO Platforms and Teleservices

2 The CIO Communication Service

The CIO Communication Service tries to offer a framework for various existing or upcoming transport protocols. Its specification resembles that of a classical transport system with a number of characteristic changes: e.g., the service provider characteristics have been extended in order to take into account its temporal behaviour and limited reliability. Special care has always been taken to provide mechanisms which allow for allocation and optimization of network resources...

2.1 Basic Modes of Operation

The service provides for conversations between two, some or many partners according to the following basic modes of operation:

- **Peer-to-Peer**   (PP, two partners)
- **MultiPeer**   (MP, small number of partners)
- **Broadcast**   (BC, large number of participants)

2.2 Topologies and Transmission Directions

From the point of view of a communication "initiator" the "peer-aware" modes (PP and MP) allow for the following transmission topologies:

- **Peer-to-Peer**   (PP)
- **Peer-to-MultiPeer**   (P-MP)
- **MultiPeer-to-MultiPeer** (MP-MP)

The P-MP topology connects a "central" peer with several participants (and, thus, fits well for lectures, etc.) while a MP-MP configuration provides for complete interconnection of all participants (which is well suited for conferences, etc.).

Transmission of user data can be

- **unidirectional**   (send-only or receive-only)
- **bidirectional**   (send and receive, half-duplex or full-duplex)

The initial choice of transmission direction(s) - to be done at connection setup - defines a set of *possible* data flows. In the course of a conversation, a participant may switch between different directions within these bounds of possibility - or "mute" to save bandwidth.

2.3 Topologies and Service Types

Instead of defining a small number of service types the CIO Communication Service distinguishes between *Context Handling Attributes, Data Transfer Mechanisms* and *Data Transmission Attributes.*

Context Handling Attributes

With respect to the amount of work that has to be done inside of Interworking Units (IWUs, e.g. routers) the following Context Handling Attributes are distinguished:

- **Connectionless-Mode**   (no connection context needed)
- **Connection-Mode**   (simple connection context)
- **Broadcasting-Mode**   (special traffic handling required, see below)

The last mode has been introduced in order to provide for conversations between a large number of participants; it assumes that special *Broadcast Routers* take on a number of management tasks freeing the sender from that work.

Data Transfer Mechanisms

In advance to a communication, it is necessary to chose a single Context Handling Attribute which may be combined with one or multiple *Data Transfer Mechanisms*:

- **Datagram**   (single message)
- **Transaction**   (request-reply message pair)
- **Monitoring**   (single request with multiple replies)

When creating a Communication Service Access Point (CSAP) it is necessary to specify a set containing all wanted mechanisms; later on, they may be freely intermixed for data transfer on the same CSAP.

Data Transmission Attributes

In addition, one or multiple *Data Transmission Attributes* may be assigned to every individual message:

- **Acknowledgment**   (asks for return of an acknowledgment)

    As the underlying transport system is assumed to use its own acknowledgments in order to realize a reliable service this one is intended to inform the sender about the actually achieved Quality-of-Service (see below);

- **Out-of-Band Transmission** (allows for expedited data transfer)
- **Retransmission**   (to a subset of participants)
- **User-defined Notification**   (allows to mix control information with normal data)

This somewhat exotic approach allows to use the same context for various operations (e.g., an application sending large amounts of data to a storage device may, from time to time, uses a transaction in order to obtain some information about the remaining disk space).

2.4 Multipeer Communication

Semantics and behaviour of multipeer communication primitives have been derived from their peer-to-peer counterparts according to a small set of rules like

- **Similarity to the corresponding PP Service** or

  A primary guideline was to issue a single primitive action for multiple recipients and - if need be - to get a single primitive event back;

- **Individual Peers**

  Logically, every single participant should still be visible (e.g., in order to take care of individual capabilities) - although the underlying transport system may perform any traffic optimization that seems to be feasible (Multicast).

These rules led to the necessity of "sum-up functions" being responsible for combining individual responses yielding a common result (e.g., the transmission of a single message may cause numerous acknowledgements. However, as only *one* primitive event is foreseen to be generated for this operation, then all these Acks have to be "summed up"). Three possible approaches are supported:

- **Pre-defined "Sum-Up" Strategies;**

  These are defined and performed by the communication system itself;

- **No "Sum-Up" by the Communication System;**

  Every single feedback message is passed on to the application which has to perform the sum-up itself and to inform the communication system about the final result in a special "proceed" primitive;

- **User-defined "Sum-Up" Strategies;**

  The communication system calls a user-defined function which performs the sum-up. Data and feedback events are not intermixed; thus, it is easier for the application programmer to separate code for communication and sum-up;

2.5 Conversation Management

Communication between more than two peers requires special conversation management. In CIO this is based on *roles* which may be assigned to individual participants [6].

- an *organizer* is responsible for conference planning and establishment, it also becomes the first manager of that conference;
- a *manager* is designated to control a conversation - e.g., to add or remove participants, to assign roles, manage tokens or to terminate a conference;
- a *talker* is allowed to send data;
- a *listener* actually receives data (this role may be temporarily given up due to a local mute for the sake of saving bandwidth);

Role assignment may change in the course of a conversation. A channel with n participants may have 1 organizer, a few managers (usually less than n), a few simultaneous talkers (usually less than n) and up to n listeners.

The protocol used for implementing conversation management has been designed to be robust, i.e. to behave well even in case of control message losses.

2.6 Access Control

Conversations between multiple participants are principally insecure: ill-behaving senders may flood the network with messages, simultaneous changes in documents by several participants may result in data loss, uncontrolled join of conferences by foreign people may disturb private meetings, a.s.o.

The CIO Communication Service provides two levels of communication control in order to get rid of that problem:

- **Role Assignment**

  The abovementioned roles are always associated with the access rights which are required in order to perform a related operation.

- **Access Control, Voting, Notification**

  It is also possible to assign the right to perform certain communication operations (like adding/removing participants, sending data, ...) on an individual basis. The right might be granted unconditionally or depending on the "vote" of other attendees. Other participants might have to be informed about (im)proper completion of an operation.

2.7 Broadcasting

Being aware of every peer in a multi-party configuration may become infeasible for a large number of participants. However, even today there is already some demand for "TV" or "radio broadcasting" over the internet [7].

A special broadcasting mode takes care of this problem as it assumes that special Broadcast Routers (BCRs) take on some management tasks: e.g., a new participant going to join an existing broadcast conversation no longer has to contact the sender itself. Instead, when constructing a path to the sender, the first BCR encountered which already processes traffic from that conversation may complete the join. The sender is still unaware of this new recipient as the "join" request is not further propagated - the broadcast mode spreads the cost of conversation management (see above) over the whole topology.

From then on, outgoing messages are branched off at that BCR (and additionally sent to the new recipient) while incoming messages (feedback) are summed-up at the same site in order to provide for some traffic in the opposite direction ("interactive TV").

Topologies and Transmission Directions

Depending on how many senders are foreseen the following topologies are supported:

- **Peer-to-World**   (P-W)
- **MultiPeer-to-World**   (MP-W)
- **World-to-World**   (W-W)

The first two topologies assume a peer belonging to the "world" to be a receiver only while the latter one allows for bidirectional data transmission between all peers. Knowledge of a particular topology may help routers in constructing their distribution trees.

From Peer-to-Peer to Broadcast

Again, a small set of rules is used to derive broadcast service primitives from their peer-to-peer counterparts. Compared to multipeer configurations, important consequences are:

- **Pre-defined "Sum-Up" Strategies** only;

  As BCRs have to perform the sum-up, user-defined strategies are infeasible;

- **Restricted Form of Replies for Transactions and Monitors**;

  As BCRs have to combine transaction and monitor replies, their contents are restricted to a few pre-defined data types for which sum-up functions can be defined;

- **Timeout Definitions necessary**;

  When collecting feedback or replies, a BCR waits for a contribution from all "neighbours" - or until a timeout occurs.

2.8 Arrangement: Bundles and Channels

Multimedia applications often handle a number of different data streams (e.g. audio and video) which - although they belong together (e.g. because of the same set of communication participants) - have different transmission characteristics.

Channels and Bundles

Actual data transmission is performed using *Channels* which model single (bidirectional) data streams between two or more participants. *Bundles* are used for grouping multiple channels (and further bundles) together in order to handle possible relationships (for the purpose of multiplexing, synchronization a.s.o.) between all bundled objects which have to be handled by the local and remote communication system entities.

Addressing

The CIO Communication System allows for two different ways of addressing:

- **CSAP Addressing**

  A certain participant may be addressed "conventionally" using the port number of the CSAP that is used to access a given channel;

- **Channel Addressing**

  If a new participant is going to join an existing channel it is not necessary to contact the current attendees individually (in a broadcasting environment this is even impossible as the complete set of participants is unknown). Instead, the application only has to supply the name of a channel (which is guaranteed to be unique) and the communication system performs the contact. Subaddressing (i.e., sending to a subset of participants) still remains possible;

2.9 Qualities-of-Service

Apart from providing an extensible set of Quality-of-Service parameters the CIO Communication Service also specifies appropriate mechanisms for QoS negotiation and control.

Quality-of-Service Parameters

The CIO Communication System defines QoS parameters with user-level semantics and maps them to the appropriate transport system counterparts, if necessary. In addition to attributes describing traffic characteristics a.s.o. a few parameters are available that can be used for special purposes (like resource optimization etc.):

- **Selective Forwarding**

  A router (or the communication system entity itself) may be instructed to selectively dismiss a certain amount of messages according to a priority scheme in order to send the same data stream to more powerful and less powerful recipients simultaneously;

- **Number of Simultaneous Talkers**

  In a multipeer configuration it can be assumed that only a limited number of talkers is sending simultaneously - and, thus, only that amount of bandwidth has to be allocated;

- **Half-Duplex Transmission**

  If a certain peer is able either to send or to receive data bandwidth requirements can be further reduced;

Separate QoS parameters exist for channels and bundles. By specifying appropriate values it is possible to optimize resource requirements, to establish (and to negotiate) multiple channels at the same time, to specify which channel should be preferred in case of resource bottlenecks, etc.

Initial QoS Negotiation vs. QoS Renegotiation

Establishment of connection-mode conversations may be combined with an initial QoS negotiation: the path between sender and receiver is chosen according to the requested QoS parameters, a negotiation failure will prevent a connection from being established.

In the course of a conversation, the initial QoS values may be renegotiated. However, the previously chosen path remains the same and negotiation failures don't affect the communication.

If certain Quality-of-Service parameters are unknown (e.g., because the initiator of a QoS negotiation is different from the sender) they may be left open and filled in by the responder.

Resource Reservation, QoS Announcement and Application

If need be, the transport provider is wanted to reserve resources according to the actually known QoS settings, at least for a certain amount of time. The application is then asked to announce the final negotiation results (which may differ due to multipeer-negotiations) and/or to inform the communication system when these settings are to be used.

Quality-of-Service Probing

As it might be tedious to find out available Qualities-of-Service between certain peers, a special service primitive exists that informs about theoretically and practically possible QoS values. The service provider is wanted to behave similar to a QoS negotiation (e.g., to chose the proper path) but not to reserve any resources.

QoS Handling

While Quality-of-Service *provision* may depend on the capabilities of the underlying transport system, it is always possible to monitor the achieved qualities. Up to five QoS values may be specified for negotiation in order to

- ask for provision of a certain quality,
- specify a narrow range for Quality-of-Service control,
- define absolute outer QoS limits from where on messages should be dismissed by the transport system.

Measurement of QoS Values

Every QoS parameter measurement is smoothed using a floating average technique which may be adjusted by the application. It is possible to specify when to start with QoS control (see figure 4) and whether to start at zero or with the first measured value.

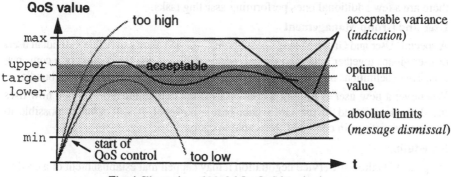

Fig. 4. Illustration of Model for QoS Monitoring

2.10 Pathological Situations

As a consequence of multipeer communication with access control a conference may encounter "pathological situations" which require a special handling in order to continue with normal conversation. Examples of pathological situations are:

- **Talker Loss;**

  The only talker in a conference might crash or get differently disconnected from all the other participants without having had the chance to pass on the right to send data.

  A special primitive allows to find out which participant actually owns certain access rights - it is up to a "manager" to detect this situation and to commit a new talker.

- **Manager Loss;**

  Similarly, the manager(s) of a conference might become unavailable.

  The abovementioned primitive can be used (by any participant) to detect this situation and to initiate a three-step approach leading to the commitment of one or multiple new managers - always trying not to jeopardize the previously fixed access control settings.

- **Conference Decomposition;**

  If the attendees of a conference get cut into multiple groups (perhaps due to the failure of an intermediate router) several fragments of that conference may continue to coexist independently until the broken link becomes available again. Then, a special algorithm starts to "recompose" the original conference - however, this procedure may fail for many reasons (f.e., because the requested Qualities-of-Service might not be available any longer).

As it can be seen, special care has been taken to avoid or to get rid of "pathological situations" by provision of appropriate primitives that allow to detect and to clear up such constellations. These facilities are intended to work hand in hand with an application (e.g., a "session manager") as it might not always be possible for the communication system to detect such a situation whereas it is necessary for it to control any repair in order to keep internal communication contexts up-to-date.

2.11 Assisting Services

Apart from the abovementioned services (which are related to communication itself) there are a few additional ones performing assisting tasks.

User and Group Management

A special "User and Group Management Service" (GMS) allows to define individual users or user groups together with their (initial) access rights outside the context of a conversation. These definitions may be independent of or bound to a concrete channel or bundle.

Whenever a new user is joining a conversation the communication service now uses the definitions found in the GMS data base - however, it still remains possible to change the settings in course of the communication.

Scheduling

In spite of Quality-of-Service negotiation it may happen that establishment of a conference fails due to missing resources. The reason might be that, although the line itself would be powerful enough, only few resources are available due to actual traffic load.

In order to get rid of that problem it is possible to plan a conference (with channels and bundles and the foreseen participants) in advance and to announce it to the transport system. Provided that PNOs support this kind of scheduling it is then possible to coordinate the conference with conversations announced by other parties and to reserve appropriate resources. Later on, when this conference is going to be established a simple reference to the previous reservation is sufficient in order to get access to the related resources.

Management, Test and Verification (MTV)

From the very beginning, people with experiences in running existing networks asked for mechanisms to manage networks and to test and verify the CIO transport services:

* **Management**

  a Management Information Base (MIB) based on SNMPv2 but with an own interface provides a way for examination and modification of status variables or other important internal parameters of the Communication System. It allows to supervise the system's operation, to restrict access to the various services or to limit the total traffic load;

* **Test**

  Provision of "Qualities-of-Service" has introduced a need for tools performing *quantitative* tests, e.g.

  * *ping* now performs a "Quality-of-Service probing";
  * *traceroute* returns information about available service qualities between neighbouring hops (perhaps, indicating which hop is responsible for a QoS degradation, as a QoS-based path selection in routers may not lead to the "best" path unless considering the *whole* path between sender and receiver);

* **Verification**

  Requested and achieved service qualities may differ for various reasons - an internal traffic generator (which may be controlled using QoS parameters) and a corresponding traffic consumer allow for verification. In contrast to measurements done by special applications, there is no need to transfer data between application and Communication System - this reduces internal bus load and the number of context switches.

3 The Communication Platform Programming Interface

A primary design goal of the CIO Programming Interface (API) has always been to offer a *common* (similar calls for every CIO hardware platform), *uniform* (similar calls and data structures for every kind of communication service) and *application oriented* (hiding details of the underlying communication system) interface to the Communication Platform. Consideration of this design goal had some important consequences:

- **Blocking and Non-Blocking IO Calls**

  Macintosh computers have a strong need for non-blocking IO calls which return to the calling program before having completed their operation;

- **Event-Oriented Control Mechanisms**

  The Macintosh Operating System is based on *Event*s and requires appropriate *Event Handling* mechanisms for processing of incoming events;

- **pre-defined QoS Parameter Sets**

  A simple mapping translates user requests like "video channel, 320x200 pixel, 8 bit, 30 frames per second, uncompressed" into the proper set of QoS values;

- **Efficient Data Transfer Mechanisms**

  With regard to smaller computers (like apple Macintosh and PC/AT systems) the API provides mechanisms for efficient data movement between application and network;

4 Work in Progress

A first prototype of the Communication Platform described herein is currently being implemented (using TCP/IP) on a Sun SPARCstation running SunOS 4.1.x, an IRIS Indigo running IRIX 4.0.5f and on apple Macintosh computers running MacOS 7.x. This prototype offers the complete interface but limited functionality (e.g., User and Group Management, MTV and Scheduling use a local database instead of a distributed one). Implementations for PC/AT computers with MS/Windows 3.11 will follow soon.

5 Related Work

The CIO Communication Platform tries to combine a number of trends in real-time multiparty and multi-media networking yielding a common service which is prepared to integrate foreseen advances in the near future without having to change its interface or principal behaviour. F.e., the following projects and protocols have been taken into account:

- **OSI 95 Enhanced Transport Services (ETS)**

  ETS provides a number of peer-to-peer transport services with good QoS support [10]. It is subject to be standardized by ISO;

- **IP/Multicast**

  This seems to be the most popular and widespread solution (MBone) [11]. However, practical experience [7] has shown some deficiencies with respect to resource allocation and admission control;

- **Multicast Transport Protocol (MTP)**

  MTP provides a reliable transport service on top of network layers with multicast capability [12]. A conference is controlled by a central "master" which is responsible for group management and token handling;

- **ST-II**

  ST-II is a network protocol providing guaranteed end-to-end bandwidth and delay together with some multicast support [13];

In addition, it has been tried to consider the aspects of routing - not to jeopardize the basic mechanisms of routing protocols:

- **Distance Vector Multicast Routing Protocol (DVMRP)**

  DVMRP is an experimental routing protocol for internetwork multicasting [14] which has become very popular with the Mbone. Problems concerning efficiency [16] ("truncated broadcasting" with "multicast pruning") have lead to development of PIM;

- **Multicast OSPF (MOSPF)**

  MOSPF provides Multicast Extensions to the OSPF (Open Shortest Path First) routing protocol [15]. A "source/destination routing" approach leads to distribution trees with "least cost paths", path commonalities may be used in order to reduce the number of datagram replications at tree branches;

- **Protocol Independent Multicast (PIM)**

  PIM is a very new approach to multicast routing: two different modes (dense and sparse mode) ensure efficiency both when members of a multicast group are situated close together as well as when they are distributed sparsely across a wide area [16].

Additionally, in [4] several ideas for MultiPeer communication and related Group and User Management are given - these ideas are already subject for being standardized.

In [5] an "API" concept is developed putting emphasis on network resource optimization by using "a priori" knowledge about traffic patterns of typical group communications such as "even in a big conference only a few people talk at the same time".

A detailed examination of multipeer multimedia communication ending up with an abstract description to be exploited for reducing the amount of resources is presented in [6].

These are just two references to work done by the "Tenet Group" - however, papers published by Ferrari et al. have played an important role for design of the CIO Communication Platform.

6 Future Plans

Four major topics outline our future plans in that area:

- **Implementation of the CIO Communication Platform**

  First of all, implementation of the CIO Communication Platform has to be completed (including integration of XTPX as additional transport protocol);

- **MBone**

  Apart from just extending the scope of the Communication Service a contact to the MBone could offer the opportunity to use the scheduling service for announcing conferences (and, perhaps, to implement a method for multicast address assignment);

- **Transport Protocols with Quality-of-Service Support**

  The Quality-of-Service concepts proposed for the CIO Communication Platform are not yet part of any transport protocol - however, it is intended to integrate (some of) them into XTPX (this includes handling of bundles);

- **Routing with proper Quality-of-Service Handling**

  While there are already some protocols with QoS support running on a single LAN proper routing algorithms still have to be developed. PIM seems to be an interesting approach - in addition, it offers mechanisms that fit well to CIO's view of multicasting and broadcasting.

In all cases, applications relying on the Communication Platform may be used in order to test and to compare different transport and routing protocols under practical conditions.

7 Conclusion

It has been tried to create a common communication platform with a service well suited to the needs of modern multimedia applications and with an interface flexible enough to keep track of the appearing advances in real-time broadband communication.

8 Acknowledgments

This work was founded by the European Commission in context of the RACE program. I would like to thank all partners responsible for CIO teleservices for their input regarding requirements of a modern transport system and the other project partners for their cooperation and for their patience when I tried to design the CIO Communication Service.

9 References

Below, you find a short list of references with a subset of papers and documents that have been taken into account when developing the CIO Communication Platform:

[1]   Edgar Ostrowski et al.
        2nd Version of Requirements to the Transport Infrastructure
        RACE Project 2060, Internal Report, 30. June 1992

[2]   R. Braudes, S. Zabele
        Requirements for Multicast Protocols
        RFC 1458, May 1993

[3]   D. Ferrari
        Client Requirements for Real-Time Communication Services
        RFC 1193, November 1990

[4]   Laurent Mathy, Guy Leduc, Olivier Bonaventure, André Danthine
        A Group Communication Framework
        University of Liège, December 1993

[5]   Amit Gupta, Mark Moran
        Channel Groups
        A Unifying Abstraction for Specifying Inter-stream Relationships
        Tenet Group, University of California, and International Computer Science Institute

265

[6] Clemens Szyperski, Giorgio Ventre
Efficient Multicasting for Interactive Multimedia Applications
Tenet Group, Computer Science Division, Department of EECS, University of
California and International Computer Science Institute, Berkeley

[7] Hans Eriksson
MBone - the Multicast Backbone (or: Multicasting considered harmful)
Proceedings INET '93

[8] Lutz Henckel, Spiridon Damaskos
Multimedia Communication Platform:
Specification of the Broadband Transport Service
RACE Project 2060, Deliverable No. 14a, 15. December 1992

[9] Bernhard Metzler, Ilka Miloucheva, Klaus Rebensburg
Multimedia Communication Platform:
Specification of the Broadband Transport Protocol XTPX
RACE Project 2060, Deliverable No. 14b, 30. September 1992

[10] Yves Baguette, Luc Leonard, Guy Leduc, Andre Danthine
The OSI95 Enhanced Transport Services
RACE Project 2060, January 27, 1993

[11] Steve Deering
Host Extensions for IP Multicasting
RFC 1112, August, 1989

[12] S. Armstrong, A. Freier, K. Marzullo
Multicast Transport Protocol
RFC 1301, February 1992

[13] C. Topolcic (ed.)
Experimental Internet Stream Protocol: Version 2 (ST-II)
RFC 1190, October 1990

[14] D. Waitzman, C. Partridge, S. Deering
Distance Vector Multicast Routing Protocol
RFC 1075, November 1988

[15] John Moy
OSPF Version 2
RFC 1583, March 1994
companion documents (RFC 1584 and RFC 1585) describe "Multicast Extensions to
OSPF" and provide an "Analysis and Experience"

[16] Steve Deering, Deborah Estrin, Dino Farinacci, Van Jacobson (et al.)
Protocol Independent Multicast (PIM): Motivation and Architecture
IETF Internet Draft, March 22, 1994
two companion documents describe the dense and sparse mode protocols of PIM

[17] P. Jones
Resource Allocation, Control and Accounting
for the Use of Network Resources
RFC 1346, June 1992

FROM REQUIREMENTS TO SERVICES: GROUP COMMUNICATION SUPPORT FOR DISTRIBUTED MULTIMEDIA SYSTEMS

Andreas Mauthe, David Hutchison, Geoff Coulson and Silvester Namuye

Computing Department,
Lancaster University,
Lancaster LA1 4YR, UK
e.mail: mpg@comp.lancs.ac.uk

Abstract. The *GCommS* (Group Communication Support for Distributed Communication Systems) project at Lancaster University is concerned with the support of group communication, especially for applications employing multimedia information. In this paper we introduce a study on requirements of multimedia group applications. Group support in existing systems is discussed. Examples to illustrate group application requirements are given and a set of characteristics to define these requirements is introduced. Based on the requirements study application level services are proposed.

1 Introduction

Group communication for distributed multimedia systems is a new area of research in computing and telecommunications. Group applications place additional requirements on the emerging multimedia architecture (including operating systems, networks, communication protocols, etc.). Group communication in multimedia systems refers to the exchange of continuous and discrete media data between multiple communication entities. Identical data units from one or more senders have to be transmitted to a group of receivers. With multicast one copy of a data unit is sent to a group of receivers, therefore a better utilisation of resources can be achieved and temporal data inconsistency is minimised. However, additional functions to maintain and administer multicast and group information are necessary.

Apart from the exchange of different kinds of data between multiple heterogeneous communication entities, multimedia group applications need support for various other tasks related to group communication. For instance, flexible, dynamic management of groups, their features and their members is required. Ordering of operations on shared objects, and synchronisation of different media and the same kind of media from different sources is needed. Further, support to handle multiple multipeer connections is required.

Application requirements also have to be considered at the design of group communication services and protocols. In general, the main characteristics of group applications employing multimedia components are multiple senders and receivers, high data volumes, high data rates and time-dependent data values. Thus there is a need to support real-time transmission of high data volumes with high data rates to multiple heterogeneous receivers.

Current systems which offer group support and group communication have been developed for special purposes such as the support of distributed file systems, distributed data bases, fault tolerant systems, etc.. These systems typically place a higher priority on reliability than on time constraints which are required in multimedia communication. Their suitability for the exchange of continuous media data is limited.

In this paper we introduce a study concerned with group application requirements and services to support group communication. To illustrate the nature and requirements of group applications, the paper offers examples of an existing co-operative system and an application scenario. Group applications can have very specific requirements. To offer general services it has to be possible to describe these requirements in a structured and systematic way. A set of characteristics to classify, order and describe group applications is therefore defined.

The paper is organised in five sections. In section two, group support of existing systems and the services they offer at different levels of the system architecture are discussed. Section three discusses examples of co-operative applications and scenarios. A set of characteristics to describe applications and their requirements is defined. Subsequently we list general requirements of group applications and propose group communication services. The final section gives an outline of our intended future work and summarises the paper.

2 Group Support in Existing Systems

Currently groups and group communication are supported by different systems and system components at different levels in the system architecture. In this section we discuss group concepts and group communication support in existing systems.

2.1 Group Support in Distributed System Toolkits and Operating Systems

Distributed system toolkits support the development of distributed applications. They are placed on top of traditional operating systems like for example UNIX. In the *ANSA* (Advanced Network Systems Architecture) *system, interface groups* are introduced as an abstraction for group communication [11]. An interface can be viewed as access point to an object (all interacting entities are treated as objects). An interface group is defined as a collection of interfaces which are accessed via a single interface. ANSA interface groups are mainly designed to provide fault tolerance and reliability and to divide tasks to exploit parallelism. With interface groups on source and sink ends of a continuous media stream object a (M : N) connection can be modelled. An interface group has, apart from a service interface, a group management interface which allows one to manage and control group behaviour. This interface offers functions to join and leave the group, to add and remove policy categories etc.. The group management interface is used by external group managers to control the interface group.

The *ISIS* toolkit was developed to support synchrony of transactions, it is based on the concept of process groups [7]. Messages are transferred by multicast in an asynchronous manner to each member, i.e. different members might receive the message at different times. ISIS tools are mainly designed to allow members to access shared or replicated information, to perform certain forms of co-ordinated distributed execution, and to tolerate and recover from failure. Group management is performed by the system.

At the operating system level application programs consist of one or more processes. A group at this level is defined as a number of processes, which act together in a system or user specific way to provide a distinctive service [18]. Groups are dynamic entities with a changing set of members. The major areas where group communication among processes is currently employed are distributed file systems, replicated program execution and distributed data bases. The client/server model for example employed in the V Kernel [3] and AMOEBA [18] is a good abstraction for such tasks. Distribution transparency provides an interface for client processes where the server group is hidden.

In the V Kernel, for example, several operating system processes on the same or different hosts can belong to a group. All processes in the group have equal rights, they communicate via the exchange of messages. The communication is based on the client/server model, a message sent by a client is indicated as a request at the server (or each member of the server group), the client blocks until it receives a reply message. Messages can also be sent as unreliable datagrams; the sender does not block because no reply is required. Processes can create, join and leave a group, and a group ceases to exist when the last member leaves. Groups are managed by the system [3].

In CHORUS co-operation among processes is based on message passing between processes. Messages are exchanged via ports, a port can be a member of a group and receive messages addressed to the group [2].

2.2 Group Support in the Transport System

In the transport system group communication is supported by some data link, network and transport protocols.

In LANs group communication depends on the ability of the underlying network to broadcast messages. The protocols standardised in IEEE 802.x (viz. Ethernet, Token Ring, DQDB) and FDDI provide multicast.

In IP a multipoint delivery of messages is possible. IP multicast groups are dynamic, a host can be member of several groups. If a multicast group spans multiple networks group membership information is communicated by the Internet Group Management Protocol (IGMP) [4]. RSVP is a companion protocol to IP which controls the packet transmission of IP. RSVP is build around the concept of multiple senders, multiple receivers [5]. ST-II is a stream oriented network protocol which offers multicast services. Multicast data is transmitted between one sender and multiple receivers. Several streams can be clustered into stream groups [5].

At the transport layer a group consists of sending and receiving processes. Most traditional transport protocols like OSI-TP4 and TCP do not support any multipeer data communication. New protocols like VMTP and XTP offer multicast services. XTP provides a one-to-many multicast. The slowest group member determines the data transmission speed. Receivers send error control messages in multicast mode and error recovery is done by go-back-n. Multiple senders can send to the same multicast address. XTP offers zero and one-reliable data transfer. There is no mechanism for the establishment and management of multicast groups [14].

New transport systems and protocols especially developed for the transmission of multimedia data offer multicast support to some extent. In HeiTS (Heidelberg Transport System) multicast is supported from the data link layer HeiDL on. HeiDL interfaces several networks. The Heidelberg Multicast Address Negotiation Protocol (HeiMAP) is an integral part of it, it manages transient multicast groups at the data link layer [19]. In HeiTP, the transport protocol, multicast is provided based on ST-II multicast services. The ST-II multicast group concept is enhanced by the concept of static and dynamic groups [6].

3 Group Applications and Characteristics

Computer Supported Co-operative Work (CSCW) is a major application area for group communication. More and more co-operative systems are being enhanced by multimedia components. Visual interaction is important in most kinds of co-operation and communication [8]. Good audio quality and short time delays in the transmission and processing of data are key factors for the acceptance of such systems [17]. Therefore we have chosen CSCW applications as the main target of this study. We looked in detail at ethnographical studies, existing CSCW systems, prototypes and system scenarios to learn more about their characteristics and requirements. In this section we introduce an existing co-authoring system and a scenario of a co-operative application for scientist using microscopes. These examples illustrate the nature, functionality and requirements of existing and future group applications.

3.1 Examples of Co-operative Group Applications

TeamWorkStation-1 (TWS-1) is a co-authoring and argumentation system. It provides a small group of users (2 - 4 members) with a shared workspace on which participants can simultaneously see, point and draw on [8, 9]. Distributed Macintosh computers are connected with a data network (LocalTalk), a voice network (telephone), a specially developed video network (NTSC & RGB) and an input device network. The shared screen shows a shared drawing window and windows with the video images of every participant inter alia. For the shared drawing window a translucent video overlay technique is employed (i.e. the hand images of the users writing on the board are overlayed). A video server controls the video network; it gathers and processes the shared screen images. After overlaying, the video images are redistributed via the video network to the shared screens. A further development of TWS-1 is TWS-2 [8]. It is designed to use narrowband ISDN (i.e. instead of four there is just one network). The video overlaying is done by a desktop overlay server. TWS-2 is just designed for two users. No group communication is supported.

In a collaborative research project between Lancaster University and ICI Chemicals & Polymers the potentials of multimedia applications have been investigated [21]. The following scenario is based on the prototype of a microscope application developed during this project [20]. High quality video, X-ray spectra and high resolution pictures have to be transmitted from a microscope to remotely sited experts examining a sample. The experts discuss the viewed sample over an audio connection. For some media (e.g. X-ray spectra) a reliable transfer to a subset of participants is required. Experts involved in a sample examination may have differing expertise on specific parts of the sample. Therefore different views on the sample by these experts might be required. Whereas one expert needs to see all movements of the sample (e.g. vesicular breathing), another one might be interested in the cell structure and therefore needs a high resolution detail of the picture[1]. The experts have different rights towards the sample. The sample owners are allowed to give instructions during the examination. Other experts are consulted for comments, passive participants are just allowed to see the video output (or parts of it) and listen to the discussion. Depending on hard and software capabilities and personal needs participants may have individual quality requirements. Further, participants might wish to change the quality of the

---

[1] Note, they are still viewing the same sample but each of them has a different view of it and hence different requirements towards the output.

presented data during a session. The set of participants is dynamic, participants can join and leave and their roles can change during a session. Like conventional cameras, microscopes in a fully automated system will be remote controllable, therefore control commands in real-time have to be transmitted. The video sequences, X-ray spectra and every other output of the microscope may be recorded for documentation and later examination. The microscope output can be incorporated in multimedia documents created by multiple authors using shared editing tools.

3.2 Group Application Characteristics

The above discussed examples show that group applications are characterised by certain distinctive features. In this section we introduce a structured list of characteristics to describe group applications comprehensively.

3.2.1 Organisational Characteristics

These features are concerned with the time, space, size and topology of the group communication.

i.) *Event Scheduling:* Refers to the degree of planning events ahead. Three categories are considered: *Unplanned* events occur at an unknown time. *Planned* events occur at an in advanced known time; reservations can be made ahead. *Regular* events are scheduled periodically in co-ordination with other regular events.

ii.) *Number of Participants:* The number of participants is expressed in integer values. Each application has a range of potential numbers of participants.

iii.) *Location:* Describes the physical location of all participants. In *co-located* systems users are in one place, *virtually co-located* systems are similar to co-located systems but users can be at different locations (e.g. conference rooms). Systems that provide high-bandwidth real-time accessibility between users are called *locally remote* systems. In *remote* systems only minimal accessibility between users exists [13].

iv.) *Communication Topology:* Refers to the way data flows between participants. We distinguish the following types of communication topology: *(N->N), (1->N), (N->1), (M->N)* (where (M < N)), *(1-> 1)* and *(1<->1)*.

Table 1 comprises the organisational characteristics and the set of parameters for each characteristic.

Characteristics	Parameters
event scheduling	unplanned, planned, regular
number of participants	integer
location	co-located, virtually co-located, locally remote, remote
communication topology	(N->N), (1->N), (N->1), (M->N), (1->1), (1<->1)

Table 1: Organisational Characteristics

3.2.2 Data Characteristics

Data characteristics are concerned with specific features related to the kind of exchanged data.

i.) *Kind of data:* We distinguish between *continuous media data* which consists of consecutive time dependent information units and *discrete media* data which consists of time independent information values [15].

ii.) *QoS*: Four QoS parameters are considered: *throughput, end-to-end delay, jitter* and *packet error rate*. Further, three classes of service commitment are distinguished, (i) *deterministic*, intended for hard real-time applications, (ii) *statistical* and (iii) *best effort* [1].

Data values of non-continuous media data are usually not time dependent but long delays can annoy users. We propose three different time classes for non-continuous media, *time-critical*, *time-sensitive* and *non-critical*.

For group communication in a heterogeneous environment different QoS requirements of individual receivers have to be considered. Each participant should be able to state individual QoS requirements.

iii.) *Data View:* Only particular parts of a data object might be of interest to individual group members, but they are interested in all changes affecting the object and their view of it.

iv.) *Reliability:* Reliability in group communication refers to a reliable transfer of data to one or more participants. We distinguish between *0-reliability* and *k-reliability* (the data transfer is deemed successful when at least k receivers receive a message correctly). Further, *det-k-reliability* might be required (k-reliability with a determined sub-set of receivers). *Atomic* delivery to a stated sub-set might be needed.

The following table comprises the proposed QoS and reliability characteristics.

QoS Characteristics	Parameters
QoS parameters	throughput, end-to-end delay, jitter, packet error rate
service commitment	deterministic, statistical, best effort
time classes	time-critical, time-sensitive, non-critical
reliability	0-reliable, k-reliable, det-k-reliable, atomic

Table 2: QoS Characteristics

v.) *Synchronisation:* Related data streams transmitted over different connections have to be synchronised before the playout. With *continuous synchronisation* two or more continuous streams are synchronised. *Event based synchronisation* is used to determine and act upon significant events [15].

Group communication requires additionally the synchronisation of information from different sources. Continuous media from different sources might be *mixed* together (e.g. audio) or *overlayed* (e.g. video).

Concurrent operations on shared objects (e.g. documents, video playouts) have to be ordered in a specific way. Different ordering semantics are known, e.g. *total ordering, source (local) ordering, partial ordering, causal ordering* [10].

3.2.3 Group Characteristics

The third class of characteristics comprises features concerned with the structure and organisation of groups, and the interrelationships between communication participants.

i.) *Group Features:* Refer to the way groups are organised; they cover attributes characterising the group as an entity. In an *active* group, communication can take place, else the group is called *passive*. Groups that allow a change in membership during their lifetime are called *dynamic;* otherwise they are *static*. In *open* groups non-group members can take part in the communication, in *closed* groups only members. *Restricted* groups (in contrast to *unrestricted* groups) restrict communication to be only between group members. Groups can be *permanent, long-term* or *short-term*. In *determined* groups individual members are known to each other. *Partially-determined groups* are groups where a sub-group of members is known to one or more other members. In *anonymous* groups all group members are unknown to all other group members.

In table 3 the group features we consider are summarised.

Features	Parameters
state	active, passive
constitutional properties	static, dynamic; open, close; restricted, unrestricted
lifetime	permanent, long-term, short-term
membership transparency	determined, partially-determined, anonymous

Table 3: Group Features

ii.) ***Interrelation:*** Participants in a group communication can be related through different features. Participants are distinguished by different roles, i.e. they are *members* and *non-members*. Further, a distinction between *initiator, master, voters* (members who have to give a vote for certain operations), *acting* (sender/receiver) and *non-acting* (receiver) participants is made. Participants can be *active*, i.e. currently participating in the group communication, or *passive*.

Table 4 shows the characteristic concerning interrelationship we consider.

Interrelationship	Participants
affiliation	group member, non-member
role	initiator, master, acting, non-acting; voter
state	active, passive

Table 4: Interrelationship

A further dimension of interrelation is the notification of operations on shared objects and of changes in the state of the group and its members.

Interaction between entities can be *asynchronous* or *synchronous* (i.e. all participants have to be present at the same time). In the synchronous case interaction can be either *uncontrolled* (unrestricted sending of data) or *controlled* (restricted data exchange). Rights towards data exchange can change during the lifetime of the application (*dynamic*) or be defined for the whole communication (*static*). If only a sub-set of participants has distinctive rights towards the data exchange we have *sub-uncontrolled* and *sub-controlled* interactions. The membership in a sub-set can be *variable* or *fixed*.

4 From Requirements to Services

Although application requirements are often very specific, certain general requirements of multipeer multimedia applications for group communication support can be identified. In this section we list the requirements retrieved from our study of group applications. Subsequently we propose general services to support multimedia group applications.

4.1 Group Application Requirements

The requirements of group applications employing multimedia components are manifold. In the following we list the main requirements.

Data Exchange Requirements

- General support for data exchange between a group of senders and a group of receivers is required. Support for data transmission in a heterogeneous environment is necessary. Data exchange does not necessarily have to be connection oriented.
- Applications need support for the exchange of different data types over various links to a changing sub-set of participants.
- Requirements towards quality of data transmissions can vary and are application and receiver dependent.

- Reliable data transfer between the group or a sub-set of participants is sometimes required. Atomicity might be needed.

Group Management Requirements
- Group creation and deletion and the assignment of group features are basic functions required by group applications. During the creation of the group an unique group address has to be assigned.
- Administration of membership and the management of the group and its features is required. This includes the support of join and leave of members, dynamic changes of membership roles and of group features.
- The management instance must be able to provide information on the group and to answer queries about the group and its members.

Synchronisation and Ordering Requirements
- Continuous and event based synchronisation of multimedia data is required.
- Concurrent operations on shared objects and devices have to be ordered.
- In group communication the same type of media from different sources have to be synchronised, viz. audio has to be mixed and video might have to be overlaid.

Additional Requirements
- In interactive applications the time requirements of all kinds of data and operations are crucial.
- Awareness is important for many group applications. Specific actions and operations from one or more participants have to be notified to all or a sub-set of participants.
- Some group applications have a demand for dynamic media scaling. Such a request causes application initiated QoS re-negotiation for all involved resources.
- Support of different data views is required. Changes affecting the view of participants have to be visible to each of them at the same time.
- For planned and regular events it should be possible to make resource reservation ahead.

4.2 Recommendations on Group Services

At the application level general services have to be offered to serve the above listed requirements. These services can be provided at different levels of the system architecture. The services we propose are general services and not connected to any specific layer.

4.2.1 Group Data Transmission Services

Group data transmission services are all services that are concerned with transmission and quality of data transfer between communication entities.

Communication Topology

A multicast service from a single sender to multiple receivers (1 -> N) is the keystone of every group communication service. This together with a unicast transport service (1 ->1), allows most other group communication topologies to be modelled. The service has to allow dynamic changes in the receiver group.

At the application level very often a (N : N) service is required, i.e. each member of the group can send data to the group addressed with a single group identifier and can receive all data addressed to the group. Moreover, a (N : N) communication means that events affecting one participant and its ability to communicate must affect the communication between all participants. One way to provide such a service over multiple (1->N) connections is to allow relations between different connections. These relations can be tight (i.e. if one connection, a receiver or sender, crashes, the transmission over each of these channels is stopped) or loose (i.e. the group is just notified of a crash).

Beside the connection oriented service there should be a connectionless asynchronous group transmission service.

Quality of Service

QoS refers to the provided quality of data transmission. The QoS requirements of continuous media can be stated in the parameters *throughput* (no. TSDU/s), *max. end-to-end delay* (ms), *max. jitter* (ms), and *max. packet* error rate. We distinguish three classes of service commitment, *deterministic* (resulting in a guaranteed QoS), *statistical* and *best effort* [1].

Whereas QoS of unicast connections is negotiated between sender, receiver and service provider, in a multicast connection multiple receivers will probably have different QoS requirements. Filters can be applied to serve theses requirements, these filters can be located anywhere between the sender and each individual receiver [12].

A further QoS aspect of group communication are QoS relations between connections. For instance a (N : N) video conversation implies that a degradation of the QoS has to affect all video connections in a similar way rather than reducing just the quality of one connection. Therefore the QoS for all these connections should be stated together and be managed as a unit[2].

Reliability

The third service aspect within data transmission is reliability. Reliability in this context refers to the number of receivers which have received a correct copy of the original message. It is generally distinguished between *0-reliability* and *k-reliability*, (k $\leq$ N), i.e. at least *k* participants have to receive a correct copy of the message. Sometimes an application must be able to state a subset of group members which have to receive a correct copy of the initial message, we call this *det-k-reliability*. Atomicity is required by some applications. Thus, a service to support reliable data exchange between the group or a sub-group of participants is required. The offered error control strategies can range from error detection and announcement over error detection and correction, to error detection, correction and announcement. For continuous media, reliability can be enhanced by making sufficient resource reservation and employing forward error correction but reliability can not be guaranteed by these means.

4.2.2 Group Management Services

Groups have to be administered by a group manager. Group management might be distributed over all layers in the system architecture. It can be centralised, with one group manager executing all functions related to the administration of groups, or decentralised.

---

[2] Note, this does not necessarily interfere with the concept of filtering for individual receivers. QoS might be reduced for all channels belonging to a set. The QoS provided by the filter might be just reduced if necessary.

The group manager is activated by the initiator of a group communication. It has to negotiate a unique group identifier which is assigned to the group. Depending on the location of the group communication this might involve other group managers, address managers and directory services. If a group is deleted the group manager has to free the group address and possibly notify other entities that the group ceased to exist.

Group management involves the administration of individual group members, their roles and current state (active or passive). In an access list all potential participants are stated. Further, group features have to be managed. For dynamic groups it has to be possible to add and remove group members, to change their roles and the group features. A majority quorum might be necessary to change specific characteristics.

The group management has to offer a directory service. Information about addresses, purpose, state and features of the group and individual group members has to be provided for group members and non-members. For determined groups information about single group members and their current state and role can be retrieved. Anonymous groups are treated as a unit, only information about the group as such can be interrogated.

The group management also has to administer a notification list. For each shared object the operations which have to be indicated and the group members which are affected are stated. Notification about joining and leaving of group members is a task of the group manager. Voting lists at application level are required for decisions with majority quorum.

4.2.3 Synchronisation Services

In a group environment concurrent operations of individual group members on shared objects have to be synchronised. The synchronisation ranges from the ordering of concurrent operations on data bases, over the ordering of messages in a computer conference, to the synchronisation of concurrent operations on shared devices and editors. In an editing scenario a copy of the original document or parts of it might be stored with each group member. Concurrent operations on the document have to be executed in the same order at every copy. The service which supports ordering has to consider time constrains as well.

A second form of synchronisation is especially concerned with the synchronisation of continuous media streams from different sources. Audio from different sources has to be mixed, video has to be overlayed. A need for a general audio mixing service especially for conferencing applications can be identified. Video overlaying is rather specific and might not be qualified for a general service.

5 Conclusion

Group communication support for distributed multimedia systems has not received much attention as a research topic in its own right. A few of today's distributed system platforms, operating systems and transport protocols offer multicast to some extent. However, most of these systems have been designed to support specific services. The group framework and the services they provide often do not serve all the needs of multipeer multimedia applications. The functions and group structure of distributed system toolkits and operating systems are in general too simple, inflexible and/or specific to support high level multimedia group communication. Provisions for multicast/group communication in some communication protocols are promising but further work has to be done to define sufficient functions and services. General group management services are not yet available.

With set of characteristics defined in the paper we are able to describe a group application and its requirements comprehensively. Apart from general requirements these characteristics have to be considered in the design of application level group services. The services we propose are general services required by multimedia group applications. Note, however, that the location of these services in the system architecture has yet to be determined.

In the GCommS project we are trying comprehensively to address the problem of group communication in distributed multimedia systems. Our main concerns are efficient data transmission services and group management services. Thus, in the design of services and protocols, performance aspects have to be as well considered as functionality and complexity of the services. Further, to optimise resource utilisation and to allow a larger number of users to participate, relations between different connections can be exploited. Relations between connections are also important to reflect dependencies between different connections. Additionally, this supports the (N : N) communication model.

Group management services are a challenging area of research. The best location for services offered by group managers has yet to be found. Further, relations between group managers at different levels have to be investigated and interfaces have to be defined. Thus, to model group services according to the object-oriented paradigm seems to be a sensible approach.

Apart from CSCW applications, other group applications such as public hearings, general meetings of public limited companies, etc. are mentioned in the literature (e.g. [16]). These have partially different characteristics and therefore different requirements. Nevertheless, we believe that they can be described with the characteristics given in the above framework.

Finally, a new group of applications are co-operative work applications employing virtual reality technology. In their current state of development these have no special requirements on group communication support. However, virtual reality is still in its infancy and further developments might make much higher demands on bandwidth, time bounds, group management, etc.. Therefore we will pay attention to new developments in this area.

Acknowledgement

We gratefully acknowledged Tom Rodden and Jonathan Trevor for the many helpful advises and the fruitful discussions. They helped us a lot to get a clearer view on CSCW.

For research support the authors are indebted to the Science and Engineering Research Council of Great Britain. The work on this project was enabled by the SERC funded CDS project GR/J47804.

References

1. A. Campbell, G. Coulson and D. Hutchison, "A Quality of Service Architecture", *ACM Computer Comunications Review*, 1994.
2. Chorus Systems, "Overview of the CHORUS Distributed Operating Systems", CS/TR-90-25.1, Chorus Systems, Saint-Quentin-En-Yvelines Cedex, 1991.
3. D. R. Cheriton and W. Zwaenepol, "Distributed Process Groups in the V Kernel", *ACM Transactions on Computer Systems*, Vol. 3, No. 2, 1985, pp. 77-107.
4. D. E. Comer, "Internetworking with TCP/IP; Principles and Architecture." Vol. I, Prentice Hall, Inc., Engelwood Cliffs, 1991.

5. L. Delgrossi, R. G. Herrtwich, C. Vogt and L. Wolf, "Reservation Protocols for Internetworks: A Comparison of ST-II and RSVP", *Proc. 4'th International Workshop on Network and Operating Systems Support for Digital Audio and Video*, Lancaster, 1993, pp. 199-208.
6. R. G. Herrtwich, "The HeiProjects: Support for Distributed Multimedia Applications", Technical Report, 43.9206, IBM European Networking Center, Heidelberg, 1992.
7. ISIS Group, "The ISIS Distributed Toolkit, Version 3.0, User Reference Manual", Ithaca, 1992, pp.
8. H. Ishii and M. Kobayashi, "ClearBoard: A Seamless Medium for the Shared Drawing and Conversation with Eye Contact." *Groupware and Computer-Supported Cooperative Work.* Editor: R. M. Backer, Morgan Kaufmann Publisher, San Mateo 1993, pp. 829-836.
9. H. Ishii and M. Ohkubo, "Design of Team Work Station: A Real-Time Shared Workspace Fusing Desktops and Computer Screens." *Multi-User Interfaces and Applications.* Editor: S. Gibbs and A. A. Stuart, Elsevier Science Publisher, 1990.
10. E. Mayer, "Multicast-Synchronisationsprotokolle f. kooperative Anwendungen", *Ph.D. Thesis*, Univ. Mannheim, 1993.
11. E. Oskiewicz, J. Warne and M. Olsen, "A Model for Interface Groups", APM/TR.009.00, Advanced Networked Systems Architecture, Cambridge, 1990.
12. J. C. Pasquale, G. C. Polyzos, E. W. Anderson and V. P. Kompella, "Filter Propagation in Dissemination Trees: Trading Off Bandwidth and Processing in Continuous Media Networks", *Proc. 4'th International Workshop on Network and Operating Systems Support for Digital Audio and Video*, Lancaster, 1993, pp. 269-278.
13. T. Rodden and G. S. Blair, "CSCW and Distributed Systems: The Problem of Control." *Groupware and Computer-Supported Cooperative Work.* Editor: R. M. Backer, Morgan Kaufmann Publisher, San Mateo 1993, pp. 389-396.
14. R. T. Sanders, B. J. Dempsey and A. C. Weaver, "XTP: The Xpress Transfer Protocol." Addison Wesly Publishing Company, Inc., Reading, 1992.
15. R. Steinmetz, "Multimedia Technologie." Springer Verlag, Heidelberg, 1993.
16. C. Szyperski and G. Ventre, "Efficient Multicasting for Interactive Multimedia Applications", Technical Report, TR-93-017, The Tenet Group, International Computer Science Institute, Berkeley, 1993.
17. J. C. Tang and E. Isaacs, "Why Do Users Like Video", *Computer Supported Cooperative Work - An International Journal*, CSCW, Vol. 1, No. 3, 1993, pp. 163-196.
18. A. S. Tanenbaum, "Modern Operating Systems." Prentice-Hall, Inc., Engelwood Cliffs, 1992.
19. B. Twachtmann and R.-G. Herrtwich, "Multicast in the Heidelberg Transport System", Technical Report, 43.9306, IBM European Networking Center, Heidelberg, 1993.
20. N. Williams, G. S. Blair, G. Coulson, N. Davies and T. Rodden, "The Impact of Distributed Multimedia Systems on Computer Support for Co-operative Work", Computing Department, Lancaster University, Lancaster, 1994.
21. N. Williams and G. S. Blair, "Distributed Multimedia Applications: A Review", *Computer Communications, Oxford*, Vol. 17, No. 2, 1994.

Harmonization of an Infrastructure for Flexible Distance Learning in Europe with CTA

Gerold Blakowski
Werner Steinbeck

IBM European Networking Center
Vangerowstraße 18
69115 Heidelberg
Mail: {gblakowski, wsk}@vnet.ibm.com

Abstract: An analysis of the DELTA research program for computer supported distance learning has shown that there is a great variety in the use of technologies and standards, even for supporting similar services for flexible distance learning. This results in incompatible applications, missing reusability of courseware material and a number of diverse requirements towards the underlying hardware and software platforms. The objective of the Common Training Architecture (CTA) is to harmonize the technologies to set up an infrastructure for an European market for learning applications. The approach of CTA is to identify the main learning scenarios, to derive the necessary services to support these scenarios and to describe technical options for the realization of these services. For the technical options, the emphasis lies on finding profiles of de-jure or de-facto standards. In the area of communication, the results show the need for integration of recently developed distributed multimedia application services and standards as well as the need for the development of new standards in these areas. This paper presents an overview of the whole CTA project from the perspective of the authors with focus on the approach and results in the area of communication.

1 Introduction

Short innovation cycles lead to increasing education and training demands. Computer supported learning may be a cost-effective solution to these increasing demands. Within the EC DELTA programme (*D*eveloping *E*uropean *L*earning through *T*echnology *A*dvance) technologies for flexible distance learning are developed to satisfy the increasing training demands in the future. A variety of efforts exist within and outside of the EC DELTA programme in the area of flexible distance learning, resulting in a number of different solutions for local and distributed learning and training systems. To overcome the diversification of ELT systems, harmonization is an important prerequisite for a successful deployment of computer based training in the future. It is the base for a wide distribution of learning applications by courseware providers and offers a user the possibility to take advantage of a large number of learning applications. It protects effort in time and money spent on the development of courses and material, on tools development and on building up an infrastructure for educational purposes.

The CTA (*Common Training Architecture*) project is the main carrier of the harmonization efforts among the projects participating in the DELTA programme. CTA is a stepwise approach towards a harmonized training architecture, see Figure 1.

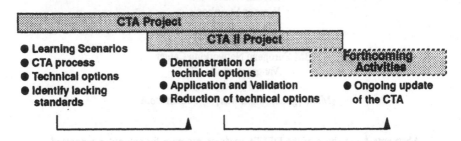

Fig. 1. Development of the CTA

Starting point of the CTA activities is the CTA project with the following objectives:

- The development of methods to describe the requirements of learning systems in the form of learning scenarios.

- The development of the CTA process to examine the requirements of learning applications and to derive services necessary to build up a suitable environment for this specific kind of applications.

- The recommendation of technical options for these services.

- The identification of missing standards.

The technical options are a contribution to the harmonization effort and, what is more, they represent knowledge about system integration. That is of great importance for application developers, even outside of the existing projects.

The CTA-II project puts the concepts of the CTA project and its results to practice. Technical options found in CTA will be demonstrated, the CTA architecture will be validated, the number of technical options will be reduced and CTA is bound to be used for the development of new applications. The work will be carried out together with DELTA projects and available prototypes within and outside the DELTA community.

Regarding the importance of future teleservices in the learning area that are offered in learning centers by training providers a harmonization enabling a standardized access to these services is very important. CTA as harmonization effort in the DELTA programme of the European Union can influence future Europe-wide efforts in that area. The CTA project takes experience of other harmonization efforts into account, like the BERKOM Reference Models [9], [10], but the CTA recommendations are mainly motivated by the training demands and strongly based on the integration experiences in the practical development of training applications and services.

Due to the great importance of hardware and software that exists now or will be shortly available, the CTA recommendations include three kinds of standards in the area of computing and telecommunications. The *de-jure* standards defined by formal international standardization bodies (e.g., ISO, ITU), *de-facto* standards agreed upon by consortia of international IT providers, universities and other contributors and, less preferable, products of single vendors that dominate the market and have to be included because of the relevance of their installation base. As a result, CTA is beside ISO and ITU standards mainly concerned with Internet and PC-related standards.

The results of the CTA project will be published in the CTA Handbook series. The focus of the Handbook series is to provide harmonization and integration recommendations to learning application developers, learning application and service providers, persons responsible for learning system procurement and users

In the following section, the organization of the CTA project and its technical work areas are described. The main focus of the paper will be the work on the *Open Communication Interface (OCI)*. The CTA OCI reference model and an example architecture is introduced in section three of this paper. Section 4 shows the use of the CTA architecture to support the user during the selection process for an ELT system. The last section summarizes the content and gives an outlook on future work in CTA.

2 The CTA Project

The CTA-Project is a 2-year-project that has been started September 1992. The second part of the harmonization effort within the DELTA programme, the *CTA - demonstration and validation* project CTA-II, started January 1994 and will last 15 months.

2.1 European partners

The CTA project consortium comprises partners from companies, public and private education, training organizations and other institutions. They are EPOS International (Rapperswil) from Switzerland, SELISA (Chilly), CCETT (Cesson Sevigne), and Citcom (Paris) from France, ICL Peritas (Old Windsor, Berkshire) from United Kingdom, Open University of the Netherlands (Heerlen), CNR - Instituto technologi didattiche (Genova) and TECNOPOLIS Cosata Novus Ortus (Bari) from Italy, the Institut für Graphische Datenverarbeitung der Fraunhofer-Gesellschaft (Darmstadt), Dornier Deutsche Aerospace GmbH (Friedrichshafen) and the IBM European Networking Center (Heidelberg) from Germany.

All partners are also involved in other DELTA projects. The selection of the participating partners guarantees a well founded coverage of all the different areas of the CTA.

2.2 Overview of the CTA working areas

The idea behind CTA is to define a framework for future ELT systems and to identify services in the main areas of computing that ELT systems have to rely on. The development of the CTA is therefore divided up into four areas.

The first area comprises enterprise modelling, learning application requirements, scenarios, and meta models.

The other three are the technical work areas. They comprise

- the Common User Interface (CUI),
- the Common Information Space (CIS), and
- the Open Communication Interface (OCI).

Each of the technical work areas deals with a specific set of services for ELT systems that are combined to area specific reference models. Figure 2 shows an overview of the service groups offered in the integrated CTA Reference Model. These service groups have been identified based on an analysis of practical existing learning scenarios. The CUI offers the user interface services, the CIS the information management services and the OCI the distributed application and networking services. These services are used during the development and the runtime of the learning applications, tools and application specific system management functions which are not part of the CTA (shaded gray in figure 2).

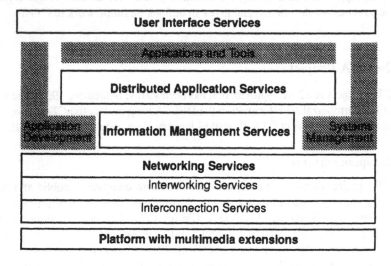

Fig. 2. Overview of the CTA Service Groups

Before discussing the Open Communication Interface in detail, we give a short overview of the enterprise modelling and learning scenario work area and of the kind of services that the CUI and CIS work areas define in their particular reference models and subarchitectures.

2.3 Enterprise Modelling, Learning Scenarios and Pedagogical Situations

The enterprise modelling identifies the objects (actors, agents artifacts) involved in the learning in an enterprise, the operations on these objects (method, actions, processes, functions) and boundaries (obligations, permissions, states) [8].

A scenario is a type of application with specific behavioral patterns of the actors involved in the learning process. The general meaning of a *scenario* is a sequence of events including interactions, information exchange etc. A series of scenarios is proposed that serve as framework for a classification of flexible distance education and training needs, settings, configurations, and components. As a result of CTA, major scenarios are identified and described.

In the *professional expertise update scenario* experts and consultants need to update their knowledge. In the *professional expertise demand scenario* professionals need to get additional or new knowledge about an area to solve a problem occurring for example in a project. In the *employee training scenario* employees need to update their knowledge and skills regularly in order to perform their tasks and jobs. The scenario of an *open distance institute offering telematic services* models a school that provide education to remotely located students. Within the *distributed production and delivery scenario* a company needs to develop, produce and deliver learning materials in a way which is adapted to frequent changes of contents, types and formats. In the *distributed collaborative authoring scenario* teams need to collaborate in producing learning materials or computer-based training materials.

Distance education scenarios include several different general pedagogical *learning situations*. These pedagogical situations are derived from existing conventional learning and training situations (e.g. teacher in a classroom).

In the *virtual classroom situation*, a teacher gives, by means of a communication system, a training course to a group of students, which is in fact constituted of several groups geographically separated. In the *distance tutoring situation* students work individually or in small groups more or less geographically separated. A tutor controls their activities and helps on request. In a *self learning situation* a learner performs a course on a personal computer without any human help during the training session.

The educational scenarios and situations are a framework for the classification of a particular ELT system. They provide a first step in a mapping process which helps the user to define his requirements and needs in means of technological components to build up a working flexible distance learning environment. This mapping process is discussed in more detail in section 4.

2.4 CTA Common User Interface (CTA CUI)

The objectives of the Common User Interface work area in CTA is to give recommendations for user interface architectures in heterogeneous distributed ELT applications based on currently and in the future available standard and to support a common look and feel for ELT application across platforms. The CUI addresses the developer of a user interface for a training system and the persons that decide on the selection of ELT software.

Requirements to the user interface at the enterprise level have already been specified in earlier CTA work [3]. Requirements have been identified including portability and multi-platform compatibility as well as further individual learning requirements.

The CTA CUI relates to the Seeheim model, a de facto standard for interactive systems. The model represents a layered architecture and is one of the recommendations made for the user interface. A goal of the Seeheim model is the separation of application programs and its user interface (*dialog independence*). The Seeheim model is shown in figure 3. The reference model for the CUI is an extension of the Seeheim model. It consists of the presentation and interaction, user interface data, hardware, user interface manager and dialog, and the application interface component. ELT applications use the application interface component.

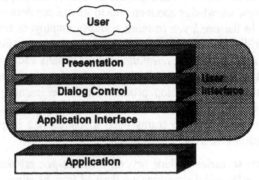

Fig. 3. Architecture of the Seeheim model

2.5 CTA Common Information Space (CTA CIS)

The Common Information Space (CIS) is a common repository addressing the information management issues of an ELT organization. It is shared by tools, services, users during the development, delivery, use and evolution of computer-based ELT material. It provides facilities for the computerized recording, storing, processing and controlling of data and data models of an ELT organization. It manages all the information related to the development, delivery and use of ELT material: units of learning material, personal information, administrative information and so on. The type of data and the rules to manage them are defined by the meta-model CIS component. All the objects contained by the CIS are accessed through the Common User Interface, the user interface of the CTA architecture. The same objects are distributed by means of the services offered by the Open Communication Interface, the communication services provider of the CTA architecture. The CIS also provides an interface to the application and tools development environment through which tool systems interact with CIS, and an interface to the system management environment.

An enterprise can use the CIS to perform the following sample activities [6]:

- define and implement a policy for ELT information environment;
- define, modify, update and obtain information describing the ELT information environment;
- locate data and information that may be distributed throughout the ELT information environment;

- control access and modification of data, information or processes in the ELT information environment;
- interchange information with software tools.

The CIS comprises five functional areas: The information models management area offers means for defining, controlling and maintaining schema of CIS contained information. The structure and representation area covers the interworking and interchange of information using predefined arrangements of the content of documents. The information storage and retrieval area provides means for the creation, the storage and the reference to information according to previously defined schema definition. The administration area includes the operations of monitoring and supervising the above three central components. The application development comprises the means to collaboratively tailor a particular CIS out of the above three central components. It includes specialised products which support these operations. These areas of the CIS are depicted in figure 4.

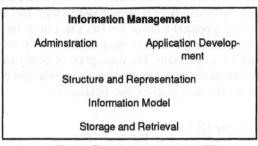

Information Management

Administration | Application Development

Structure and Representation

Information Model

Storage and Retrieval

Fig. 4. Functional areas of the CIS

The CIS offers the following APIs: the CIS interface, the database service processor interface, the operating system input/output processor interface and interfaces to CUI and OCI.

3 CTA Open Communication Interface (CTA OCI)

A major objective of the Open Communication Interface (OCI) is to offer a set of services for learning applications according to the requirements specified in the *CTA Communications Requirements Specification* [1].

The following approach has been chosen for the development of technical options for the OCI. As a first step, a *reference model* has been developed that integrates the services identified as required in the OCI. As a second step, standards are identified that can be used to realize these services.

3.1 Introduction to the CTA OCI Reference Model

To identify the necessary services several DELTA projects as well as projects outside

DELTA have been analyzed [1]. These projects cover the main learning situations Virtual classroom, Distance Tutoring, Self Learning and, in addition, a Distributed Production situation. The basic communication patterns, like conversation, conferencing, sending, distribution, retrieving and collection, used in these projects have been identified and the service structure chosen by the projects was used for the development of the OCI *reference model* [30]. Further requirements like management and security services which are important in a real open learning environment have been taken into consideration. These requirements are only partially addressed in the project prototypes.

The applicable standards for a service in the Reference Model are the *technical options* for this service. A set of coherent standards that cover all services and that is typically based on a standard family is called an *architecture*. For the OCI reference model, several architectures are available. An important aspect of the architecture concept is the support of interoperability. Applications based on an architecture of the same standard-family have a stable base for interoperability. A standard for a service together with the needed underlying standards is called a *standard profile*. Standard profiles are important because they represent a set of services that can be integrated to provide a service. The concrete products that implement an architecture on a specific platform (e.g. Unix) build up a *configuration*. The description of configurations is beyond the scope of the CTA specification, which gives recommendations down to the level of technical options, their standard profiles, and architectures.

3.2 Architectures in the OCI

The focus of interest for the OCI lies on existing or upcoming de-facto and de-jure standards and on relevant products that are in wide use today. As we regard CTA not as a "one-solution-architecture", and for a wide applicability, it is necessary to take existing environments into account. Three main types of architectures are envisaged:

- Architectures based on *ISO* standards, i.e. an ISO/OSI-based protocol stack with application layer services (e.g. FTAM, MHS/CCITT X.400) and management and security services (e.g. ISO Management and Security Framework).

- Architectures based on *Internet* standards, i.e. the Internet protocol family for communication, applications on top of them (e.g. NFS, FTP, SMTP, OSF DCE) and related management services (e.g. OSF DME).

- Architectures based on (de-facto) standards in the *PC*-environment, i.e. NetBIOS-based protocol stacks, application services (e.g. de facto file server and mail standards) and corresponding management and security functions.

These architectures reflect the three main lines of development in the IT industry.

CTA is currently understood to be in a state of transition, as a dynamic architecture that needs continuous update especially in the areas with a high intrinsic dynamic (combination of television, telecommunication, and multimedia). Therefore also upcoming standards in the area of distributed processing, management and multimedia are considered in CTA.

The reason for selecting a specific standard for an OCI service is its wide availability - now or expected for the future - and the degree it supports the specific FDL-requirements. In general, one standard will not completely fullfil the FDL requirements so often a decision for a standard implies a trade-off between the advantages/disadvantages of all selectable standards. Therefore, the OCI handbook series does not only list the standard profiles but it includes an evaluation of the standards considering the CTA-related qualities. These qualities are availability, extensibility to new demands, ease-of-use, reusability of learning material, and portability of the learning application. Not each of the quality aspects is relevant or applicable for each standard.

3.3 The OCI Reference Model

The OCI Reference Model is shown in figure 5 as a refinement of the OCI service groups (see figure 2). To give an impression of which standards are part of the technical options for the CTA OCI services, some selected standard are added to the following description of the service groups. These standards have been analyzed and evaluated in the OCI development [2], [4].

- *Interconnections services*: This service group comprises the asynchronous and isochronous data transport based on telecommunication and broadcasting networks. Regarded standards are the OSI communication protocols, TCP/IP, UDP [11] for asynchronous communication in the Internet, RFC 1190 (ST-II) [29] for isochronous communication in the Internet, X/OPEN-XTI [23] as abstract transport protocol interface hiding the usage of TCP and OSI Transport Protocols [13], [14], SPX/IPX [24] as transport protocol in the PC area, and DIGICAST and VSAT for broadcasting. The asynchronous data transport is the base for any type of asynchronous communication in the learning application, e.g. for exchanging messages between learner and tutor and to exchange non real-time learning material. The isochronous communication is the base for the learning-related audio and video communication like audio/video conferencing or the real-time delivery of distributed stored audio/video sequences. The broadcasting mechanisms are mainly needed in learning situations with high bandwidth communication to a large audience and smaller bandwidth communication to the teacher. An example is a virtual classroom in which a video showing the teacher is transmitted to the learners.

- *Network Applications*: The network applications service group comprises remote file access, file transfer and electronic mail. In the OSI area FTAM for file access and transfer [14] and MHS/CCITT X.400 [7] as electronic mail service has been selected. In the Internet world, NFS and AFS have been considered. For mail in an Internet environment, SMTP and MIME are relevant standards. Proprietary system like Novell Netware for remote file access and Lotus :cc mail play an important role in the PC environment. But also implementations of the Internet related standards are today available for this environment. Also gateways from proprietary mail systems to existing standardized electronic mail systems are available. Basic file transfer is used for the non real-time delivery of courseware, for example for self-study. Electronic mail can be used for deferred time communications between learners or within the learner/tutor relationship, for bulletin boards and also for the delivery of courseware material. Of special interest in the learning environment is the support of multimedia in the electronic mail systems.

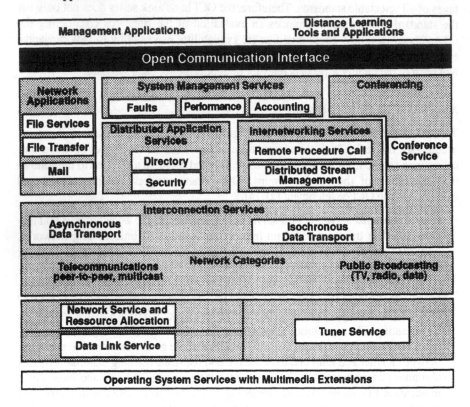

Fig. 5. Service groups of the OCI

- *System Management Services*: Fault management, performance management and accounting belongs to this service group. The ISO Management Framework and the corresponding standards as de-jure and the upcoming OSF/DME as de-facto standards are part of CTA. Some of the services of these service group can be found in proprietary systems for PCs. An example is accounting in the Novell Netware environment. The management services are the base for accounting, monitoring and maintaining learning services in open distributed learning environments.

- *Distributed Application Services*: Directory and security services are part of this service group. ISO/CCITT X.500 [18] with the ISO/CCITT recommendation X.509 (authentication) [17] and the ISO Security Architecture for OSI with its related standards realize these services in the ISO world. Also the OSF/DCE Distributed Directory Service [25] together with the OSF DCE User Registry/Authentification Service [25] are selected. The authentication and protection of the learning material against unauthorized access and the protection of copyright are important functions provided by these services.

- *Internetworking Services*: The Remote Procedure Call and services for distributed multimedia data streams are located in this service group. RPC is supported by ECMA-127 RPC and OSF DCE RPC [25]. Also OMG Corba is considered in CTA. RPC and stream mechanisms do not directly affect the end user of an ELT system, but are of major concern for application developers. They are important for the interoperability between learning systems.

- *Conferencing Service:* The conferencing service comprises computer conferences as well as other forms of electronic conferencing systems. It includes any kind of conferencing system used in the area of flexible distance learning. They are based on PSTN, ISDN, B-ISDN, satellite or cable networks. Important standards are H.242 [21] or H.221 [22]. Motion-JPEG is a popular compression method used for computer conferencing [26].

- *Multimedia data and document encoding and exchange standards* (e.g. JPEG [28], MPEG [28], MHEG [19], [20], HyTime [13], HyperODA [28]) make another important point especially regarding the reusability of learning material and the data exchange between heterogeneous platforms.

The integration of traditional communication services, multimedia communication services and support for distributed applications and their management is one central issue which was identified and considered during the work on the OCI. Multimedia data streams, interchange of complex multimedia documents and standardized methods for copyright protection were found as areas where standards lack or not yet established. Even within standard families, the integration advanced services and the existing standards is not completed. Management and security services and the support for distributed applications are not yet able to deal with multimedia requirements in a coherent way.

3.4 The OCI Service Groups with an Internet-based Architecture

Depending on the platform to be used, CTA examines several types of architectures. To demonstrate how the service groups of the CTA can be used, an *Internet-based architecture* is chosen as an example. Figure 6 shows the architecure that includes new support services for multimedia data, distributed application services and management. This architecture brings together a well-known environment and standards that are not yet stable. As a matter of fact, the new services are not yet fully harmonized with the existing services.

For the interconnection services, TCP/IP is a classical pair of protocols while ST-II and HeiTS or RTP are not very widespread and widely used, but offer new and necessary services for distributed multimedia communications on digital networks. OSF DCE and DME gain more and more importance for the Internet domain, IMA streams are on the other hand not yet implemented although the importance of the stream concept for distribued multimedia applications is generally recognized [15], [16]. The network applications include basic file services (FTP, NFS) and message handling systems (SMTP) that appear to be "part of the system" while later services with advanced features (Andrew file system (AFS), Multimedia Mail (MIME) still have to be brought into motion.

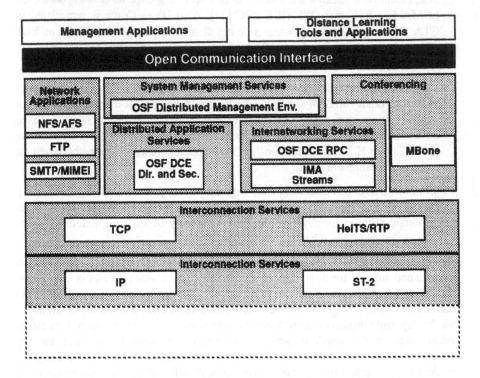

Fig. 6. Internet-based Architecture

The Internet-architecture can typically be found on Unix machines. The family of Unix operating systems is a typical platform for all the DoD-based communication protocols and its applications. On the other hand, implementations of ISO standards also exist on this platform. For the Unix environment, worldwide standards exist that are more or less independent of a single vendor while in the PC environment, Microsoft and Novell dominate the share of de-facto standards.

An architecture should offer a strong coherence of all the services it provides. But for all the three architectures examined in CTA, lack of standards were found within the architecture and a missing interoperabilty and integration of services of different architectures. In the Internet environment, the isochronous data transport service is not yet filled with a widely accepted standard. The PC platform offers very few de jure standards and has a very small base of internertworking services, system management services and distributed application services. To provide interoperabilty between platforms, gateway functions are not a completely satisfying solution. The quality of service offered over a gateway is the common subset of the services supported. If for example a MIME mail is sent over a MHS/CCITT X.400 gateway, the multimedia information will be lost behind the gateway.

To provide a seamless and transparent access on learning material cross-platform regardless of location, format or hardware environment, the main effort has to focus on integration issues and on introducing and applying existing and upcoming standards on all platforms.

4 Mapping of application requirements to technical options

Bearing in mind the educational settings of a concrete learning environment, the user of the Common Training Architecture recommendations should be able to select technical options depending on his specific requirements. The CTA provides a mapping process to guide the user of ELT systems through the selection process. It allows to select technical options specified within the CTA recommendations that are the most suitable for the pedagogical needs. The technical constraints of the target platform also determine the selection of technical options.

4.1 Six layers of the mapping process

The mapping process runs through six layers which guide the user from general aspects of an ELT situation to the technical options required for the particular system. The layers are shown in figure .

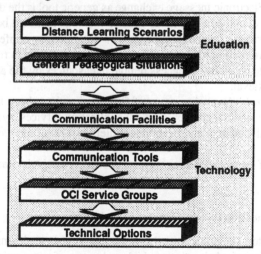

Fig. 7. The 6 layers of the mapping process

The first two layers reflect the educational point of view, the four lower layers represent the use of technology within the selection process. Each of the layers consists of a set of components. During the mapping process, the user starts at the top with a learning scenario and is guided through the selection process using tables and matrices. The layers the process runs through are:

- The *distance learning scenarios* level with six distance learning scenarios identified in earlier work [1] within the DELTA and CTA activities: Professional expertise update, professional expertise demand, employee training, open distance institute telematic services, distributed production and delivery, distributed collaborative authoring.
- The *general pedagogical situation* level includes three fundamental learning environments, namely virtual classroom, distance tutoring, and self-study (see [27]).
- The *communication facility* level offers six generic communication services that have been defined according to [1]. They are conversation, conferencing, sending, distribution, retrieving, and collection.
- At the *tools* level, twelve generic communication tools were found during the requirements analysis (see [1]). All these tools are being used the one or other way in the DELTA projects. The tools listed now are generic tools that are named accordding to the function they fulfill. The tools are: Television-like communication facilities including broadcast and cable TV, telephone, videophone as a video conference based on anlog technology, audio conference as analog solution based on conventional technology. The video conferencing tool is in most of the cases based on analog technoiogy with digital components. A means of application sharing tool is the possibilty to jointly work with one application sharing resources and data. A tool that provides remote control as generic tool is the sharing of applications and views with an asymmetrical relationship (e.g. teacher controls learner application). Electronic message exchange as generic tool is the deferred time communication with computers. Fax as generic tool was included because of its wide use. The generic tool "electronic forum" is a deferred time conference with contributions of a group of participants. File exchange and access to remote information resources is the possibilty to access remote data sources by file access or data base access and retrieval.
- At the *OCI service* level six groups of OCI services have been identified (see [1], [26]): Interconnection services, network application services, system management services, distributed application services, internetworking services, and conference services.
- At the level of *technical options*, several alternatives exist for each of the OCI services.

4.2 Mapping educational requirements to technical options

Starting from the top, the process works as follows: A distance learning scenario comprises general pedagogical situations, which makes use of communication facilities. A communication tool is supplied for a communication facility and a OCI service is implemented for a communication tool. For a specific OCI service, a technical option is chosen.

Step 1: Educational Scenarios and pedagogical situation

The CTA OCI process enables a CTA user to determine for his own case what are the most satisfactory technical options. Nevertheless, it is assumed that the user is able on the one hand, to express his requirements by conforming to the scenario and situation decomposition suggested by the CTA. The derivation of the communication facilities can be done with the help of a classification (described in [1]). The basic idea is to subdivide the pedagogical situations in several basic steps that require specific communication activities. These activities are classified in generic communication services GS(i). The services GS(i) are mapped to the communication facilities (see figure 1).

GS(I)	Conversation	Sending	Retrieval	Distribution	Collection	Conferencing
Transfer of Multimedia Resources		▨		▨		
Cooperative Work	▨					▨
...	▨		▨	▨		▨
Supervision					▨	

Table 1: Identification of required communication facilities

As a result, the user started e.g. with the idea of employee training within a virtual classroom scenario. According to [1] he finds the subdivision of the virtual classroom situation in steps and the necessary generic communication services to fulfill the steps. With a matrix like the one in figure 1 he derives "Sending" as one necessary communication facility. Then, the process described here is aimed at conducting him smoothly towards a set of technical options compatible to a technology deployment which may be available in his enterprise. It is important to recognize that computer networks rely on heterogeneous technology. Consequently, CTA will not determine one solution as the preferrable one but suggest a set of solutions in relation to the standard families.

Step 2: From communication facilities to tools

If one or several communication facilities are determined, a set of matrices will be used to identify what are the generic communication tools satisfactorily meeting the user expectation. This expectation is defined under a finite set of characteristics or qualities coping with various aspects such as audio, image and video quality, delay acceptance, confidentiality. There is one matrix for each communication facility. In each matrix, communication tools are matched to characteristics. Here, the discussion of the communication facility *sending* as an example is continued.
In the appropriate matrice for the communication facility "sending" the user finds the communication tools audio conference, fax, electronic mail, deferred time computer

conferencing, and file transfer in this group of tools. Each of the tools has got a certain set of characteristics (media, time behavior, data format etc.). In table 1 the time characteristics of selected tools are given.

Audio conference needs real time, electronic message exchange seconds up to days may is the suitable range. For file transfer a time range of seconds up to minutes is expected. The service characteristics have significant impact on the selection of the tools and standards. Depending on the service quality the application requires, the number of technical options that fulfill the requirements goes down.

		Tools			
		Audio Conference	Electronic message exchange	File Transfer	...
Characteristics	Real time				
	Seconds				
	Minutes				
	Hours				
	Days				

Table 2: Example matrix for communication tools and their characteristics (delay)

Step 3: From communication tools to OCI Services

Once the CTA OCI user has identified one or several tools, more matrices will help him to determine which OCI services need to be implemented. For example, communication facility "sending" using an electronic message exchange tool will need the implementation of both e-mail and asynchronous data transport OCI services. This information will be given by defining matrices matching OCI services against the same previous set of characteristics. There is one matrix per couple (facility, usable tool). As a result, boxes will be ticked with numbers, each number corresponding to a constraint applied to the OCI service (for example, 1: binary transfer mode is requested.

Step 4: From OCI Services to Technical Options

Once OCI services are determined, the user's problem is solved at the functional level. Indeed, the OCI services level are independent of the available technology. In this respect, it avoids all the process performed so far to become obsolete due to technological changes. It also guarantees that any new technical option can be included into the process by being plugged in at the last level. It also provides sufficient extensibility of the process in the future. Achieving the process will lead the CTA OCI user to turn in OCI services into a choice of technical options. All the technical options recommended by the CTA project have been included into the annex of the OCI part of the CTA handbook series. As an example, the description of NFS can be found in the annex of this contribution.

A last set of matrices, in fact one per OCI service, will match the OCI services to the technical constraints enumerated in the previous matrix. This last procedure shows which of the technical options are satisfactory. Now the user has got with finishing the selection process, a number of techncial options that will as an ensemble meet the requirements specified in the beginning. The families of architectures and the standard profiles give help to do so. The architectural families are a guideline when selecting for a specific platform or if interoperability with a platform is requested. It is recommended to use the standard families for a coherent solution.

The example illustrates the e-mail service for which the technical options are Internet electronic mail, MHS ISO 10021/CCITT X.400 and MIME [5]. Several numbers are attached to each tool giving indications about the technical constraints so that the OCI user is sure to choose one or several tools corresponding to his ELT requirements. The numbers mentioned above appear here again as the quality of the characteristics of the implementations of the generic OCI services.

		OCI Services		
		MIME (RFC 1341)	Internet Mail (RFC 822)	MHS ISO 10021/ CCITT X.400
Characteristics	1			
	2			
	...			
	7			
	8			

Table 3: OCI Service "E-Mail" list of technical options with supported constraints

At the end of the process the user selects a technical option (e.g. CCITT X.400) and with the option a number of standards that are interrelated with it (the standard profile). A more illustrative example of a technical option is given in the annex.

5 Summary and outlook on future work

The objective of the CTA is the harmonization of technologies underlying flexible distance learning applications. In a first step, learning scenarios have been identified. Based on these scenarios, services have been derived that have to be provided by the underlying environment. The technical options for the implementation of the services are given as recommendations and are based on standard profiles. Also CTA offers a process and method to identify the learning application requirements and to map them to the technical options. The results of the CTA project will be published in the CTA Handbook series.

The CTA-II project as successor of the CTA project will demonstrate in cooperation with other DELTA projects recommendations and will perform validations. The result will be an updated CTA Handbook series.

The work on CTA indicated that integration of existing and distributed multimedia applications and management services within an architecture is an open issue. The interoperability of the lines of development on the different platforms requires at least a common estabilhed exchange format of all kinds of multimedia material. Lacking standards in the multimedia area must be filled, but the dynamic process in this area does not allow for a fixed solution within a short time scale. The solution is a continous update of changing parts of the CTA and an approval of what is considered as existing and well established. Further standardization and integration of services will help to reduce the number of technical options.

Acknowledgments

This paper is based on work of all participants of the CTA project: B. Béchon, N. Benamou, J. van Bruggen, M. Capurso, A. Chioccariello, J. R. Collins, J. Fromont, C. Hornung, W. John, U. Kohler, J.F. Maudet, H. Mispelkamp, J. M. Pratt, R. Price, D. Ray, H. Schmutz, M. Vafa, W. A. Verreck, A. Villemin, H. G. Weges and Peter Zorkoczy.

References

[1] B. Béchon, J. Fromont, JF. Maudet, D. Ray, H. Schmutz, *CTA Communications Requirements Specification*, DELTA CTA Deliverable 7, Cesson, 1993

[2] N. Benamou, C. Hornung, U. Kohler, M. Vafa, *CTA Common User Interface - Requirement Specification*, DELTA CTA Deliverable No. 9, Friedrichshafen, 1993

[3] N. Benamou, C. Hornung, W. John, U. Kohler, G. Paquot, M. Vafa, *CTA Common User Interface - Technical Options*, DELTA CTA Deliverable 13, Friedrichshafen, 1994

[4] G. Blakowski, J. Fromont, D. Ray, W. Steinbeck, T. Schütt, *CTA Survey of Existing Solutions for an OCI*, DELTA CTA Working Paper, Feb. 1994

[5] N. S. Borenstein, *MIME: A portable and robust multimedia format for internet mail*, in: Multimedia Systems (1993) 1, p. 29-36

[6] M. Capurso, J. Collins, A. Chioccariello, M. Malerba, J. Pratt, P. Romanazzi, L. Stijnen, W. Verreck, *CTA Common Information Space Technical Options Specification*, CTA Deliverable 12, Valenzano, 1994

[7] B. Plattner. C. Lanz, H. Lubich, M. Müller, T. Walter, *X.400 Message Handling: Standards, Interworking, Applications*, Addison-Wesley, Wokingham, 1991

[8] J. R. Collins, U. Kohler, J. M. Pratt, W. A. Verreck, *CTA Description and Definition*, DELTA CTA Deliverable 6, Heerlen, 1993

[9] DETECON and GMD-FOKUS, *BERKOM Reference Model - Lower Layers*, DETECON Technisches Zentrum Berlin, Germany, Feb. 1991

[10] DETECON, GMD-Fokus (ed.), *BERKOM Reference Model II - Application-Oriented Recommendations*, BERKOM Dokumentation Band V (in German)

[11] Douglas E. Comer, *Internetworking with TCP/IP, Vol. 1: Principles, protocols, and architecture*, Prentice Hall, 2nd ed., 1991

[12] C. F. Goldfarb, *HyTime: A standard for structured hypermedia interchange*, IEEE Computer, August 1991, p. 81 - 84

[13] J. Henshall, S. Shaw, *OSI explained*, Ellis Horwood, 2nd ed., 1990

[14] J. Henshall, *Opening up OSI*, Ellis Horwood, 1992

[15] Interactive Multimedia Association, *Multimedia System Services*, Version 1.0, June 1993

[16] International Business Maschines, *Multimedia Presentation Manager /2*, Programming Guide, 1992

[17] ISO/IEC, *Information technology - Open systems interconnection - Basic reference model, Part 2: Security architecture*, International Standard ISO/IEC 7498-2, 1989

[18] ISO/IEC, *Information technology - Open systems interconnection - The Directory: Overview of concepts, models, and services*, International Standard ISO/IEC 9594-1, 1992

[19] ISO/IEC, *Information technology -Coded representation of multimedia and hypermedia information objects (MHEG)*, Committee draft 13522-1, June 1993

[20] ISO/IEC, *Information technology -Coded representation of multimedia and hypermedia information objects (MHEG)*, Update of the MHEG press release, August 1993

[21] International Telecommunication Union, *CCITT recommendation H.242*, 1990

[22] International Telecommunication Union, *CCITT recommendation H.221*, 1990

[23] Object Management Group, X/Open (Ed.), *The common object request broker: Architecture and specification*, OMG document number 91.12.1, Revision 1.1, 1991

[24] Novell, *NetWare system interface technical overview*, Addison-Wesley, 1990

[25] Alexander Schill, *DCE, Das OSF Computing Environment*, Springer, 1993

[26] Werner Steinbeck (ed.), CTA OCI Technical Options Specification, DELTA D2023 CTA Deliverable No. 10, Heidelberg, 1994

[27] Werner Steinbeck et al., A multimedia environment for distributed learning, in: Robin Mason, Paul Bacsich, ISDN applications in education and training, Herts, 1994

[28] Ralf Steinmetz, *Multimedia-Technologie: Einführung und Grundlagen*, Springer, 1993

[29] C. Topolcic: *Experimental Internet STream Protocol, Version 2 (ST-II)*, Internet RFC 1190, 1990.

[30] W.A. Verreck, H.G.Weges, J. van Bruggen, D. Ray, J. Collins, *Models and scenarios for education and training*, DELTA D2023 Common Training Architecture, Deliverable No. 4, Heerlen, 1993

6 Annex A: Definitions

Within the CTA project and its work packages, several terms are used with special meaning for the project and its topics. To clarify the ideas behind the terms, definitions are given of the most important terms used in this document.

6.1 Reference Model

A reference model is an abstract model of functional components and their relationships, used to partition functionality so interfaces can be identified. A service reference model is a set of services that work together to fulfil the interface requirements of a technical system. An example of a service on the level of the reference model is a file transfer service.

6.2 Architecture

An architecture defines de-jure, de-facto or industry standards to use for the services of the reference model protocols for the communication between implementations of the architecture or with other systems, the necessary integration information, that shows how the components of the architecture are working together.

Several architectures may exist for one reference model. To derive the architectures from the reference model it is necessary to identify components that are instances of the reference model services. For example, the selected international standard for the file transfer may be the OSI standard FTAM.

6.3 Configuration

A configuration of an architecture is a set of implementations (products, public available solutions) that realise the architecture on a specific platform A configuration comprises also integration information that is specific for this configuration. Several configurations may exist for an architecture. In the example, a configuration of the architecture on a PC platform comprises a PC-specific implementation of the FTAM standard.

7 Annex B: Technical Option Specification of the NFS service

Criteria	Description of the technical option
Service	Network and name transparent access of a distributed file system. On-line shared file access that is transparent and integrated. User can execute an arbitrary application program and use any file for input and output (invisible network access). The file names do not show whether the files are local or remote.
Interoperability	NFS has become an important part of the UNIX file system and the transparent access to remote file systems. Applications running on heterogeneous platforms can exchange data transparently by accessing files on common servers. NFS supports interoperability in a very high degree.
Reusability	NFS offers name transparent file access to applications. If an application runs on one platform accessing local disks, it also can use NFS transparently and hence access remote file systems. NFS poses no restriction on the applications. The network transparency allow to store the learning material on one or several file servers in a network and make the material accessible via NFS. A specific user will be able to access transparently the file systems he needs, but no other file system he is not authorised to use.
Ease-of-use	NFS is during most of the time invisible for the user and therefore as easy at it can be. At a lower level (operator level, API) NFS is very good integrated in the UNIX naming and conventions and therefore not easier or more difficult than the same task for the local case. Network transparency supports the ease of use of learning material because the data is stored on a central file server. Updates and new versions can be done on the fly, the end-user will not be disturbed and it takes no changes in the system configuration or the file system of the learner.
Portability	NFS is available on all major UNIX derivatives, for IBM OS/2 and PC. NFS itself is transparent for the application program and there are no restrictions concerning portability for the application program, e.g. learning applications written on the base of a local file system can easily be ported to NFS.
Availability	NFS is available as a product for all important UNIX systems, IBM OS/2, and DOS environments.
Cost	NFS is integrated in the UNIX environment (file system). For OS/2, it is integrated in the IBM TCP/IP for OS/2 product, It is also available as a special product at 500 ECU.
Communic. Scenarios	NFS itself is independent of the underlying network infrastructure. It runs on all networks that support the use of TCP/IP or UDP/IP.
Dependencies	NFS relies on TCP/IP transport service or UDP/IP datagram service.

Table 4: Technical Options Specification for NFS

8 Annex C: Standard profile for the NFS file service

Message format according to RFC 822	MIME (RFC 1341)	SMNP (RFC 1098)	NFS	AFS	FTP (RFC 959)
SMTP (RFC 821)		External Data Representation (XDR)		ASCII or binary format	
Sun RPC		NULL		TELNET (RFC 764)	
TCP (RFC 793)		UDP (RFC 768)		RTP	
IP (RFC 791)			ST-II (RFC 1190)		
Data link services (IEEE 802.3, X.25/T1, NetBIOS, FDDI)					
Physical layer services					

Table 5: Standard profile for NFS

DEDICATED - MODULAR TRAINING SYSTEM
DELTA Project D2014

José M. Velez[1], Mário Rui Gomes[2]
Instituto de Engenharia de Sistemas e Computadores (INESC) [1,2]
Rua Alves Redol 9 - Apt 13069 1000 LISBOA PORTUGAL
Instituto Superior Técnico, Departamento de Engenharia Electrotécnica e de
Computadores[2]
E-mail: jmv@inesc.pt, mrg@inesc.pt

B. Tritsch, C. Hornung
Fraunhofer Institut for Computer Graphics
Wilhelminenstrasse 7, 6100 Darmstadt Germany
E-mail: tritsch@igd.fhg.de, hornung@igd.fhg.de

Abstract: This paper describes the architecture and implementation of
the Dedicated's MTS - Modular training System, a Computer Based
Teaching system. It is a distributed, multiplatform and multimedia system
that relies strongly in the comunication features to add a new dimension to
the area of the CBT. A big emphasis is put in the student's evaluation,
course material reusability and it's automatic distribution throughout the
entire European Training Center's network.

1 Introduction

The DEDICATED (DEvelopment of a new DImension in european Computer
Aided TEaching) is an European project of the DELTA program. It aims both the
development of a *computer based teaching* (CBT) system (the MTS - Modular
Training System) and the creation of an European network of *local training centers*
(LTCs). The *LTCs* will be available in Portugal, Germany, France and Greece and
will use DEDICATED's CBT system to deliver the courses.

In order to promote the interchange of the courses developed in the LTCs, they
will be interconnected, creating an European wide educational network, using the
international broad or narrow band ISDN networks. Thus, a student in a given LTC
has the possibility to access and use the courses available in any of the other centers
and a teacher, while developing a course, may integrate in it a module already
developed elsewhere.

All the machines in the LTC's will be connected with a LAN allowing the
students to interact with each other or with a teacher. The teacher that supports a
given course, or chapter of a course, may be contacted by the student independently

of being in the same or in a remote LTC. This way, the student can access the know how of the best researchers in Europe without the need of expensive travels. However, they still benefit from the advantages of a personalized contact.

Another goal of this system is platform portability. This system (MTS) is being developed as machine independent as possible and the platforms supported in the moment are both UNIX machines (SUN, SGI and HP) with X and MOTIF and PC-compatible machines running DOS and WINDOWS. The LTCs provide both types of machines and the MTS not only runs in all the above machines but it can also operate in a distributed way between them. This is a truly distributed system, using a message-based protocol to exchange multimedia information between the different modules that build the system. The MTS supports not only multimedia courses (as most of the other systems already do) but also allows multimedia communication between students and between students and teachers in the form of teleconferencing. Cooperative work and remote control of the system (allowing the teacher to show something to the student) are also supported.

This paper will describe the architecture and the main features of the DEDICATED's Modular Training System. It was developed by a group of Portuguese (INESC e IST-DEM) and German (IGD-FhG and CLS) partners.

2 Functionality

This system is supposed to be used by two very different kinds of users: the teachers, who develop the courses, and the students, who use the courses. Both have their very specific needs and the MTS must be designed to meet them.

As far as the teacher is concerned, he must be able to specify the contents of a course. He can do that using a script language that describes the contents of the course as well as the interaction with the student. A great emphasis is put on the reuse of the previously written material and on the student's evaluation and guidance throughout the course.

To achieve the first goal, the reusability of the developed material, the course is divided in several modules. Some of the modules are specific to that course but others may be generic modules that are already written for previous courses. In that case, it doesn't make sense to waste time writing it again. A course is described by a table of contents where the teacher details the several sections. Those sections are called *modules*.

The main course has dynamic references to these modules which are solved only at run-time. Therefore, a teacher may not know exactly which module is going to be used to deliver his course to a given student: the system may choose different modules according to the student's expertise (beginner or expert), language preference (the student may prefer a module in English but, if not available, the system will give him the second preferred language), etc. This way, all the courseware available in the educational network can be easily reused, integrating in the course the most recent modules available in the network.

Special care was also put in the student's evaluation. This will not only give a mark to the student at the end of the course but will also guide the student through the entire course, detailing the subjects where the system feels the student has difficulties and speeding the explanations in the other case. To ease the implementation of these evaluation features, the teacher has several evaluation primitives that may be used to design tests, like true or false questions, multiple choice questions and two different test models.

The student can specify his favorite parameters to the system (as his preferred language, his degree of knowledge in the field, etc.) and, with that information, the system will choose at run time the modules that belong to that course and that best match his preferences. Some of those parameters (the degree of expertise, for instance) may change during the lesson, and the system will automatically choose again the correct modules.

The student also has the possibility to interact with other students using the communication facilities embedded in the MTS. If some doubt about something that is not clear in a given course arises, he can contact the teacher that developed the course or that is in charge of it's support. As the courses are available on an European scale, it is possible that the teacher is not present in the local LTC. Once again, the communication infrastructures may be used by the student to reach the teacher, wherever he may be, for a quick explanation.

3 Architecture

The MTS is a distributed, object oriented system based on a server-client architecture. The course server is the entity that provides the courses, after the student request, and interprets the courses scripts. When it interprets the courses, it sends requests to the interaction client (that interacts with the student: the Generic Learning Support - GLS) and receives the messages resultant of the student actions (mouse commands, typing at the keyboard, etc.).

The GLS, must connect to a course server. All the interaction with this system is processed by the GLS and then sent to the course server, which reacts to this input according to what is specified in the course script, currently being interpreted. The next figure will detail the overall system architecture.

Fig. 1. MTS Architecture

In this diagram we can identify the client process (the GLS) and the Course Server. These components will be further detailed in the next sections.

3.1 Course Server

The course server has a layered structure. Each of the three layers is supposed to offer to the upper layers high level services, hiding the protocol and implementation details.

Fig. 2. Course server architecture

The services implemented in the several layers are the following:

- *API* The Application Programming Interface is, as the name suggests, an interface to both the communication services and to the operating system in use. The communication part of the API is intended to hide the communication protocol between the server and the client, offering a transparent way of filling the message structures and sending and receiving those messages to and from the GLS. In this way we have the freedom to change completely the communication protocol without changing the upper layers.

   The API also has an important role in achieving the operating system independence of the MTS, because it eliminates the differences found in each of the supported operating Systems (UNIX/Motif and DOS/Windows). Among these services we can find file access, low level system calls like timers, etc.

   Adding to all these features, the API also defines the programming model, implementing a callback based system to deliver the messages to the upper layers.

- *LMI* (Learning Material Layer) is a library of high-level objects that may be used, by the teacher, to define the contents of a course. As an example of these objects, we can mention test objects that are used to evaluate the level of knowledge of the student, or objects as simple as text or bit map images.

- *CI* (Course Interpreter) controls the flow of the course and creates all the objects needed to present the course. The complete structure of the course, as well as the definition of the interaction with the student, is described using a script language (CDL - Course Description Language). This script can be created using a plain text editor or a graphical authoring tool.

3.2 The GLS

The GLS is a stand alone process that may even run in a different machine from the course server. It is supposed to handle the user interaction and all the graphical display required to present a course. All the service requests are received from the network in the form of messages and the user input is also translated in messages that are sent to the course server.

The services offered by the GLS range from the simple vector graphics primitives, like line and circle drawing, to display of images as well as timers, menus, dialog, video, sound and other interaction.

The GLS is machine dependent and it is usually necessary to program a specific implementation for each machine.

4 Communication

As one of the aims of this project is to create a network of LTCs, this project always had a very strong emphasis on communication issues. There are two different types of network in this system: the Local Area Network (LAN) that interconnects the computers inside a LTC and the Wide Area Network (WAN) that interconnects the several LTCs. The LAN allows the students to access local courses, communicate with the local teachers and colleagues and reach the gateways that provide the same functionality but on the remote LTCs. This way a student can put questions to specialized remote teachers, if there is no local expertise in that field, or interact with other students in order to exchange opinions. The communication may be done by exchanging written messages or e-mails but it can also be a multimedia link, using video and audio. This approaches the functionality of a video phone or even the teleconference, because the communication can be established between more than two persons.

The LAN is implemented using Ethernet over coaxial cable and/or twisted pair, depending of the sites.

This infrastructure is used to communicate between machines in the same LTC. The communication is message based, using the Berkeley sockets, allowing an easy port of the MTS to other platforms. All that is required, with this type of architecture, in order to use a different machine, is to provide another GLS, because the course server can run on a already supported machine.

All the multimedia communication is also done using messages. In order to use the hardware specific to each machine, the sound is coded into a subset of the wave format and, in every machine, there is a process that converts this generic format to the native one of each machine (AIFF in SGI, Sun's format, WAVE for the PC's, etc.). The video communication between computers uses a software-only compress algorithm in order to lower the price of the machine configuration needed to the LTCs. Even so, the video has a reasonable frame rate. This subject will be detailed further ahead.

In what concerns the WAN communications, in the first phase of the project we used the resources of the Internet to provide a data link between the several LTCs.

However, if this is the easiest and faster approach to have a communication link up and running, the available bandwidth is not only limited but also unpredictable, as we are using a non dedicated link. This makes the synchronization of data and the interaction of the system very poor. So, an alternative solution for the interconnection of the LTCs was considered. It should provide a relatively large bandwidth for data communication, assure a minimum delay, and be cost effective. Our choice fell on the narrow-band ISDN for long distance communication and, eventually, broad-band ISDN between near-by LTCs. As this technology began to mature we started the first tests. We used INESC's hardware and software to establish a link between Coimbra and Lisbon, using a 64 Kbps ISDN link. The ISDN adapter card was a data and voice PC-card called "PC-BIT". The software (the TCP/IP stack) was also developed at INESC. This combination allows a PC to be connected to an ISDN link.

We can summarize the architecture described above with the following figure, where we can also see some of the communication features:

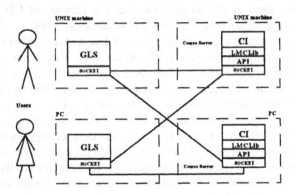

Fig. 3. MTS architecture and communication features

4.1 VIDEO COMMUNICATION

The video we are used to seeing on our televisions and VCRs is an analog signal. To capture the video signal and transmit it over a computer network as a Tele-Medium, the video data must be digitized and this requires special purpose hardware. Digital video is based on computer technology and is visual information coded as a sequence of discrete numbers. Therefore it can be read into a computer's memory.

Capturing and transmitting a continuous video sequence requires up to 216 Mbits/sec, which exceeds most systems buses, LANs or Wands. The peak rate of 216 Mbits/sec is the data rate that is equivalent to the digitized PAL video signal following the CCIR 601 specification (PAL video norm: 625 lines, 833 columns, 25 frames/sec, 2 bytes color/pixel). The vast amount of data required to represent a digital image and the hardware requirements for the capture are the key obstacles for most of the applications being developed in this field. If we reduce the amount of data to reproduce an image then the transmission speed will increase. Consequently, on-line video capture for Tele-Media purposes generally also include on-line video CoDec to reduce the bandwidth required (see figure 4).

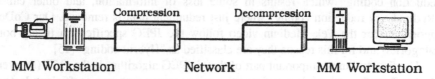

MM Workstation   Network   MM Workstation

Fig. 4. The video connection between multimedia workstations

Our prototype uses VideoPix hardware on SUN workstations, Indigo Video on SGI workstations, and miroMovie and ScreenMachine capture hardware on PCs. VideoPix, IndigoVideo, miroMovie and ScreenMachine are commercial frame grabbers.

The original resolution of our individual digital video image is given by the PAL video standard. These images are then reduced to 176x144 pixels with 16 grey scales, following the H.261 (Px64) specification [13, 14]. Other operation modes supported by our prototype use 352x288 pixels (resolution of the Common Intermediate Format - CIF) or 176x144 pixels (resolution of the Quarter Common Intermediate Format - QCIF) with 16 or 128 grey scales, or 160 colors (see *Video Compression, video Scalability* and [15]).

Video Compression. The development of digital video technology in the last years made possible not just the transmission of sequences of video images over computer networks but also the use of digital video compression for a variety of telecommunication applications [15].

There are two different compression techniques: The first does a pure intraframe compression which means that it compresses a single image at a time, without regard to the previous or succeeding image. The *JPEG* (Joint Photographic Expert Group) standard belongs to this first compression technique. The second does intraframe compression and then additional inter frame compression which takes previous and succeeding images into account. This technique provides a very high quality sequence of images. However, is very costly to encode, does not provide easy access to single images of the sequence and it is not useful for applications which need to be symmetric. By symmetric CoDec applications we refer those applications that require essentially equal use of the compression and the decompression process.

The *MPEG* (Moving Pictures Expert Group) standard for full motion video compression belongs to the second category of compression techniques and could have been used in our application [16]. However, for the real-time video part of our system we needed a fast compression and decompression algorithm that could be performed even on low-end hardware or in software. Additional requirements were easy access to single images and symmetric processing of the CoDec. The real-time issues for video as a communication channel turned out to be more important than high quality pixel and color resolution. So we decided to use the JPEG CoDec standard for our video frames, calling it *M-JPEG* (Motion-JPEG) [10].

Two categories of data compression algorithms can be used. One called "Entropy reduction coding" which results in some loss of information, and other called "Redundancy reduction coding" where just redundant data is removed. Our CoDec algorithms for the Tele-Medium video follow the JPEG specification (using both categories) and for this reason they are classified as "Hybrid coding" [15].

The first and most important part of the M-JPEG algorithm is the *DCT* (Discrete Cosine Transformation). It transforms the source image data to a coefficient domain space. This sub-process is specified for lossy compression because after this step a great part of the resulting coefficients are considered to be without any relevant information. For that reason they are discarded. To the not discarded coefficients we apply a vector quantisation process terminating the entropy reduction of the image data. The following steps are applied in order to reduce data redundancy. For some samples (DC coefficients), we use one of the most well known techniques designated by DPCM (Differential Pulse Code Modulation) [15]. Finally, we apply a Huffman coding to the coefficients increasing the compression rate factor without losing more information. Figure 5 shows these three steps of the CoDec algorithm.

For the DEDICATED project, video communication functionality was implemented on UNIX workstations. The CoDec algorithm is realized as a software solution for Sun Sparc 2/10 and Silicon Graphics Indigo workstations. The frame grabber hardware is machine-dependent (VideoPix on Sun, IndigoVideo on Silicon Graphics).

There are a number of other research efforts which are focused on developing generic platform-level technology for video communication and CoDec [8, 17, 18]. But none of them seems to address both cross-platform portability and low-bandwidth networks (< 1 Mbit/s) like we do with our M-JPEG approach.

Video Scalability. One of the main features of our prototype is the manipulation of a set of CoDec parameters. The goal was to enable the application (or the user) to work in a flexible space, selecting parameters for image quality, frame rate, compression and decompression speed, color mode, image size and so on. All this scalability is implemented and it has been possible to observe interesting results and new ideas for future developments.

With the Motif-based software-only solution, the frame rate is approximately 4 to 5 frames per second on a SUN Sparc 2. Interpolating additional frames from existing, decompressed ones and inserting them between the original frames increases the frame rate on the workstation and allows cross-platform adaptation (see 2 - Tele-Media). This mechanism is called Intermediate Image Interpolation (I^3) and is represented in figure 5. As a result of the I^3 mechanism, the frame rate is doubled at very reasonable computing costs without additional network usage [8].

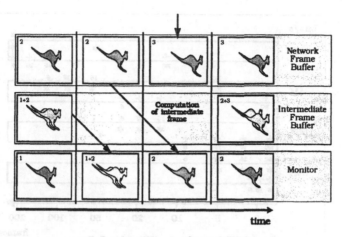

Fig. 5. The Intermediate Image Interpolation (I^3) scheme. Time scale: left to right; Path of the frames from network to monitor: top to bottom.

With our prototype we give the user the possibility to choose three different levels of image quality. The lowest quality can be reached with 6 DCT coefficients, the medium quality with 10 and the highest quality with 15. The higher the quality is, the lower the frame rate reached and, as a consequence, the transmission rate is affected (see *Video Transmission and Time Rates* and figure 6). For grey scale images and quite good light conditions, the lowest quality can be enough to produce a very good image quality. Even if the images are compressed with a high quality on the sender side it is possible to decompress them with a lower quality on the receiver side. This means that just a part of the image data received is handled to reconstruct the image. The advantage is the increasing speed the decompression process due to the smaller amount of processed data. If the network used to transmit the video images is fast enough (e.g. LAN), it may not be necessary to perform all steps of the CoDec algorithm. For example, the compression achieved with just the first CoDec step - DCT transformation - results in a lower frame rate. But on the other hand the outcome is less information loss (better image quality) and faster CoDec algorithms.

Taking color into account, we can work with 16 and 128 grey scale images and with 160 color images. For grey scale images CoDec operations typically produce a very good image quality with a reasonably good frame rate and CoDec time. Color image compression can be regarded as compression of multiple grey scale images. However, the use of color substantially limits the system performance.

Due to the fact that we are following the specifications of the M-JPEG CoDec, the only restriction on the dimension of the images is that the size in pixels must be dividable by 8, both for width and for height [15]. Within this restriction all image dimensions can be used. As default we use the QCIF standard size (see 4.1- Video Communication).

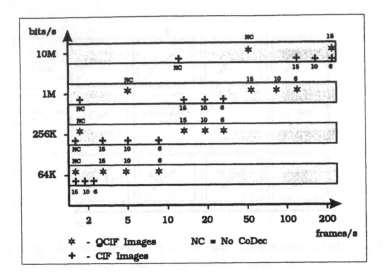

Fig. 6. Frame rate table

Video Transmission and Time Rates. In our prototype tested on a Sun Sparc 2, the currently most significant limiting factor is the access time to the frame grabber hardware. With the VideoPix hardware, it takes about three times longer to grab the video data than to compress them. Therefore, within the current prototype, the video performance is not higher than 4 to 5 images/sec.

On our LAN (Ethernet), at a resolution of 176x144 and 16 grey scales (128 grey scales) per uncompressed image, we measured a peak video transmission rate of approximately 80 (40) images per second. Each individual uncompressed image has a size of about 12 Kbytes (16 grey scales) or about 25 Kbytes (128 grey scales), what results in a peak transmission rate of approximately 1 Mbyte/sec. Applying the CoDec algorithms, the image data size is dependent of the quality demanded. We implemented three different qualities of image (see section 3.2) and collected the following data where the CoDec rate gives us an indication of the data reduction factor. (A CoDec rate of 20 means: the volume of the data is reduced by the factor of 20 when compressed).

The CoDec rate of 20 to 30 (6 DCT coefficients) is the fastest (10 to 14 images/sec) but gives the lowest image quality. A CoDec rate of 9 to 12 (15 DCT coefficients) is the slowest (5 to 7 images/sec), but gives the highest quality. This applies for both resolutions, 16 and 128 grey scales. Therefore, when using a CoDec rate of 9 (about 2800 bytes/image), it is theoretically possible to transmit up to 500 images/sec using the same peak transmission rate. This shows that with compressed images the LAN is not the limiting factor but the frame grabber hardware and the CoDec algorithms. We also measured the performances using images with the same size but with 160 RGB colors. With 6 (15) DCT coefficients we have a CoDec rate of 7 to 10 (3 to 4) what results in 3 to 5 (about 2) images/sec.

The presented results are measured with our test environment. If we ignore the limitations due to the speed of the frame grabber hardware then for series of different network bandwidths some interesting theoretical results can be seen. Together with a variation on the image quality, figure 6 shows how the transmission frame rate can change. Two standard image sizes were considered for these results: CIF and QCIF (see 4.1 - *Video Communication*).

As a conclusion we can state that, by using fast frame grabber hardware and new-generation workstations, reasonable video frame rates could be reached in practice. This is true for both the sending and receiving of video images simultaneously because the CoDec algorithms are symmetric.

4.2 AUDIO COMMUNICATION

Different multimedia platforms provide different types of audio devices and services. Nevertheless, all of them offer a common set of functionalities, like record, store and playback. We realized on-line Tele-Media speech communication, using these basic functionalities. In order to provide cross-platform usability, we defined a common, intermediate exchange format and realized a number of on-line bi-directional converters supporting different audio formats (see figure 7).

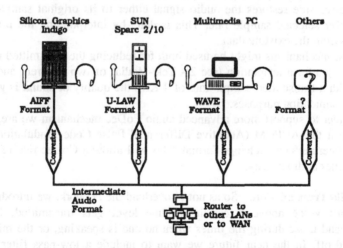

Fig. 7. The architecture for the audio communication using a heterogeneous audio environment.

Our intermediate exchange format used for all platforms is a subset of the WAVE format. The reason for the choice of the WAVE format was its simplicity and the superior processing speed of the converters on Sun and Silicon Graphic workstations in comparison to converters on the PC. On the workstations the converters only produce delays of approximately 12.5 milliseconds each. The

Multimedia PC accepts the intermediate WAVE format without any conversion which results in reduced computational costs. The WAVE format as used on our prototype is based on the Pulse Code Modulated (PCM) data format of 8-bit mono at a sample rate of 8 KHz (voice-quality). It can reproduce a limited dynamic range of the human voice. The audio transfer rate sums up to 64 Kbits/sec, what is precisely the data rate of one ISDN channel. In combination with audio CoDec mechanisms, we intend to reach data rates of about one quarter of an ISDN channel.

Audio Compression. On occasions it might be useful to record the exchanged audio data during a Tele-conferencing session, for later playback, for archiving issues or for editing. But even voice-quality audio is data intensive; one minute of voice quality audio takes almost half a Mbyte of storage space. Therefore, on-line CoDec algorithms to minimize storage space and network bandwidth are needed, covering both the raw data and precise timing information such as sample rate, recording time and synchronization markers.

We realized mechanisms called "up-sampling" and "down-sampling". They allow very fast changes of the sample rate of the transmitted and processed audio data without modifying the used audio hardware. Down-sampling means that a part of the audio samples on the sender side are discarded before transmitting the audio stream and broadcasting the "new" sample rate. If down-sampling from 8 KHz to 4 KHz is selected every second sample is discarded. The up-sampling algorithm on the receiver side restores the audio signal either to its original sample rate or to some other selected sample rate. This is done by interpolating the missing audio samples from the existing data.

These mechanisms might be used both for reducing the transmitted data and for the cross-platform adaptation (see 2 - Tele-Media) of two different audio systems. Even with a transmitted sample rate of 4 KHz the quality of sound is good enough for communication purposes.

In order to support more advanced audio CoDec mechanism we are planning to implement the ADPCM (Adaptive Differential Pulse Code Modulation) algorithm for our intermediate exchange format. This will allow a CoDec rate of 2 to 3 with much better audio quality.

Audio Transmission. So as not to overload the network, we introduced a noise gate. Just voice above a certain volume level gets transmitted, but not the background noise during the times when no one is speaking, or the microphone is switched off. In the near future we want to include a low-pass filter in order to reduce audio aliasing effects.

As with the on-line converters, the noise gate is realized using software on the workstations. The delay during the audio data transmission produced by the noise gate is less than 7.5 milliseconds on Sun or Silicon Graphics workstations, dependent on the processing load. The on-line converters consume less than 12.5 milliseconds each. This totals a worst case delay of approximately 30 milliseconds for the entire conversion, not taking into account the delays due to the physical network. The audio data rate is always less than 64 Kbits/sec which can be easily provided by LANs or ISDN.

5 Implementation

The architecture here described was implemented in C++, using an object oriented approach, both in UNIX and Windows. There are parts of this system that are dependent of the machine, so they were implemented differently for each system. That's the case of the GLS and of the API. However, the rest of the software layers in the course server are exactly the same for all the supported systems, thanks to the API, that contains and encapsulates all the system dependencies.

For the interprocess communication the sockets streams were chosen. They offer a versatile way of communication and meet all the requirements we need in terms of reliability.

For the Unix implementation the X graphical system with the Motif tool kit was chosen because of their wide acceptance as a *de facto* standart. This eases the port of the GLS among the several *flavors* of the UNIX systems and machines. The Unix machines currently supported are Sun sparcStations, SGI Indigos and HPs, running Unix System V (with sockets extensions) and BSD 4.3.

As the PC compatible machines are getting more and more importance they were included in the list of machines supported from the begining of the project. In these machines Windows was chosen as the operating system and graphical interface. A third party TCP/IP communication stack is also necessary because Windows doesn't provide interprocess comunication using sockets.

A different aproach from the one we used was considered in the begining of the project. This involved the use of one machine independent graphical toolkit to provide graphical support in multivendor platforms. However, that was just one of our problems, and if we had addopted that solution we would still have many of the other problems unsolved, such as different system calls for each operating system and the support for the distribution of the system. This way we achieved a more integrated product, where we can have total control over each module to tune it to obtain the best performance of the complete system.

6 First Field Tests

In our test, we had a PC at Coimbra that accessed a course server at INESC, in Lisbon. This course server was running on a Unix machine, a Sun Sparc station, and could be acessed through a ISDN to Ethernet PC gateway. Both PCs (the gateway at Lisbon and the one in Coimbra) where using INESC's ISDN board PC-BIT (voice + data version) and software.

The first test assumed a course server at Lisbon and a student trying to connect from Coimbra, using a PC and the ISDN line. Everything worked correctly and the student was able to establish a successfull connection and see the simulated course.

To test the connection in the other direction we established a course server in the PC at Coimbra and tried to connect from the Unix workstation from Lisbon. To be able to visualize the remote student's screen at Coimbra, we used a X server running in the PC (at Coimbra) and redirected the X output in the workstation to that remote server, using the same ISDN link. Now things were a litle bit slower, because of the extra X data sharing the same ISDN data link without compression, but everything worked as planned.

REFERENCES

1. J. M. Velez, M. R. Gomes, J. M. B. Lopes, "Sistema de Formação Modular", *6° Encontro Português de Computação Gráfica*, February 1994

2. J. M. Velez; B. Tritsch, "Revised MTS Specification", *DELTA Deliverable 11, Project D2014 DEDICATED*, DELTA Commission of the European Community, August 1993.

3. B. Tritsch, Ch. Hornung, "DEDICATED - Learning on Networked Multimedia Platforms", IFIP WG 3.2 Working Conference, Visualisation in Scientific Computing: Uses in University Education, July 1993

4. Brisson Lopes, Gomes, Graf, Knierriem, Lindner, Tritsch, Velez, Zysk, "Functional Specification of the Modular Training System", *DELTA Deliverable 3, Project D2014 DEDICATED*, DELTA Commission of the European Community, August 1992.

5. W. Stallings, "Handbook of Computer Communications standarts", volume 3,Howard W. Sams & Company,1989

6. M.Hall, M. Towfiq, G. Arnold, D. Treadwell, H. Sanders, "Windows Sockets - An Open Interface for Network Programming under Microsoft Windows", January 1993

7. A. Banerjea, B. A. Mah, "The Real-Time Channel Administration Protocol", The Tenet Group, *Proc. 2nd Intl. Workshop on Network and Operating System Support for Digital Audio and Video, Lecture Notes in Computer Science, No. 614, pp. 160-170*, Heidelberg, Nov. 1991

8. R. Steinmetz, Th. Meyer, "Modelling Distributed Multimedia Applications", IBM European Networking Center, Heidelberg

9. Ch. Hornung, M. Jaeger, A. Santos, B. Tritsch, "Cooperative HyperMedia - An Enabling Paradigm for Cooperative Work", *The Visual Computing Special Issue on "Techniques and Applications of Computer Graphics in the Context of Telecommunications"*, 1993 (in press)

10.   B. Tritsch, Ch. Hornung, "Cooperative Multimedia on Heterogeneous Platforms", *Proceedings of the Dagstuhl Workshop on Multimedia System Architectures and Applications*, November 1992

11.   A. Santos, B. Tritsch, "Using Multimedia to Support Cooperative Editing", as accepted at the *Eurographics '93*, Barcelona, September 1993

12.   M. Jaeger, B. Tritsch, "Multimedia in a Tele-Cooperative Environment - Performing Video on an ISDN-PC", *Proc. of the 2nd Eurographics Workshop on Multimedia, pp. 127-138*, May 1992

13.   "Video Codec for Audiovisual Services at px64 kBits/s", *CCITT Rec. H.261, CDM XV-R 37E*, CCITT, August 1990

14.   M. Liou, "Overview of the px64 Kbit/s Video Coding Standard"; *Communications of the ACM 34, 4 (1991), pp. 59-63*, 1991

15.   "JPEG Technical Specification", *Joined Photographic Expert Group ISO/IEC, JTC1/SC2/WG8*, CCITT SGVIII, August 1989

16.   D. Le Gall, "MPEG: A Video Compression Standard for Multimedia Applications", *Communications of the ACM, Vol. 34, No. 4, pp 46-58*, 1991

17.   J. Hanko, D. Berry, Th. Jacobs, D. Steinberg, "Integrated Multimedia at Sun Microsystems", *Proc. 2nd Intl. Workshop on Network and Operating System Support for Digital Audio and Video, Lecture Notes in Computer Science, No. 614, pp. 300-313*, Heidelberg, Nov. 1991

18.   A. Hopper, "Pandora - An Experimental System for Multimedia Applications", *ACM Operating Systems Review, Vol. 20, No. 2, pp. 19-34*, April 1990

Multimedia Teletutoring over a Trans-European ATM Network

Yu-Hong Pusztaszeri*, Maurice Alou*, Ernst W. Biersack**, Philippe Dubois**,
Jean-Paul Gaspoz*, Pascal Gros**, Jean-Pierre Hubaux*

*Department of Electrical Engineering
Swiss Federal Institute of Technology, Lausanne (EPFL)
CH-1015 Lausanne, Switzerland

**Institut Eurecom,
2229 Route des Crêtes, B.P. 193,
F-06904 Sophia Antipolis, France

Abstract. We describe the multimedia teletutoring application jointly
developed by EPFL and Eurecom in the context of the first 34 Mbits/s Trans-
European ATM network interconnecting sites in France and Switzerland.
This network was called the Broadband Exchange over Trans-European
Links (BETEL). The aim of this paper is to describe the BETEL teletutoring
platform and its building blocks, together with performance evaluation,
limitations and future enhancements of this prototype. Focus is placed on
the interactive audio and video communication part of the application.

1 Introduction

The trend in today's telecommunication networks is a migration towards Broadband
Integrated Service Digital Networks (B-ISDN) to support integrated high-speed data,
voice and video communications. Asynchronous Transfer Mode (ATM) is the packet
switching and multiplexing technique chosen for B-ISDN to provide services with
different Quality of Service (QoS) requirements. Meanwhile, new video and audio
coding standards are emerging, and many commercial products, both hardware and
software, are now available to integrate audio and video with conventional digital data
communications.

With this in mind, the European Parliament launched the DIVON program
(Demonstration of Interworking Via Optical Networks) in 1992 to prepare and
promote ATM technology and new B-ISDN services in Europe. The BETEL project,
funded by the European Commission and the Swiss Federal Office for Science and
Education, was one of the four projects in this program. The aim of BETEL was to
run user driven applications over one of the first 34 Mbits/s international ATM
networks in the World.

A teletutoring application was demonstrated between EPFL and Eurecom. It used videoconferencing and shared workspace tools to allow interactions between a teacher at EPFL and a group of students located at Eurecom. The teacher at one site could teach a class or supervise students in their individual work in another site while the students could seek assistance from their teacher located several hundred kilometers away.

The goal of this paper is to give an overview of the BETEL teletutoring application. Section 2 describes the BETEL teletutoring platform and its building blocks, while Section 3 is devoted to an interactive audio and video communication tool developed for this prototype and Section 4 contains a brief description of echo cancellers built for this experiment. Section 5 gives detailed performance studies of this prototype. Finally, Section 6 discusses the limitations and future enhancements of the system.

2 An Overview of the BETEL Teletutoring Application

2.1 The BETEL Network

The BETEL network infrastructure shown in Figure 1, is based on ATM technology, and supports FDDI LAN interconnection, ATM transfer service, and AAL 3/4 data service [1]. The FDDI LANs at EPFL and Eurecom were interconnected to the BETEL network by means of Cisco routers.

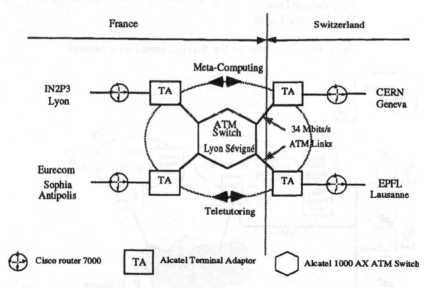

Fig. 1. BETEL: Europe's first operational ATM network

The end system protocol stack was imposed by the Cisco routers. Standard Internet protocol suite and FDDI protocols were used. The protocol stacks at the interworking units and workstations are given in Figure 2. The Switched Multimegabit Data

Service (SMDS) and point-to-point ATM connections were implemented but multiplexing of several Virtual Channel Connections (VC) over a Virtual Path Connection (VP) was not available in the BETEL network. A separate VP connection was thus established between each BETEL user sites, forming a fully meshed VP network.

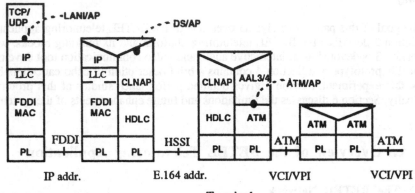

CLNAP : ConnectionLess Network Access Protocol
HSSI : High Speed Serial Interface
HDLC : High-level Data Link Control
AP : Access Point
PL : Physical Layer

Fig. 2. Protocol stacks in the BETEL teletutoring network

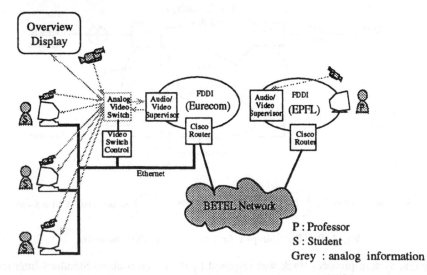

P : Professor
S : Student
Grey : analog information

Fig. 3. Teletutoring network infrastructure at EPFL and Eurecom

2.2 Teletutoring Network Infrastructure at EPFL and Eurecom

The teletutoring network infrastructure at EPFL and Eurecom is shown in Figure 3. The network topology at Eurecom is more complex than at EPFL. At Eurecom, an FDDI ring was dedicated for high speed audio and video transmission, and an Ethernet was used to connect student workstations to the BETEL platform to support connection control and shared workspace data communications. Because the shared sessions generated relatively low data rate, it was not required to connect student workstations to the FDDI ring.

Moreover, the distribution of audio and video signals at the student site at Eurecom used an analog audio/video switch. All cameras, microphones, monitors, and loudspeakers were connected to the switch, which was controlled by dedicated software to establish and release audio, video and data connections. Using the existing analog infrastructure at the site provided a cheap implementation strategy since video compression hardware was not required for each student workstation.

2.3 Application Building Blocks

Commercial workstations were used to build this prototype. Sun Sparc10 stations equipped with Parallax boards were used for audio and video acquisition and transmission, while Hewlett-Packard (HP) workstations were used as workspaces which could be shared between the teacher and the students using SharedX, a Shared Workspace Manager provided by HP.

Connection Control and User Interface modules [2] provided an intuitive and user friendly access to the Audio-Video Supervisor and Shared Workspace Manager. Moreover, echo cancellers were used to reduce echoes generated in the BETEL teletutoring network. The building blocks were as follows:

- Hardware:
  - HP 9000/700 workstations
  - SUN Sparc10 stations
  - Parallax video acquisition board
  - Echo canceller

- Software Modules:
  - User Interface
  - Connection Control
  - Shared Workspace Manager
  - Audio - Video Supervisor

3 Audio - Video Supervisor (AVS)

AVS provided real-time audio and video acquisition and end-to-end transmission. Under the supervision of AVS, audio and video signals from analog sources were digitized and encoded, and then were transmitted to a remote station via the BETEL network. At the receiver end, the data were decoded, and video images were reconstructed and displayed while audio was being replayed. Audio and video signals were handled in a similar fashion (see Figure 4).

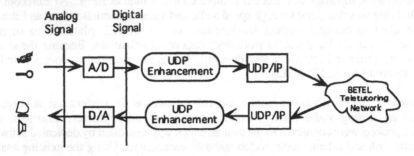

Fig. 4. AVS processing pipeline

3.1 Video Acquisition

The current AVS implementation used the Parallax XVideo board. This board was the only hardware available on the market which permitted real-time video compression and decompression at a reasonable frame rate. Details about this board and its performance are given in [3].

The Parallax board can handle analog video input and output in various standards (PAL and NTSC) and formats (YUV, RGB, super VHS, and Composite). The video signals are first digitized, then compressed by the XVideo board based on the JPEG standard before being sent to the network. On the receiver side, the digital video signals are decompressed, converted to the analog signals and displayed on the receiver's monitor.

The programming interface that came with this board uses an extension of the X11 library called XVideoToolkit. Its functionality is exploited by AVS through an extended X-Window server, which provides access to Parallax's graphical accelerator and frame buffer, and supervises video digitization and compression / decompression.

One of the major drawbacks of this board resides in digitized video images, which have to be first stored in XVideo's frame buffer then compressed by the JPEG Image Compressor on the XVideo board. Hence, video images cannot be compressed without being first displayed locally.

3.2 Audio Acquisition

Audio streams are digitized, recorded and played by a SpeakerBox on the Sun Sparc10 station. The SpeakerBox audio peripheral provides an integral monaural speaker and microphone, stereo line in/out and headphone connections. This SpeakerBox supports different audio qualities and encoding techniques. Moreover, it has a programmable audio device interface.

3.3 Networking Issues for Audio and Video Transmission

Real-time audio and video transport service imposes several performance requirements on the network. Since both audio and video sources produce continuous data streams, not only do their temporal relationships have to be satisfied, but they also require a large network bandwidth. In summary, interactive audio and video data generated by the teletutoring application impose the following network requirements:

- guaranteed high throughputs
- bounded end-to-end delay and delay jitter
- low loss and error rates
- connection-oriented service, i.e., in sequence delivery
- support for real-time data service
    - higher priority for real-time data
    - selective discarding of data according to their priority in case of congestion
- synchronization
    - intra-medium
    - inter-media
- adaptive and preventive rate-based flow control

The transport layer protocol was restricted to TCP and UDP since IP was imposed by the Cisco router in the BETEL network. TCP provides reliable end-to-end connection oriented transport service while UDP and IP support best effort services based on connectionless techniques. The Internet protocol suite was designed for point-to-point non-real-time data service and has many difficulties to meet the network performance requirements demanded by the teletutoring application.

TCP/IP is unsuited for a network with a large bandwidth-latency product. The BETEL network is one of such networks. The sliding-window flow control with credit allocation does not allow to use the full bandwidth of the BETEL network. Retransmission in TCP significantly increases end-to-end delays and is unsuitable for interactive audio and video data transport service. Hence, window based flow control and error control mechanisms found in TCP create problems for real-time audio and video transmission, as in the BETEL context.

On the other hand, the lack of retransmission mechanism makes UDP a better candidate to transport real-time audio and video data. Since UDP does not guarantee in sequence delivery and does not have any error or flow control, some end system enhancements added to UDP are needed. For instance, video frames in general are larger than the UDP datagram limit (9 Kilobytes), thus video frames need to be segmented into smaller frames before being sent to a UDP socket and be reassembled together at the receiver end. In order to make the reassembly process efficient, missing frames and out of order frames have to be detected. Minimum UDP enhancements are loss detection and packetization (including segmentation and reassembly for video frames). Therefore, UDP/IP protocols with these enhancements at the end system were used to transport audio and video data in AVS.

3.4 Implementation and Performance Issues

AVS was designed to guarantee optimal performance. This was done with as little data movements and copying as possible. Only minimum UDP enhancements were implemented. Data were packetized (segmented if necessary) and sent to a UDP socket without any buffering and copying. A small header was added to each packet. The frame sequence information (for loss sequence detection and sequence check) was put in the header. For video frames, the sequence number and the total number of segments were also needed in the header. A typical video packet size was about 4 Kilobytes, and so was the MTU in the BETEL network (excluding the Ethernet segment since no audio and video information was distributed across Ethernet using UDP). On the other hand, audio frames had 128 audio samples (one byte per sample in this context) to ensure low delay and loss rate.

The audio quality was the most important factor in the design of this teletutoring prototype. Voice is the most common and effective means of human communication, although eye contact and other facial information are also important. Some steps were thus taken to ensure optimal audio quality. The use of smaller audio frames was one. Another was the use of high quality sound equipment. The Sun microphone did not give as good an audio quality as did the semi-professional microphones, so the latter was used (connected to line in port of the SpeakerBox). In addition, echo cancellers were necessary to reduce echoes generated by the large round trip delay in the BETEL network. Further, it was impossible to use CD quality audio supported by the SpeakerBox because the echo canceller used could not treat audio with sampling frequency higher than 8KHz. Therefore, audio was restricted to telephony quality. Audio data in AVS were not compressed to reduce processing delay to ensure good audio quality, as performance studies in Section 5 show that the performance bottleneck of this prototype is not in the network.

On the other hand, one may ask why teletutoring applications and other interactive multimedia applications need broadband networks nowadays? The most obvious answer is video quality. High quality digital video signals demand a large bandwidth. Since the XVideo board is one of the most preferment real-time compression hardware available, it was used to have the optimal video performance. In addition, using a pair of Sun Sparc10 stations to send and receive video frames did not obtain high video frame rates and video bit rates and cannot fully exploit the 34 Mbits/s high-speed link for this teletutoring application. In order to overcome this problem, a Sun Sparc10 station was dedicated to either transmit or receive video sequences. As AVS consists of four independent processes, each process is used for receiving or sending video and audio data. Transmission of audio and video is hence independent. Therefore, four Sparc10 stations equipped with Parallax boards were involved in the video acquisition and transmission of this prototype.

4 Echo Cancellation Through Adaptive Filtering

The adaptive filtering technique [4] used the Least Mean Square (LMS) algorithm, which has the advantage that no prior knowledge of the room impulse response is required. The adaptive filtering shown in Figure 5, consists of two distinct steps:

- estimating an filtering error

$$e(k+1) = y(k+1) - \vec{g}^T(k) \cdot \vec{x}(k+1) \qquad (1)$$

- updating the coefficients $\vec{g}(k)$, using the error e(k+1).

$$\vec{g}(k+1) = \vec{g}(k) + Ke(k+1)\vec{x}(k+1) \qquad (2)$$

Where

$$\vec{x}(k) = \begin{bmatrix} x(k) \\ x(k-1) \\ ... \\ x(k-N+1) \end{bmatrix} \qquad \vec{g}(k) = \begin{bmatrix} g_0(k) \\ g_1(k) \\ ... \\ g_{N-1}(k) \end{bmatrix}$$

x(k) represents the sample at instant k,

$\vec{x}(k)$ represents the vector of the N most recent samples at time k

$\vec{g}(k)$ is the vector of the N filter coefficients at time k

y(k) is the sample coming from the microphone at time k

e(k) is the error at time k

K represents the adaptation step

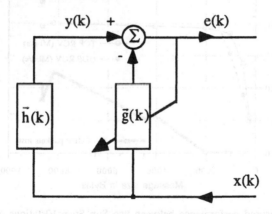

Fig. 5. Feedback of e(k) on the filter coefficients with
emphasis on the adaptation process

The convergence time and stability of this system depend on the value of the adaptation step K, which depends on the length of the filter N and on the input signal power [2].

323

The measured echo attenuation was 20 dB in a given test room and the signal decreased by 20 dB in 102.5 milliseconds. The performance of this echo canceller is closely related to the room acoustics.

5 Performance Evaluation of the Teletutoring Platform

5.1 Measured raw TCP/IP and UDP/IP Performance Without Applications

The purpose of this study is to estimate the upper bound in performance which is available for applications running on top of TCP and UDP. The TCP/IP and UDP/IP throughputs were measured (using ttcp tool) both in the local FDDI environment and on the BETEL teletutoring platform [2].

The measured performance in the local FDDI LAN environment is shown in Figure 6. UDP/IP (without UDP checksum) could achieve 58.6 Mbits/s throughput for a message size of 4K bytes but with a maximum of 20% losses while TCP/IP had a maximum throughput of 40 Mbits/s for the same message size.

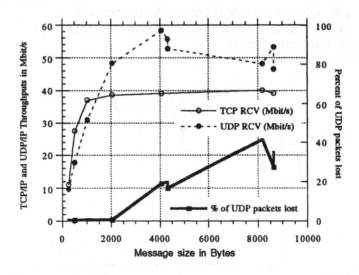

Fig. 6. Measured performance between two Sun Sparc10stations over FDDI

On the other hand, TCP/IP throughputs stabilize at around 8.4 Mbits/s in BETEL teletutoring network for message sizes above 2000 bytes. The Round Trip Time (workstation back to back) of the EPFL-Eurecom BETEL link is about 12 milliseconds for 64 byte messages and 17 milliseconds for messages of 1024 bytes.

These results show that the BETEL teletutoring network has a large bandwidth-latency product and that UDP is unreliable and suffers from losses but UDP can attain higher throughputs than TCP. Hence, at least a 8.4 Mbits/s of bandwidth on top of UDP/IP is available for real-time audio and video data communications between EPFL and Eurecom using the BETEL teletutoring network.

5.2 Measured Video Performance

Video communications demand a large bandwidth and their peak performance is likely to be limited either by the network or by the end system. Understanding the parameters influencing video performances can help us to obtain the best video quality. By measuring video bit rates and frame rates, we can gain an objective insight into video performance and hence identify the bottleneck of this prototype. The performance measurements in this study were taken between two Sun Sparc10 stations in an FDDI LAN at EPFL.

First, video bit rate depends on the Q factor (ranging from 25 to 1000) which is used by the XVideo board to determine the quantization level and to control the compression factor. The higher the Q factor and the larger the compression factor, the lower the video bit rate and video quality. Another important factor is the video frame rate, defined as the number of video frames captured and played per second. Figure 7 shows the relationship between the measured unidirectional video bit rates using tcpdump [5] and the number of frames captured per second when only a quarter of PAL resolution was used and the Q factor was at 50. The video bit rate increases proportionally to video frame rate.

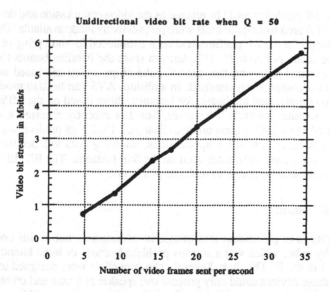

Fig. 7. Measured unidirectional video bit rate between two Sun Sparc10 stations over FDDI using UDP/IP

When we pushed the system to its maximum capability, we can reach 34 video frames captured and sent per second at a video bit rate of 5.7 Mbits/s. Similarly, transmitting the full PAL resolution video images of with the same Q factor, the system can reach only 12 video frames per second and can generate a video bit rate of 5.8 Mbits/s. As a consequence, the measurements from section 5.1 show that the bandwidth available for video communications in BETEL teletutoring platform is larger than the maximum video bit rate which the end system can deliver. It is clear that the performance bottleneck lies in the end system and not in the network.

6 Limitation and future enhancements

The implementation philosophy of BETEL was to integrate currently available technology and build a demonstrator within a year. The teletutoring prototype inherited the limitations of the current technology. This teletutoring experiment used a hardware dependent, point-to-point configuration (i.e., EPFL-Eurecom), and used the UNIX operating system and the Internet protocol stacks. There was no built-in synchronization and rate-control mechanisms implemented in the videoconferencing system. Therefore, enhancements will be needed in the following areas: multipoint and multiplatform teletutoring configurations, and system support for interactive real-time teletutoring applications.

6.1 Hardware Dependency

The hardware dependency could be relaxed as the video compression and decompression hardware and shared workspace tools were progressively made available. The release of a Parallax board for the HP platform had been announced for the Spring of 1994. AVS may then be easily ported to the HP platform since the Parallax boards (both for Sun and HP platforms) are using the same C-cube chips which are based on the JPEG compression/decompression standard. In addition, AVS can be also modified to use other video compression hardware, for instance, those based on the MPEG standard when they become available. Moreover, Sun has recently released a commercial product called ShowMe2, a competitor to SharedX. Unlike its predecessor (ShowMe), ShowMe2 can be used to share applications, allowing both videoconferencing and shared workspace tools to be integrated on the Sun platform. The BETEL teletutoring prototype could then be ported to multiple platforms.

6.2 Audio Quality

Audio quality is of paramount importance in teletutoring, but in this context it was hampered by echo, which was a serious problem because of large latency that audio experienced in the BETEL link. Several echo cancellers were designed to cancel this effect, but these devices could only process one speaker at a time and created problems when used with audio mixing devices. An audio enhancement would be to design new echo cancellation algorithms which support CD quality audio and can be used with mixing devices. This would benefit mostly a teletutoring scenario with geographically dispersed students.

6.3 Scalability

The current prototype is limited to point-to-point communications. A teacher can interact with one student at a time, although his image and voice can be broadcast to everyone in the classroom. Students may not engage in a discussion with each other. All video and audio signals have to be transported via point-to-point audio and video connections. It would be useful if either the teacher could simultaneously supervise several students from different sites, or several teachers from different sites could interact together. Thus, a fully meshed digital multipoint videoconferencing is needed.

6.4 System Support for Teletutoring

One of the long term solutions to the performance problem is to implement end system support and network support for interactive multimedia applications, such as teletutoring. A better multimedia workstation architecture is needed to sustain transmission of large amounts of data through the system buses and to minimize data movement and copying.

The UNIX operating system is not adequate to support real-time services. The scheduler granularity of Sun's OS 4.1.3 illustrate this point. Since its clock resolution is not higher than 20 milliseconds [5], it is difficult to implement any efficient audio-video synchronization mechanisms. In addition, the scheduler cannot give a higher priority to real-time data.

The Internet protocol suite cannot guarantee a high throughputs and a bounded delay and delay jitter which are required for teletutoring applications. Thus, network protocols supporting QoS requirements are needed here. Moreover, the current standard network protocols do not support a multicast service which is essential in the multipoint teletutoring configurations. Furthermore, end systems should at least support audio-video synchronization and adaptive rate-based flow control.

7 Conclusion

With the aid of high quality videoconferencing and shared workspace tools, the BETEL multimedia teletutoring application was successfully demonstrated at the end of 1993 over the first 34 Mbits/s Trans-European ATM network. Echo cancellers were essential in ensuring high quality audio in this experiment. This paper showed that it is feasible to build a high quality teletutoring prototype using the technology available in 1993. The performance bottleneck of this prototype was at the end system level, particularly in the video acquisition. The UNIX operating system and BETEL protocol stacks provided best effort services which were not ideal but satisfactory to support the BETEL teletutoring application. UDP/IP were used to transport real-time audio and video data, without any explicit audio-video synchronization and adaptive rate-based flow control mechanisms. More robust and realistic teletutoring scenarii will be realized in a multipoint and multiplatform environment in the framework of the Pan-European ATM pilot experiment starting in July 1994. Examples are distributed classrooms, teleseminars and archiving and retrieval of multimedia documents.

Acknowledgments

Sincere thanks to the BETEL teletutoring project team, in particular, Mireille Goud, Marc Furrer, Xavier Garcia, Bruno Dufresne, Andrea Basso and Marco Mattavelli from EPFL and Morris Goldberg, Antonio Fernando Vaquer Mestre and Revital Marom from Eurecom.

References

[1]   Le Moan, Y., "Data Transfer Service Specification", BETEL internal document, CIT-5, April 1993.

[2]   Pusztaszeri, Y. H., "Teletutoring over a Trans-European Broadband Network", EPFL-Eurecom Technical Report, June, 1994.

[3]   XVideo Technical Overview, Parallax Graphics, Inc., Release 1.0, 1991.

[4]   Kunt, M. Techniques modernes de traitement numérique des signaux, Collection Electricité, Presse Polytechniques et Universitaire Romandes, 1991, pp. 175-190.

[5]   Jacobson, V., Leres, C., and McCanne, S., TCPDUMP(1), 1989-1993.

A Framework for Synchronous Tele-Cooperation

Thomas Schmidt

Siemens AG

Corporate R&D Division

Stuhlsatzenhausweg 3

D-66123 Saarbrücken

Jean Schweitzer

Siemens AG

Corporate R&D Division

Stuhlsatzenhausweg 3

D-66123 Saarbrücken

Michael Weber

Distributed Systems Dept.

University of Ulm

Oberer Eselsberg

D-89069 Ulm

Abstract. Computers are ever more thought of to support cooperative work over distances. Thus tele-cooperation tools and environments are needed which bring the meeting and joint-working onto everyone's desktop. The following paper introduces a framework for such tele-cooperation. The required cooperation-awareness is achieved by a session model around which the building blocks of the entire service are configured. A session management service controls the underlying services application sharing and telepointing, annotation and sketching, and audio/video conferencing.

1 Introduction

Information processing and data transmission have made the world smaller since the first images were drawn on cave walls. Today, the technical environment for high speed data interchange enables humans around the world to interact with each other. While a face-to-face meeting involves all human senses and all means of expressing, currently the most technically based interaction is restricted to a few ways of communication only. But the rapid technical development enables the integration of all important means of communication among humans into a computer based environment.

The face-to-face meeting situation is mirrored by so-called "synchronous tele-cooperation tools" with the users sitting in their rooms in front of their computers. The applications realizing these tools can be divided into cooperation-aware applications especially designed to be used within groups, and into coordinating tools allowing existing cooperation-unaware single-user applications to be run in a group context [1], [2]. These coordination tools are necessarily realized as cooperation-aware applications. Thus, looking from a higher level of abstraction, all tools supporting synchronous group work can be seen as cooperation-aware.

However, the inflation in the development of single features makes it necessary to structure the whole bunch of currently available techniques. This paper presents a framework for synchronous tele-cooperation which enables synchronous joint working in heterogeneous workstation environments comprising different hardware platforms.

Specifically, SNI RW420 workstations with IRIX 5.x, SNI's Pentium PC with SINIX 5.x and Sun Sparcstations with SunOS/Solaris will be supported. Most of the experiences which form the base of this paper are gained by R&D projects carried out in the frame of DELTA, RACE and BERKOM II, especially ARAMIS [3], CIO [4], ECOLE [5] and BERKOM-MMC [6].

2 The Session Model

Working in a group context requires the awareness of the ongoing cooperation on two main levels, the system level and the user interface level. This paper focuses on the system level. The system has to possess and process information concerning the involved machines and users. This information may comprise dedicated user roles. The role may be associated with rights like read or read/write access on data objects.

The user relevant information has to be extracted from the system and presented on the user level by an appropriate interface realizing the group-awareness for the user, e.g. different user inputs may be mapped onto different visualizations of cursors or modified objects.

The overall frame within cooperative work takes place is described by the *session model* shown in Fig. 1. This model is derived from the application sharing specific model [7] by extending it to the general class of cooperation-aware applications. The *model* describes cooperation in an abstract way and serves as the basis for the design and implementation of the components on the system level and the user interface level. The left-hand side of the diagram is based on an abstraction hierarchy, whereas a corresponding hierarchy is reflected in terms of user participation on the right-hand side.

The abstraction hierarchy consists of three levels: The top-level abstraction is the *world*, a virtual space in which *sessions* take place. The potential participants of sessions are *users*.

The users attending a session are called *participants*. One user out of these, the *chairman*, leads the session. This chairman usually is the initiator of the session and he decides on the session participation policy. According to that policy, users can freely join, they can request the chairman to be allowed to join, or they can be invited to join the session.

The session forms the frame synchronous cooperative work is embedded in. This frame, the central component of synchronous tele-cooperation, is realized by the *Session Management Service*.

A session comprises several *cooperation-aware services* reflecting the resources and tools also brought into action in ordinary meetings commonly used by the participants. The following services are administrated by the Session Management:

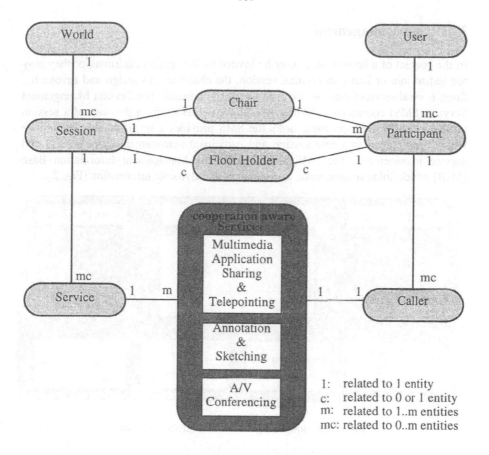

Fig. 1. Session Model

- *Multimedia Application Sharing* performing the common synchronous use of single-user applications together with *Telepointing* performing joint pointing,

- *Annotation and Sketching* providing whiteboard layers,

- *A/V conferencing* providing audio/video communication between the participants.

Moreover the Session Management Service is open to the integration of further cooperation-aware services.

Services can be involved in a session depending on the session policy only by the chairman or any other session participant, who is referred to *caller*. Some services may require a sequential user input. The assignment of the right to provide this input is subject to *floor control*. At any one time there is one participant at the most holding the floor.

3 Session Management

In the context of a session users may be invited by the session chairman, or they may request to join or leave an existing session, the chairman can assign and revoke the floor, several services may be involved by the participants. The Session Management Service (SMS) coordinates all these operations necessary from the start of a session until its end. To perform these tasks the SMS provides a set of generic operations which are grouped into core session and participant management, floor control and service management. The SMS also includes the Management Information Base (MIB) which holds session-static as well as session-dynamic information (Fig. 2.).

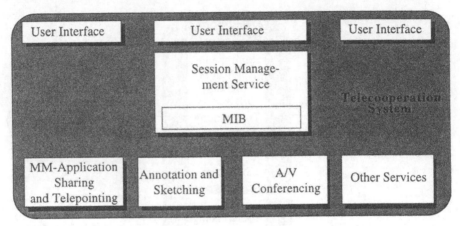

Fig. 2. Session Management and controlled services

3.1 Core Session and Participant Management

The SMS provides the following operations for session and user management:

open: An empty session is created. The issuer becomes the chairman and decides on the participation policy.

close: A session is terminated. Only the chairman can close a session.

invite: The chairman invites a user to join the session. The invited user may join the session.

join: A user requests to join the session. It depends on the policy whether he joins straight away or whether the session chairman decides to allow joining.

drop: A participant is removed from the session by the chairman.

leave: A participant leaves the session.

shift_chair: The role of the chairman is shifted to another session participant. The operation can be issued by the session chairman only.

set_policy: The chairman sets the current session policy which lays down the rules of joining a session.

3.2 Floor Control

In a session there have to be means of assigning the floor to one participant at a time. This user is designated with the right to provide input to the application currently shared. We distinguish between floor policies and floor mechanisms [8]. Floor policies describe how participants request the floor, how it is assigned and released. Floor mechanisms are the low-level means to implement the floor policies. Different floor control policies are supported:

-        The session chairman assigns the floor to a session contributor which returns it to the chairman after use.

-        The current floor holder passes the floor to another session contributor who requested it.

-        Floor requests are queued. As soon as the current floor holder releases the floor, it is assigned to the participant heading the request queue.

At a time each session follows a specific floor policy which might be changed by the chairman. The Floor Control Service is embedded into the SMS. Four primitives are required as floor mechanisms:

request: A participant requests to get the floor. Depending on the current policy the floor is granted or not.

release: The current floor holder releases the floor.

assign: The floor is assigned to a participant.

revoke: The floor is taken from the current floor holder.

The information who is currently holding the floor is passed on from the SMS to underlying services.

3.3 Service Management

The SMS controls the other involved services. Such control does not only comprise the start and termination of the respective service, but also the changing of the number of session participants and the change of the floor holder.

open: A participant requests to start a service.

close: A service is terminated.

add_user: The session management requests the service to expand the list of participants by a new one.

remove_user: The session management requests the service to shorten the list of participants by the leaving one.

assign_floor: The token is assigned to a dedicated participant. This allows him to provide input.

3.4 Management Information Base

The MIB holds session-static and session-dynamic information. Static information relates to users, such as their addresses and names which might be stored exploiting an existing directory service. Dynamic information relates to session-specific data, such as the current set of participants and their roles, or the current floor holder. The MIB can be accessed by all services to get and retrieve relevant information.Thus the MIB offers a facility for inter-service data exchange. The following operations are provided to enter and update information stored in the MIB:

create/destroy: A data object is created/destroyed in the MIB.

change: Attributes of the data object are changed.

un/register: A service is un/registered by the MIB in order to be informed about changes of a data object

indicate: A service is informed about changes of a data object

4 Multimedia Application Sharing and Telepointing

To share an application means to provide its output to all session participants and to grant the input right to one of them at a time. The basic idea of application sharing is to intercept the window protocol traffic between the application and its terminal server. The Multimedia Application Sharing Service allows the running of single-user cooperation-unaware multimedia applications in a group context. The Multimedia Application Sharing Service can be divided into the Application Sharing Component [9] and the Audio/Video Sharing Component. Application output is distributed to all session participants by multiplexing the output stream (see Fig. 3.). The application executes on one machine and its user interface is displayed onto several machines. I.e. the requests from the application to the window server are multiplexed and sent to the remote window servers as well. The remote window servers then produce the same graphical output on their terminals. Only input from the current floor holder is passed to the application. The input of other participants is filtered to maintain the single-user behaviour of the application being shared.

The Telepointing Service provides each user with a set of globally visible pointers. As soon as a telepointer is moved into a shared window it becomes visible for all other participants. Therefore the Telepointing Service tracks the movements of locally used telepointers and distributes these movements to the remote sites. Coordinate transformations have to be performed to let remote telepointers appear in the same position in

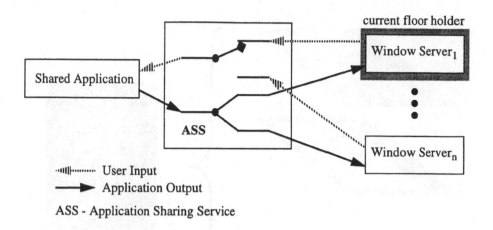

current floor holder

Shared Application

Window Server₁

ASS

Window Serverₙ

·····‖······· User Input
——▶ Application Output
ASS - Application Sharing Service

Fig. 3. Application Sharing by Distributing the Application's User Interface

the shared window as on the source site. An adjustable update frequency on remote screens avoids telepointers turning into rubbers when being moved. Telepointers also have to be synchronized with related audio streams coming from the Audio/Video Conferencing. This can be achieved by delaying the presentation time of telepointers until the corresponding audio is received, or by dropping intermediate positions of the telepointer when audio is faster.

5 Annotation and Sketching

The Annotation and Sketching Service is a generic annotation utility, to provide sophisticated functionality for annotating shared documents. Additional to the joint editing facilities the Annotation and Sketching Service provides comfortable means to annotate arbitrary applications, especially documents with text and graphics. In order to provide sophisticated annotation facilities and keep the original application unchanged, the working sheets of the Annotation and Sketching Service are transparent. A user of the system has the impression, as if the output of his document is covered by a transparency containing his annotations. Using this technique, it is possible to see usual output of the application and the annotations on the 'transparency' at the same time (see Fig. 4.). The annotations are not inserted in a shared document, but are visible to all participants of a conference. The document can be edited in the usual way, after switching from the annotation mode to normal input mode. In order to provide a homogeneous integration and to keep the user interface as simple as possible, only the most important functions are integrated into the Session Management's user interface.

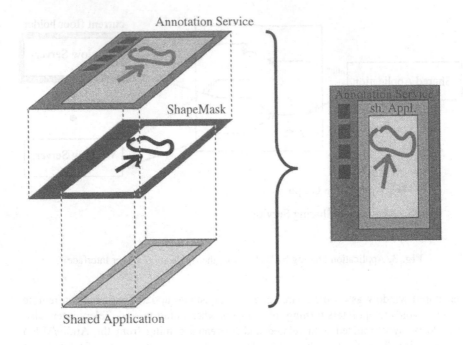

Fig. 4. Annotation service using X-Shape extension

The current version of the Annotation and Sketching Service does not support semantically linked annotations. Annotations are called semantically linked, if they are adapted to changes made in the original document automatically. For example a user has marked one sentence in a paragraph of a document under review. If additional text is inserted in this paragraph, the paragraph is moved to another location on the page or even the page has been changed, the annotation will be visible at a wrong place. It is possible to adjust the annotations automatically, if such situations occur. Future versions of the Annotation and Sketching Service will support semantically linked annotations.

6 A/V Conferencing

Cooperative work requires support by audiovisual communication. Therefore the system is enhanced by the Audio/Video Conferencing Service. The Audio/Video Conferencing Service is based on an A/V communication service which allows to multiplex and distribute audio and video sources to remote participants (see Fig. 5.). It also includes audio and video processing capabilities such as compressing and decompressing A/V data, or mixing audio channels. It can be used for a separate video conferencing session or in conjunction with other services to enhance communication. For audio and video data transport services providing isochronous delivery between the participants are required. Thus a negotiation of suitable Quality-of-Service parameters with such a transport system has to be performed.

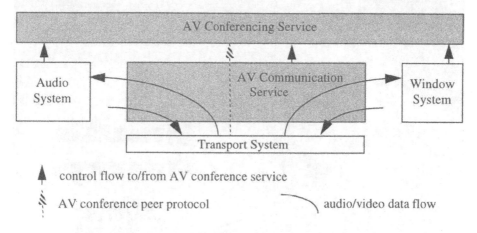

Fig. 5. AV Conference Service Architecture

7 Conclusions and Future Work

The paper introduces the session as a framework within synchronous cooperative work takes place. In the context of a session users may be invited by the session chairman, or they may request to join or leave an existing session, the chairman can assign and revoke the floor, or participants may involve several services.

The framework is realized by the Session Management Service (SMS). The SMS provides a set of generic operations to coordinate all activities necessary from the start of a session until its end. These operations are grouped into core session (open/close a session) and participant management (e. g. invite a participant), floor control (e.g. assign/revoke the floor) and service management (e.g. start/terminate a service). The SMS also includes the Management Information Base which holds session-static (e.g. users addresses) as well as session-dynamic information (e.g. participant role). The MIB can also be accessed by all services to store and retrieve relevant information.

In the next phase a conference level will be installed above the session level. The conference service maintains short- and long-term conferences and deals with the subjects of storing and recovering session contexts, address management which includes group support by a name service, a cooperative scheduler and voting support.

Furthermore, group-awareness will be implanted into the data level, e.g. by a distributed virtual group-filesystem, document managers and databases, and into the presentation level, i.e. into the window manager.

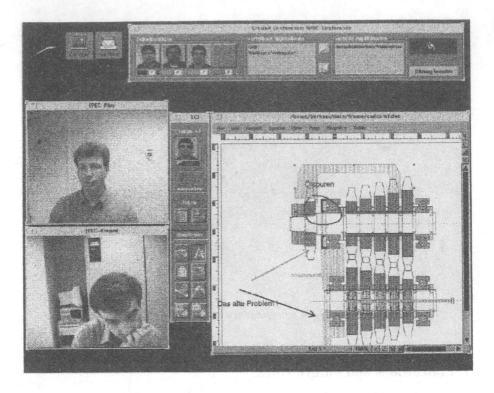

Fig. 6. Screenshot of the telecooperation system

8 Acknowledgements

We thank all our internal and external colleagues being involved in the projects BERKOM II, CIO, ARAMIS and ECOLE for valuable contributions and discussions.

9 References

[1]   Reinhard, W., Schweitzer, J., Völksen, G. Weber, M.:"Ways to a CSCW Framework: Concepts and Architectures"; IEEE Computer, May 1994.

[2]   Cronjäger,S., Reinhard, W. , Schweitzer, J.: "Functional Components for Multimedia Services". *Proceedings, International Conference on Communications (ICC '93)*. Geneva, 1993

[3]   Armbrüster, H., Humer-Hager, T., Pütz, K. J.: "Pilotprojekte mit ATM-Netzen"; telcom report, Siemens AG, März/April 1994.

[4]   Dermler, G., Gutekunst, T., Ostrowski, E., Pires, N., Schmidt, T., Weber, M., Wolf, H.: "JVTOS - A Multimedia Telecooperation Service Bridging Heterogeneous Platforms"; Proceedings, International Conference on Broadband Islands, Hamburg, 1994.

[5]   Dietel, C. et al: "A multimedia environment for distributed learning". ISDN applications in education and training, IEEE, London, 1994.

[6]   Herrtwich, R. et al.: "The BERKOM MMC Service". Proceedings ACM Multimedia '93 Conference, Anaheim, 1993.

[7]   Gutekunst, T., Schmidt, T., Schulze, G., Schweitzer, J., Weber, M.: "A Distributed Multimedia Joint Viewing and Tele-Operation Service for Heterogeneous Workstation Environments"; Proceedings, GI/ITG Workshop on Distributed Multimedia Systems, pp. 145 - 159. Edited by W. Effelsberg, K. Rothermel. Stuttgart, 1993.

[8]   Crowley, T., Milazzo, P., Baker, E., Forsdick, H., Tomlinson, R.: "MMConf: An Infrastructure for Building Shared Multimedia Applications". Proceedings CSCW '90. Los Angeles, 1990.

[9]   Minenko, W.: "Transparentes Application-Sharing unter X Window". in german; Internal Report, DFKI, Siemens AG, 1993.

[10]   Dermler, G., Gutekunst, T., Ostroswski, E., Ruge, F.: "Sharing Audio/Video Applications among Heterogeneous Platforms"; 5th IEEE Comsoc Workshop Multimedia '94, Kyoto, Japan, May 1994

Multimedia Conferencing Services in an Open Distributed Environment

Peter Leydekkers
PTT Research, P. O. Box 15000, 9700 CD Groningen - The Netherlands
E-mail: P.Leydekkers@research.ptt.nl

Valérie Gay
Université Paris VI, Laboratoire MASI, 4, place Jussieu, 75252 Paris Cedex 05 - France
E-mail: gay@masi.ibp.fr

Abstract. Distributed Processing Environments (DPE) or platforms are regarded as the future telecommunication architecture [1] on which distributed applications such as Multimedia Conferencing (MMC) services will operate. Using the DPE as target telecommunication platform, an important issue for MMC designers is the specification of generic MMC interfaces. This enables MMC users to access, in a distribution transparent way, a wide range of MMC services, that conform to these interface specifications. This paper proposes a classification of MMC services by means of three generic interface templates. These interface templates describe in an implementation independent way the functionality and management related to MMC services. The paper discusses also a possible implementation of these interfaces. Techniques and concepts are used as proposed by RM-ODP [2] and TINA-C [1].

Keywords. Multimedia Conferencing Services, RM-ODP, TINA-C, Distributed Processing Environment (DPE)

1 Introduction

Distributed applications like multimedia conferencing services (MMC) are emerging applications that are of interest for both the residential and business environment. Several MMC prototypes have been developed that operate in a heterogeneous workstation environment consisting of different hardware and software platforms. These implementations show that equipment heterogeneity can be solved which is important factor for a universal availability of distributed telecommunication services. Another important requirement for commercial success is the interoperability aspect. Interoperability in this context, allows application software located in different domains to interoperate in a consistent manner for the execution of services and management. Interoperability implies, for example, that users using a specific MMC service (e.g. JVTOS [3]) would like to work together with users using another MMC implementation (e.g. BERKOM MMC [4]).

The diversity of emerging MMC services (e.g. [3],[4],[5],[6],[7],[8]) requires homogenisation in order to make a chance in the commercial market. Standardisation of MMC services will be necessary to solve the interoperability problem which enables commercial usage and exploitation of conferencing services. Currently, this is blocked due to the fact there are almost no standards or recommendations for MMC services.

An important aspect for the standardisation of MMC services is the description of the interfaces supported by MMC services. A basic property of objects, used to model a particular service in a distributed environment, is that they are only accessible via the interfaces supported by the service. This paper classifies distributed MMC services by means of three generic interface templates. The interface templates are described using techniques and concepts as proposed by RM-ODP and TINA-C. The interface specifications are described from an ODP computational viewpoint which results in a software and hardware independent description. Users can access a MMC service via those interfaces and this paper discusses a possible engineering and technology support for this access.

The paper is structured as follows: Section 2 describes the future telecommunication architecture that will be developed in the TINA-C project. This architecture is the target platform for our MMC service. Section 3 classifies and specifies computational generic interface templates for the MMC service. Section 4 shows the possible engineering and technology support for distribution transparent access of MMC services. Finally, section 5 presents some conclusions.

2 Future telecommunication architecture

The telecommunication architecture of today causes many problems to meet the requirements of an environment where different stakeholders[1] play a role. To develop, introduce, maintain and extend telecommunication services is very expensive and costs a lot of effort. Furthermore, those services are developed for vendor specific equipment and cause interworking problems in a multi-vendor environment.

To solve these problems, the TINA-C project is created which is a world-wide initiative that consists of members from the computer industry, telecommunication network operators and service providers. Its goal is, amongst others, to define an open distributed environment that incorporates the integration of the telecommunication, computer, and television domains. TINA-C provides an architecture that enables the construction of building blocks (these can be entire services like MMC) that should operate in a heterogeneous environment. For the construction of objects, the object oriented approach is used which enables, amongst other advantages, reusability of application software. This reduces the costs associated with the development of new services.

TINA [1] propagates a distributed processing environment (DPE) that facilitates the interworking between the different stakeholders and meets the requirements of

---

1 A Stakeholder denotes a person or organisation that has an interest in distributed telecommunication services. Stakeholders are classified by the role(s) they play. In general the following roles are identified: *user* or *customer* represents a person or organisation that uses services offered by a network provider or third party service providers. The *service provider* represents an organisation that uses the network infrastructure to add value to communication services offered by network providers. The *network provider* role provides communication services to interconnect the stakeholders that are geographically separated. It is the owner of the transport infrastructure and provides end-to-end basic communication facilities.

heterogeneity and interoperability. The TINA DPE distinguishes itself from other developments in the sense that it combines developments in the IT-industry, in the area of distributed computer platforms, with specific telecommunication issues like connection management.

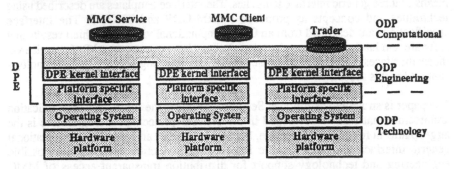

Fig.1. Distributed Processing Environment (DPE) and concerned ODP viewpoints

The DPE can be regarded as the infrastructure or platform on which distributed applications such as the MMC service will operate. The DPE provides important advantages such as heterogeneity and distribution transparency. A distributed environment heterogeneity may include: equipment heterogeneity due to a multi-vendor environment, operating system heterogeneity due to different operational contexts (office, factory), and authority heterogeneity (e.g. co-operation between distinct network providers). Distribution transparency implies that applications are not bothered with communication aspects which enables location independent applications and provides the advantage of software interoperability.

The tools that will be constructed by TINA-C to develop services like MMC, allow application designers to specify their applications in the computational language as proposed by RM-ODP. The computational specification describes how distributed applications are structured in a distribution transparent way. For the TINA architecture the computational language is refined with additional concepts and specific specification techniques (e.g. OMG-IDL) that are also suitable for the telecommunication environment. This implies that services to be built on the DPE can be viewed from the computational viewpoint (figure 1).

The DPE runtime environment consists of the DPE kernel, which resides on every node, and DPE servers. The DPE kernel is the software infrastructure that provides a set of basic functions such as communication, storage and processing capabilities. The DPE kernel provides the mechanisms to solve distribution transparency and hides the infrastructure from the application designer. This is similar to the ODP engineering view on a distributed system. The ODP engineering language describes how to structure the computational objects in order to execute them on the infrastructure. The engineering specification describes the functionality of objects supporting distribution transparencies used by the application.

DPE servers use the DPE kernel services to provide additional services to applications being built on the DPE. An important DPE server which is available on fewer nodes, is the trader. To access a service, a trader is necessary to obtain the reference of a service. The trader is a kind of broker between clients and services ([9], [10]). In a distribution transparent environment the user (e.g. MMC client) is not aware of the location of a certain service. Therefore, a service registers its interface references and properties to the trader. A request from a client for a specific service is matched with the properties of the offered services and upon agreement, a reference of the service is sent to the client. The client may then contact the service.

The ODP technology view on a distributed system expresses how the application specifications can be implemented. This viewpoint is of interest for those responsible for the hardware and software of a distributed system. From the technology view, local operating systems, input-output devices, storage, communication protocols etc. are visible. Related to the TINA DPE, the specific hardware platforms and local operating systems are addressed in this viewpoint (figure 1).

Using the MMC generic interface templates that are described in this paper, requires some software and hardware (e.g. camera, windows, DPE kernel) available on the client's machine before a MMC service can be used. We assume that the client has the generic MMC templates available in an executable form or downloads them from a server. Also the software to access the trader should be available. The user can then access a known MMC server or ask the trader for a MMC service satisfying certain properties.

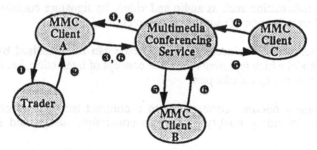

Fig.2. Interaction model to initiate a MMC service

Figure 2 shows a simplified interaction model for the set-up of a conference between three users. Initially, user A requests the trader for a reference of one or more MMC services that conform to the properties user A requires (❶). The properties describe specific requests such as 'desired QoS of audio/video connections', 'maximum price per minute' etc. The trader will respond with a list of references that conform to the request (❷). User A will choose a specific MMC service and contacts the MMC service (❸). If user A and the MMC service agree upon the provided service, a confirmation is sent to user A (❹). In this scenario, the MMC service contacts user B and user C for the establishment of a three party conference (❺). If both users agree they will join the conference (❻).

At binding time between the MMC clients and MMC server, the MMC server indicates which MMC operations are available for the MMC clients. This is reflected in the user interface of the MMC clients. For example, in a window GUI, the possible operations are in bold, and the disabled one in grey or they are even not shown in the menu.

3 Computational specification of MMC generic interfaces

Having identified the environment in which MMC services operate, it is evident that some standardisation is required to allow a wide range of end users to access these kind of telecommunication services. Focusing on the description of generic MMC interface templates provides the advantage that a collection of interface templates exists that assure interoperability when used for the specification of a MMC service.

This paper specifies the interface descriptions used between the MMC service and MMC clients. A similar approach can be found in [11] where interface templates are identified between the MMC service and different stakeholders such as the network provider and other service providers (e.g. trader). We use the interface template as defined in the computational model of RM-ODP [2] which consists of three components i.e. the signature, behaviour and the environment contract.

For a particular interface signature, RM-ODP defines three different types: an operational interface, stream interface or signal interface. We use only the operational and stream interface types for the MMC service. The operational interface is an interface whose signature describes a series of interrogations and announcements. The stream interface describes behaviour which is a single non-atomic action that continues throughout the lifetime of the interface. It can be characterised as isochronous information such as audio and video. Its signature contains the type of the flow and an indication of causality (e.g. direction of flow).

The behaviour which can occur at the interface can be visualised by a dynamic interaction model which represents the ODP concept of interaction, i.e. the actions in which two or more objects take part.

The environment contract corresponds to a contract between an object and its environment, including quality of service constraints, usage and management constraints.

This paper specifies the signatures of the MMC server interfaces. To have a complete interface description the behaviour and environment contract should also be described.

3.1. Three generic interfaces

The functionalities supported by the MMC server can be classified according to three generic interfaces, as shown in figure 3. This classification is based on the different kinds of operations identified in several MMC prototypes.

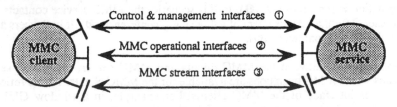

Fig.3. MMC client/server interfaces

The *MMC control & management interface (①)* is used to support operations that deal with the control & management of the MMC service. The operations described in this template are exchanged between the MMC service and MMC clients.

Interface sub-template	Operations
Administration of conferences	Conference (register, unregister, list,...) Group(register, change, unregister, list,...) Policies(user roles, token, QoS,...) User(register, change, unregister, list,...)
Conference negotiation	user role, token, QoS,...
Conference session	open, close, assign-chair, query, new-chair,...
Conference participation	join, leave, new-user, user-left, join_request,...
User role administration	change, list,...
Token passing	request, revoke, release, list, assign, ...
Application sharing control	share, unshare, query,...
Flow control	remove, add, set audio video mixing, synchronise-play-out, ...

Table 1. Interface templates for MMC management & control interface

The operations defined in these sub-templates are used to initialise and control a conferencing session. Operations to register MMC users, and policies can be negotiated among the MMC users, such as token policies and user roles. MMC users can join or leave ongoing conferencing sessions and facilities are provided to control shared applications. For the audio-visual communication, operations are defined that allow to add and remove flows.

The *MMC operational interface (②)* is used to support the functionalities provided by the MMC service (e.g. telepointer). This interface template describes the functionalities not dealing with control & management or streams. Operations related to shared applications (e.g. cut & paste actions for a joint editor) are an important group described in this template The description of this template depends on the applications that can be shared between the MMC clients which will result in sub-templates such as joint-editor, telepointer, joint agenda.

These shared applications relate to the concept of workspace [8]. A workspace contains objects that are created and manipulated by one or more MMC users. The behaviour of the shared application depends on its workspace status. The workspace can be *common* which implies that all conference actions are visible among all MMC users. It can be *closed* which indicates that the MMC service offers sub conference services between some users. Finally, the workspace can be *local* which indicates that MMC users do not share information with other MMC users.

Interface sub-template	Operations
Shared application name (e.g. Telepointer, Joint editor)	actions of the application (e.g. move telepointer, cut & paste text)

Table 2. Interface template for MMC operational interface

The telepointer application is an illustration of a shared application part of this interface template. Operations such as move-pointer are described together with the behaviour of the telepointer application.

The *MMC stream interface (③)* is used for the exchange of flows between all MMC users but also between one MMC user and the MMC service (e.g. video to be stored in the multimedia conferencing database). The ODP stream interface concept is applied to specify continuous data flows that have specific QoS requirements. This interface does not contain operations but is characterised by the properties of the flows it supports. Operations to control the flows are described in the MMC control & management interface template.

Interface sub-template	Characteristics
Video Flow	Direction (in, out) Coding (e.g. H.242, G.721, G.711) QoS parameters (e.g. HDTV, TV, Frame rate)
Audio Flow	Direction (in, out) Coding (e.g. JPEG, PCM) QoS parameters (e.g. HiFi, CD, Sample rate)
Composite Flow	Direction (in, out) Coding (e.g. MPEG)

Table 3. Interface templates for MMC stream interface

3.2. Example of interface specification in OMG-IDL

RM-ODP describes a computational model applicable for distributed applications, but it does not address particular specification techniques for computational objects and interfaces. Therefore, an additional specification language, OMG-IDL [9], is used in this section to derive a computational specification of the generic interfaces. OMG-IDL is also adopted by TINA-C, and provides the means to express telecommunication and multimedia oriented computational specifications.

```
interface template Conference-Session;
 /* sub-template of the control & management interface ① */
typedef .... UserInfo;          /* structure containing info about a MMC user */
typedef .... Tokenpolicy;     /* description of the possible token policies*/

operations
OpenConference(in UserInfo participants, in Tokenpolicy token, in UserRoles users,
....);
CloseConference(in ConferenceId confId,...);
AssignChair(in UserInfo NewChair, out Status result);
Query(in ConferenceId confId, out ConferenceInfo, info);

behaviour 'an instance of this interface template contains operations to control  the
conference session.'
```

Table 4. OMG-IDL specification of control & management interface.

The StreamInterface template consists of several type definitions of continuous flow types. For multimedia conferencing services, audio and video flows will be often applied but other flow types (e.g. graphics) can be added when necessary.

```
interface template StreamInterface
typedef enum CodingType{NoCoding, MPEG, JPEG, PCM, ASCII,..}
typedef struct {
       Boolean        data_loss; /* If the flow handles data loss */
       Integer        Jitter;
       Integer        Delay;
       Integer        Throughput;
       ....
              } QoS
typedef enum Direction {In, Out}
typedef struct {
       CodingType     coding;
       Integer        samplesize;
       Integer        samplerate;
       Qos            qos;
                    } AudioFlowType

interface template AudioInterface;
uses StreamInterface /* inherits the definitions of StreamInterface */
/* One or more audio flows, each with the direction in or out */

sequence <Direction dir, AudioFlowType audio> AudioFlow;

behaviour 'An instance of this interface template provides the possibility to exchange
one or more audio flows having certain QoS properties.'
```

Table 5. OMG IDL specification of stream interface.

3.3. Objects to support the generic MMC interfaces

The ODP computational specification describes how distributed applications and their components are structured in a distribution transparent way. This implies that the structuring of applications is independent of computers and networks on which they run. In the computational viewpoint, a MMC service consists of a collection of computational objects that provide the functionality and mechanisms required to support the generic interfaces described in the previous section.

The functionalities provided by multimedia conferencing services are extensively described in the literature. These prototypes show some similarities and several functionalities are identified that characterise a MMC service. Figure 4 shows some of the most important MMC functionalities grouped in several computational objects, as well as, the generic interfaces ①, ② and ③.

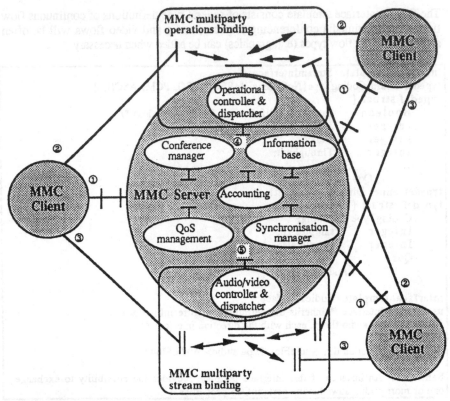

Fig.4. Objects supporting the generic interfaces

The *Conference manager* creates and manages conference sessions. It includes operations that control the flow of actions in a conference session. Furthermore, it administrates conferences and controls the assignment of roles to conference participants. This object provides the mechanisms to support many operations of the generic control & management interface (①).

The *Information base* models a kind of database where conference related information about users and user groups are stored. It deals with static (e.g. user names) and dynamic information (e.g. who has the floor).

The *QoS manager* manages the end-to-end QoS between the MMC service and MMC clients. The MMC clients have negotiated certain QoS with respect to a conferencing session which is reflected in a 'QoS contract' between the MMC service and MMC client. This contract specifies, for instance, the allowed audio sample rate, video frame rate and cost per minute. The QoS manager is in charge of controlling the negotiated QoS and to perform corrective actions when necessary. This object provides the mechanisms for the QoS operations of the control & management interface template.

The *Accounting manager* is responsible for the end-to-end billing. The MMC service uses the so-called one-stop-shopping principle which means that the MMC user will get one bill including the costs for using the transport capabilities provided by the network provider. The accounting manager will collect tariffing information from the network provider and add some costs corresponding to the use of the MMC service.

The *Synchronisation manager* manages three types of synchronisation that are required for multimedia conferencing services. The first one is intra-stream synchronisation that defines the required relation between the samples of a single stream (audio, video or composite) to ensure satisfactory presentation quality. Inter-stream synchronisation is used for the co-ordination of two or more different media types. E.g. lip-synchronisation where voice samples should be synchronised with video samples for correct presentation. The third type of synchronisation is spatial-synchronisation or multi-destination synchronisation. It controls the synchronisation of the presentation of media, at the same time, in geographically distributed sites.

MMC multiparty binding objects manage the interactions between the interfaces it encompasses. They have a control interface through which operations are provided to control its functioning. The two binding objects in figure 4 are rather similar but one handles continuous media and the other one discrete operations. An ODP description of the *multiparty audio/video stream binding* can be found in [13].

The *Operational controller & dispatcher* manages the multiparty operations of the shared applications specified in the MMC operation interface template (②). It receives all operations of the involved participants and rebounds them (after possible manipulation) to all MMC clients. It has a control interface (④) that receives from the MMC server operations such as add new client, change floor control.

The *Audio/video controller & dispatcher* receives all audio/video flows of the involved MMC clients and rebounds the flows (after possible manipulation) to all clients. It has a control interface (⑤) that receives operations such as add new interface, change QoS.

4 Engineering support for the access to generic interfaces

This section describes the engineering support for the interactions between the client and the server via the generic interfaces. First, the mapping of the generic computational interfaces onto the engineering concept of channel is described. Then, it details each of the required engineering channels.

4.1 Computational to engineering mapping

The computational specification of the generic interfaces has to be mapped onto an engineering specification to be executed. Figure 5 illustrates the mapping of the generic interfaces onto the ODP concept of channel. This concept provides the engineering mechanisms to assure distribution transparent interactions of the MMC server and its clients, as illustrated in figure 4.

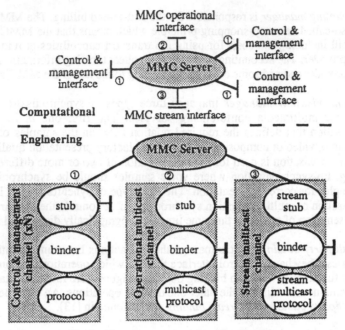

Fig.5. Mapping computational/engineering

Each instance of the *MMC Server control & management interface (①)* is reflected in the engineering specification as one control & management channel (figure 5). The environment constraints specific for the interfaces (e.g. security constraints) are taken into account while establishing the channels between the concerned objects.

The *MMC Server operational interface (②)* is reflected in the engineering specification as multiple client-server channels. The environment constraints specific for the interfaces are taken into account while establishing a channel between the concerned objects.

The engineering transformation of the *MMC Server stream interface (③)* leads to the creation of one multicast producer/consumers stream channel specialised for continuous flows (figure 5). The QoS parameters associated with the flow defined in the computational specification influence the choice of the stream multicast channel components.

As shown in figure 5, a channel consists of three engineering objects i.e. protocol, binder, and stub objects.

Stub objects provide conversion functions. A stub object has a presentation interface used by the object that is bound to the channel and a control interface for e.g. QoS management.

Binder objects interact with one another to maintain the integrity of the binding. Information is maintained about the status of the channel. Binder objects are also responsible for validating the interface reference and to maintain consistent information about the interface location in case of a binding error. Information is preserved with respect to the required QoS. A binder has a control interface which enables changes in the configuration of the channel and destruction of all, or part, of the channel.

Protocol objects communicate with other protocol objects to achieve interaction between objects. They can also interact with objects outside a channel to obtain the information they need (e.g. access to a trader).

The stub, binder and protocol objects can be specialised for the information it transports (e.g. multicast protocol, stream stub).

4.2. Channels between the MMC server and MMC clients

Different kinds of information are conveyed between the client and the server. Therefore, the channels are divided into 3 different channels each with their own characteristics (figure 5). The three channels are created to convey operations of the MMC control & management interface, MMC operational interface, and to transport the flows of the stream interface.

The *control & management channel* conveys the MMC control & management operations supporting the generic interface ① (e.g. floor control, QoS negotiation). This channel should support both low traffic volume and bulk data transfers that need high throughput and should be reliable. E.g. distribution of graphic libraries and conference minutes. In case of bulk transfer visible delays are tolerable. The stub object provides marshalling/unmarshalling of operation parameters to enable access transparent interactions. Its protocol object assures that computational objects can interact remotely with each other. The RPC mechanism is used for interaction. For the interaction with the network management the CMIP/SNMP protocols [15] may be used.

The *operational multicast channel* supports the non-real time interactions ② (e.g. text updates) that need low traffic volume but high reliability. Its stub object provides marshalling/unmarshalling of operation parameters to enable access transparent interactions. VMTP and XTP [14] are two lightweight multicast transport protocols of interest for the engineering protocol object of this channel.

The *stream multicast channel* transports real-time interactions of the MMC stream interface ③ (e.g. voice, and video data) that needs high throughput, bounded delay jitter, but tolerates transmission errors. Streams require different functionality of the stream stub object due to the different nature of information that is exchanged. It should provide the mechanisms to encode and decode video/audio information. Furthermore, data available for the consumers should be notified, and the stream stub object provides operations to control local resources (e.g. increase buffer-size). Notifications of events concerning the stream are supported (e.g. QoS change, no buffer space available, data drop out).

For the exchange of continuous flows, a stream multicast protocol without the RPC mechanism is necessary. RPC requires that each buffer of data to be transferred is treated as a separate action with no particular relationship between previous and future

RPC calls. Continuous flows require relationships between calls and a stream protocol is applied which creates a virtual channel between two protocol objects for the duration of flow exchange. In this case, relations between data can be defined specifically. XTP is a multicast transport protocol of interest for the protocol object of this channel.

5   Conclusions

A MMC service as described in this paper will operate in an environment which involves many stakeholders such as the network provider, service provider and end-users. To operate a telecommunication service in such an environment, the TINA-C initiative proposes a distributed processing environment (DPE) that supports a multi-vendor environment, co-operation between separate network providers and provides distribution transparency to the users.

For distributed applications like the MMC service, standardisation is required to operate successfully in this environment. The most appropriate way to standardise MMC services is the standardisation of the interfaces supported by a MMC service. Standardisation increases the usage of MMC services by clients since the clients do not depend on one specific MMC implementation. Furthermore, it creates the possibility to compare different implementations of MMC services. Various MMC implementations will provide different subsets of the operations defined in the generic interface templates.

We identified three main categories that should be standardised. This classification is based on existing MMC prototypes. First, the *MMC control & management interface* to describe operations dealing with the control and management of MMC services. Second, *MMC operational interface* which contains operations to support shared applications such as a joint editor. Third, *stream interfaces* which provide the audio-visual communication functions essential for multimedia services.

RM-ODP and TINA-C were used as a basis to specify these interfaces in an implementation independent way. ODP supports the definition of the generic MMC interfaces in a standardised way. The OMG-IDL language, as used in TINA-C, was introduced for the computational specification to describe computational objects and interfaces more precisely.

6   References

1.   M. Hoshi: 'Telecommunications Application Requirements'. TINA-C deliverable, TB_A0.MH.002_1.0_93, restricted distribution, December 1993.

2.   Basic Reference Model of Open Distributed Processing:
Part 2: Descriptive Model (DIS), ITU/T X.902-ISO 107046-2.
Part 3: Prescriptive Model (DIS), ITU/T X.903-ISO 107046-3. February 1994.

3.   B. Plattner, G. Dermler, K. Froitzheim, T. Gutenkunst, F. Ruge, M. Vodslon, 'JVTOS: Requirements on a Joint Viewing and Tele-Operation Service'. Deliverable RACE 2060, 1992.

4.   T. Käppner, K. Werner et al., 'The BERKOM Multimedia Teleservices Volume II Multimedia Collaboration'. 1993.

5.   L. Aguilar et al., 'Architecture for a multimedia Teleconferencing system'. ACM, 1986

6.   T. Crowley, P. Millazzo et. al., 'MMconf: an Infrastructure for Building Shared Multimedia Applications'. Proceedings of CSCW, 1990.

7.   T. Ohmori, K. Maeno, S. Sakata, H. Fukuoka, and K. Watabe, 'Distributed Cooperative Control for Sharing Applications Based on Multiparty and Multimedia Desktop Conferencing System: MERMAID'. C&C Systems Research Laboratories, pp 539-546, 1992.

8.   T. Hiroya, A. Tomohiko, M. Shigeki, S. Kazunori, 'Personal Multimedia-Multipoint Teleconferencing System'. IEEE INFOCOM, 1991.

9.   Basic Reference Model of Open Distributed Processing, Draft ODP Trading functions, November 1994.

10.   A.J. MacCartney, G.S. Blair, 'Flexible trading in distributed multimedia systems', Computer Networks and ISDN Systems 25, pp145-157, 1992.

11..   I. Kwaaitaal, P. Leydekkers and B.Teunissen, 'The Good, the Bad and the Ugly about MultiMedia Conferencing Services'. Submitted to the COST 237 Conference on Multimedia Transport and Teleservices , Vienna, Austria, November 1994.

12.   N. Natarajan et al., 'Computational Modelling Concepts', TINA-C deliverable, TP-A-2.-NAT-.002_5.0_93 edition, restricted distribution, October 1993.

13.   V. Gay, P. Leydekkers and R. Huis in 't Veld, 'Specification of Multiparty Audio and Video Interaction Based on the Reference Model of Open Distributed Processing'. Computer Networks and ISDN Systems - Special issue on RM-ODP, 1994.

14.   H.Santoso and S.Fdida, 'Transport Layer Multicast: an enhancement for XTP Bucket Error Control', Proceedings High Performance Networking A.Danthine, O spaniol Editors, 1992.

15.   ISO/IEC 9596-1, 'Common Management Information Protocol Specification'.

ISABEL
Experimental Distributed Cooperative Work Application over Broadband Networks

Tomás P. De Miguel[1], Santiago Pavón[1], Joaquín Salvachua[1], Juan Quemada Vives[1].
Pedro Luís Chas Alonso[2], Javier Fernandez-Amigo[2],
Carlos Acuña[2], Lidia Rodriguez Yamamoto[2].
Vasco Lagarto[3], Joao Vastos[3].

[1] DIT/UPM, España
[2] Telefónica I+D, España.
[3] CET, Portugal.

Abstract. Users are looking towards ATM technology as a suitable solution for specific applications in the new field of distributed multimedia. The aim of ISABEL is to take benefit of the new broadband technology in order to provide a good access to new distributed multimedia facilities. The application has been developed to cover two main fields: distance learning activities between two or more real conference rooms and a flexible framework to configure many different computer support cooperative work (CSCW) scenarios. The paper describes the functionality of ISABEL and its use to support real experiments.

1 Introduction

Users are looking towards ATM technology as a suitable solution for specific applications in the new field of distributed multimedia. Operators are preparing marketing plans in order to offer advanced telecomunications services with a completely different commercial approach. Telecom manufacturers are now launching new products. The overall environment is now under evaluation, the customers are few and fibre based infrastructure are under development.

The demand for broadband facilities is technologically driven by the emergence of new usages such as distributed high performance computing and multimedia services to support new kinds of interactive and distributed applications over LAN interconnections to access remote places.

Broadband islands are now under experimentation in several European countries in order to come up with general requirements and spécifications for paneuropean and international broadband techniques [13].

ISABEL has been a one year project devoted to put together two different ATM developments; RECIBA from Telefonica in Spain, and RIA from Telecom Portugal in Portugal. ISABEL is also the name of the cooperative work application developed in the top of ISABEL network infrastructure, to demonstrate the real usage of new ATM-based emerging technology.

The aim of ISABEL application is to take benefit of the broadband technology in order to provide a good support to new distributed multimedia services. The application has been developed to cover two main different fields:

1. Distance Learning between real conference rooms. ISABEL has been developed to manage full interaction between two or more real conference rooms. It integrates speaker presentation with audio, video, slides projection, remote execution and pointer facilities from an auditorium to the rest of sites connected to the conference. The application has been defined to provide some different views in each moment (the talk, questions, discussion time, etc).
2. A flexible Computer Support Cooperative Work (CSCW) application. The core of ISABEL is devoted to support distributed cooperative work. It allows to maintain full interaction between two or more users in order to exchange any type of information (audio, video, documents, etc.), edit documents in collaboration, distribute the execution of local applications, etc.

The paper describes the functionality of ISABEL and its use in some real environments. ISABEL has been used to demonstrate the real usage of new ATM-based emerging technology, working in the top of link between two ATM developments (RECIBA from Telefónica in Madrid, Spain and RIA from Telecom Portugal in Aveiro, Portugal).

The main use experience of the tool has been obtained during the First RACE Summer School (SS93): *Towards IBC*. During that event, ISABEL has been used to distribute the lectures between Spain and Portugal auditoriums and to perform CSCW activities within the Syndicate Sessions. The experience shows that the application is usable and helps participants to produce a document in group. Actually, a new version is under development in order to improve voice and video quality, to support the dynamic reconfiguration of scenarios during execution and integrate the manipulation of other advanced facilities. This new development will be used to support the Second RACE Summer School (SS94) to be held during July 1994.

1.1 Cooperative Systems

Throughout its brief history, computing has become increasingly concerned with supporting its human users [2]. Now massive numbers of personal computers and workstations are being networked together. This is resulting in a growing awareness of the possibilities for individuals to work together via networks, and of the need for specialized software to support specific group activities.

In parallel with these developments, academics in areas such as decision support systems, coordination systems, and office automation procedures, have been investigating the possibilities for providing computer support for groups in both face to face meetings and via networked systems.

In 1986 an international conference brought these various facets together. The aim was to discuss human group working preferences and characteristics, and to explore how they could be supported by computers. Participants immediately realized that the potent mix of disciplines and ideas had enormous potential, and since then the field of Computer Supported Cooperative Work (CSCW) has expanded rapidly.

This kind of application addresses new fields of investigation as computer supported collaborative work (CSCW) and allows to focus on more challenging technical and psychological issues.

There are three fundamental aspects of collaborative systems: common task, shared environment and time/space [2]. The first measures the extent to which the members of a workgroup can work on the same task. If the system allows many people to work on the same task, it ranks high on the common-task spectrum.

The concept of *shared environment* requires the ability to the user and to the expert to see and talk to each other as well as to have the same view and understanding of the current problem. By definition, Shared Workspace facilities provide an area which two or more participants can see and work in. A shared screen facility enables parts of an individual screen to be reproduced on more screens is an example of this concept.

A strong shared-environment system keeps you informed of what a projects condition is, what participants are doing, and what atmosphere or setting is supported. An E-mail system is a low shared-environment but an electronic classroom rates much higher.

Time/space collaborative systems focus on the item and place of the interaction. Synchronous interaction is at the same time and place. Asynchronous interaction occurs at one place and at different time. And interactions at different places and different times are distributed asynchronous interactions.

CSCW contains a myriad of components, technologies and concerns. Consequently it is difficult to provide a precise definition. However two major concerns are apparent: the support of human groups, and the technology which can be used for that purpose.

ISABEL application aims to provide both a good infrastructure and flexible software and the best support to distribute multimedia information.

2 Physical Infrastructure

The main objective of ISABEL project is to compare different technologies to demonstrate the real usage of the new ATM technology. Therefore, a set of different platforms have been used to distribute multimedia information.

First of all, a videocodec system has been selected to test such a system with respect to the videoconference system inside the workstation. Figure 2 shows interconnection protocols used to compare multimedia distribution systems. The videocodec system is based on the H.261 standard and work using a point to point 2 Mbps link. Therefore, only two places can be interconnected simultaneously. If more than two participants should be connected, a videocodec mixer to switch connections dynamically must be included in the network configuration.

The second user terminal used in ISABEL is a power multimedia workstation [1]. The system is composed of the following components:

- A SUN SPARCstation 10/30, with 32 Mb memory and hard disk (1Gb).
- A Parallax board with input, output and video compression facilities.
- Camera.
- Speaker.
- Microphone.

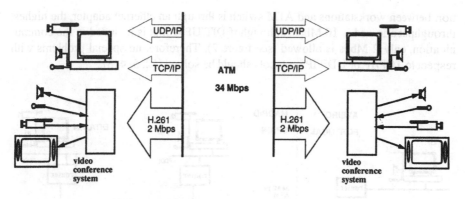

Fig. 1. Interconnection protocols

ISABEL application used a video compression board (the Parallax Xvideo) in order to get a better performance. This board allows to decompress several video sources without allocate all cpu resources. Parallax board used a variant of JPEG standard called MJPEG to allow the compression of real time video. Therefore, with this hardware facilities, the simultaneous participation of more than two partners is allowed.

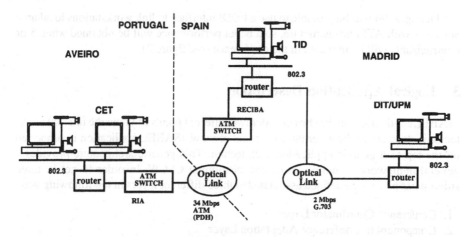

Fig. 2. SS93 network

UDP protocol is used to transfer voice and video. This protocol has not any checking facility. Therefore, if one or more packets of a single image were lost, the complete image is lost. However, over a good quality links good quality voice and video are obtained.

Workstations use IP as a network protocol to solve the gap between application network interface and basic ATM communication infrastructure. Because the connection between workstations and ATM switch is through an ethernet adapter, the highest throughput allowed is 10 Mbps, although if DIT/UPM link is included in the communication, only 2 Mbps is allowed (see figure 2). Therefore, no special problems with respect to TCP/IP or UDP/IP protocols should be solved [12, 5, 4, 3].

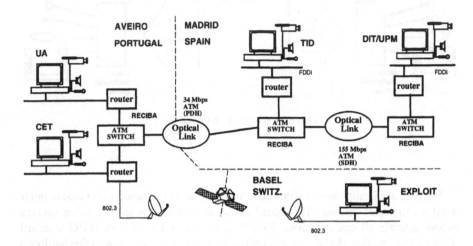

Fig. 3. SS94 network

During SS94 will be possible to use a FDDI interface to link workstations local area networks with ATM infrastructure, so a better performance will be obtained when 5 or 6 participants will be included in the conference (see figure 2).

3 Logical Application Description

ISABEL application can be shown under two different points of view: the internal architectonic structure or the external users point of view. ISABEL application is supported on the top of a generic application architecture. This generic architecture provides a general framework to build almost any distributed CSCW. In order to get it, three subsystems have been defined associated to three different layers in the following way:

1. Conference Coordinator Layer
2. Component to Conference Adaptation Layer
3. Cooperative Components Layer

Conference Coordinator Layer consists of a daemon activated by the originator of each particular conference. A conference is defined by an IP port number and a host name. This information help users to identify the conference and it is used to configure the application daemon in order to coordinate the conference. On the other hand, each participant is associated with a client process to interact with three application elements: the conference daemon, the conference participant and all components activated by the participant. The figure 4 shows an example of interconnection of daemon and conference participant clients and their components.

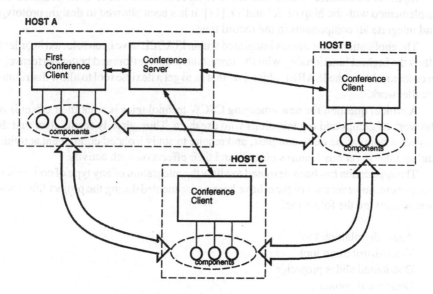

Fig. 4. Conference interconnection components

Each component handles typically two groups of connections. The first is a single connection devoted to exchange information with the conference client associated to each participant and used to obtain information of the conference status (the list of participants, connections and disconnections, etc). The second is a set of connections, one per component participant with that component activated. These links are used to simulate multicast with a set of unicast connections.

The Conference Adaptation Layer consist on a set of functions (one per component type) devoted to transfer different types of control information between the common conference subsystem and each component integrated in the ISABEL Cooperative Work Application.

The Conference System Daemon use a particular protocol to communicate information to and from each component. In order to standarize this protocol, ISABEL defines an Adaptation Interface associated to each different component integrated in the Cooperative Work Application.

The Adaptation Layer is mainly devoted to perform the following functions:

1. Send the set of participants linked to the component when it starts the execution.
2. Send the name of the new participant connected to the component.
3. Send the name of the old participant that abandons the conference.
4. Force the termination of the components execution.

Application components are the real programs executed by the users to perform certain type of interaction within the conference. ISABEL application has been designed using X11 windows system [8, 9, 10, 7] because it has been selected as the industry standard user interface software system to develop and integrate graphical applications.

ISABEL conference system and most of its components have been designed and implemented with the help of Tcl and Tk [11]. It has been allowed to design, prototype and integrate all components in the record time.

The application components integrated within ISABEL have been selected in order to allow the highest functionality with the lowest integration effort and better performance in order to address the ISABEL objective, that is, to get a flexible tool to allow distributed parallel work.

A critical question for new emerging CSCW technologies is what their effect is on the work of groups and organizations who use them. Thus, already seems clear that the answer is going to be quite complex, and require to study isolated component activities and looking at various features of tools and their effects on each activity.

The application has been designed to allow the integration of any type of cooperative component, however a set of them have been experimented during the project life. These components are the following:

- Audio distribution tool
- Video distribution tool
- Distributed slides projector
- Distributed pointer
- Distributed white-board
- Cooperative text editor
- Distributed display

3.1 Teleconference

Meetings involving remote attendees accomplish via telecommunication means. This covers audio-video-phony, audio and video conferencing, etc. where participants deal with information without any specific, predefined structure. It can also include meetings with jointly and simultaneously executing other components.

Within ISABEL, we have two ways to perform that functionality: use the commercial videoconference system, or integrate audio and video facilities within multimedia workstation. The first approach is easy to install, work properly, but it is only valid to communicate two points. When more than to participants should be connected the workstation tool must be used.

Audio and video tools work exactly in the same way as the rest of components of the ISABEL CSCW System. When a participant want to offer his video he starts the video component and after his image is enable to be distributed to the rest of participants. When a participant want to see the video of another one, he only need to connect to that activated video signal.

This allows to include in the same session hosts with very different communication facilities. If a participant has enough bandwidth, he can select many video signals (for example, between TID and CET), however if he has not enough bandwidth, only a subset of video signals should be selected (for instances, between DIT/UPM and TID).

3.2 Slides Projection

One of the most important components in the distance learning field is the tool for slides and images projection. ISABEL has a tool to update the slide display. In such a tool one station acts as a teachers host and is the only allowed to define which and when slide should be displayed.

Slide projector is based in the display of GIF image standard files format. When another format image is demanded, a tool to transform from this format to GIF should be provided. ISABEL uses ghostcript tool transformation functionalities to address such kind of facilities, so it is possible to make many standard image formats are allowed as input to the slides projector.

3.3 Distributed Pointer

It is a tool to allow the simultaneous control of a group of workstation pointers. With the help of X protocol, the tool receives all events associated to the local pointer and distribute these events to all workstations included within the group.

The first workstation acts as a controller and the rest of group members only should send initially a message in order to be included in the group. After it, the controller sends X protocol messages directly to the X server of all group workstations.

3.4 Cooperative Editor

Current CSCW systems do not adequately support the distributed creation of multi-author documents. One of the main drawbacks of the full asynchronous editing tools is the absence of an user to play the role of coordinator.

ISABEL cooperative editor is mainly devoted to help the production of a document within a meeting (a conference in the group ware terminology). Such kind of tool is designed to write a text document by two or more authors.

A coordinator of edition is always defined. When tools starts execution the first participant is selected as coordinator. He will be allays the coordinator of edition unless another participant demands the attribute. During edition only the coordinator has permission to modified the document.

The functionality of the tool is defined in order to be similar to a typical meeting session. All meetings have a person participanting as a secretary. He or she writes minutes and allows other participants to include their comments and documents in the final edition.

Within ISABEL we have been design a very easy and intuitive editor, to free users to study a new editor with new commands and facilities. Therefore, the time to be familiar with the tool commands is negligible.

3.5 Distributed White-board

A distributed white-board is a tool that allows users to write text and draw graphics in asynchronous mode. Systems such as WSCRAWL and SKETCHPAD are systems geared entirely towards shared editing sessions among two or more users. The absence of a real coordinator, the mixture of text and graphics on any order and the transition from single person editing to multiple user editing very abrupt, are the mayor shortcomings of these systems.

Within ISABEL, we have been integrated the WSCRAWL tool, only as a white-board to draw drafts and schematas, but no to be used as a coauthoring text editor.

3.6 Distributed Display

A distributor display is an X multiplexor that allows a single X client to be displayed over several X servers simultaneously. We have been tested some of such components and finally XMX developed at Brown Univ. has been integrated [6].

XMX is very simple to use and enough robust in monochrome mode to support a wide range of graphic applications. XMX has been fully integrated within the second version of ISABEL cooperative work environment.

4 Experience summary

The main experience with the use of the tool has been obtained during the First RACE Summer School SS93 *Towards IBC* [13]. During this event the ISABEL application has been used within some Syndicate Sessions in order to produce a document with the results of syndicate work.

The experience shows that the application is usable and helps participants to produce a document in group. This experience revealed the advantages and disadvantages of the application. So, during the second phase of the project were improved some aspect of the application, like its facility of use, the simplicity to coordinate with remote participants (redesign to be used by three or more persons in the same place using the same workstation), the intuition, the power of the components (quality of audio and video, new functionalities), removing some bugs, etc.

The second part of the development has been also devoted to support a real broadband ATM interconnection. It allows to support both, a higher bandwidth applications and a higher number of participants collaborating in a single cooperative work conference. In summary, the performance of previous developed applications has been increased.

However, we cannot take a full advantage of ATM technology, because the link is only 34 Mbps and, more important, the applications are based on workstations connected to LANs. Therefore, the real usable bandwidth accessible by ISABEL applications during this second project phase is only 10 Mbps.

Nevertheless, during 1994 a complete new version (the second one) has been developed to increase the quality of audio and video components, to allow the participation of more users in the same cooperative work conference and mainly to define a real flexible environment to support a wide range of CSCW applications. It has been proved to be stable and robust in comparison to the other applications tested. The plan for the near future is to integrate virtual reality facilities within the ISABEL environment.

References

1. SPARCstation 10 System Architecture. 1992.
2. J. Hsu and T. Lockwood. Collaborative Computing. *BYTE*, 18(3):113–120, Mar. 1993.
3. Jacobson. Extensions for High Performance. (RFC 1323).
4. Jacobson. Extensions for High-Speed Paths. (RFC 1185).
5. Jacobson. Extensions for Long-Delay Paths. (RFC 1072).
6. T. G. John E. Baldeschwieler and B. Plattner. A Survey of X Protocol Multiplexors. *ACM SIGCOMM*, 23(2):16–24, 1993.
7. A. Nye, editor. *X Tollkit Intrinsics Reference Manual*, volume Five. O'Reilly & Associates, Inc., 1992.
8. A. Nye. *Xlib Programming Manual*, volume One. O'Reilly & Associates, Inc., 1992.
9. A. Nye, editor. *Xlib Reference Manual*, volume Two. O'Reilly & Associates, Inc., 1992.
10. A. Nye and T. O'Reilly. *X Tollkit Intrinsics Programming Manual*, volume Four. O'Reilly & Associates, Inc., 1992.
11. J. K. Ousterhout. *An Introduction to Tcl and Tk*. Addison-Wesley. To be published in 1993.
12. Poster. Transmission Control Protocol. (RFC 793).
13. Universidad de Aveiro. *First International Summer School on Advanced Broadband Communications, Towards IBC*, Portugal, Julu 11-16 1993. RACE R.2095 BRAIN.

Demonstrating Image Communication within Open Distributed Environments[1]

Rüdiger Strack
Fraunhofer Institute for Computer Graphics,
Wilhelminenstraße 7, D–64283 Darmstadt, Germany

Ralf Cordes
Bosch Telecom,
Kleyerstraße 94, D–60326 Frankfurt/Main, Germany

Dale C. Sutcliffe
Rutherford Appleton Laboratory,
Chilton, Didcot, Oxon OX11 0QX, United Kingdom

Abstract. A wide variety of image interchange and communication (de facto) standards are employed today in different systems, applications and environments. Within the AMICS project a framework, called the *Image Communication Open Architecture* (ICOA), was defined to enable the various standards and standardization activities in the broad area of imaging and image communication to be related and the necessary support tools to be identified. Based on the ICOA, software tools to support the framework were developed focusing on different requirements for image communication. One requirement was perceived to underlie all the others, that of providing uniform access to whole images and parts of images whether they are stored locally or remotely. Such uniform access is provided through the *ICOA Image Handling Interface* (IHI). The IHI is realized by means of the *ICOA Image Handler* that is modelled as an *Open Distributed Processing* (ODP) object. The Image Handler encompasses the support of various compression schemes and (image) data formats, as well as different conversion facilities.

To demonstrate the concepts of the ICOA and the ICOA Software Tools, a remote teaching scenario was chosen. The teaching scenario illustrates the accessibility of any image storage with any kind of digital image format on it within a multimedia communication environment. The access to the media is provided by a multimedia communication service. Within this service, the Image Handler is used to retrieve still and moving images and those parts of audio–visual information that are covered by the ICOA.

1 Introduction

The communication, storage, and manipulation of the information type "digital image" is one of the most challenging tasks within the development of multimedia systems and telecommunication services. If the technological issues on image storage, image communication, and image manipulation (including image compression, image conversion and the synchronization of images with other information types) are solved then the most serious obstacles towards the usage of multimedia technology in open, distributed environments will be removed.

[1]The work has been performed within the RACE II project *Advanced Multimedia and Image Communication Services* (AMICS) (R2056) partially supported by the Commission of the European Communities (CEC).

A wide variety of image interchange and communication (de facto) standards are employed today in different systems, applications and environments. Most often, these standards have been developed separately and in isolation from each other, each addressing particular needs. Nevertheless, for the development of systems for open image communication, the standardization of compression schemata, data formats and communication protocols turns out to be a key issue. Within the RACE II project *Advanced Multimedia and Image Communication Services* (AMICS) a framework, called the *Image Communication Open Architecture* (ICOA), was defined to illuminate the broad area of imaging and image communication. Although the framework sets a clear emphasis on the handling of images, the ICOA fulfills the image communication requirements from a wide range of application areas including multimedia.

Software tools to support the ICOA framework were developed. They focus on different requirements for image communication. Within AMICS, one requirement was perceived to underlie all the others, that of providing uniform access to whole images and parts of images whether they are stored locally or remotely. Such uniform access is provided through the *ICOA Image Handling Interface* (IHI). The IHI is realized by means of the *ICOA Image Handler* that is modelled as an *Open Distributed Processing* (ODP) object. The Image Handler encompasses the support of various compression schemes and (image) data formats, as well as different conversion facilities.

To demonstrate the concepts of the ICOA and the ICOA Software Tools, a remote teaching scenario was chosen. The teaching scenario illustrates the accessibility of any image storage with any kind of digital image format on it within a multimedia communication environment. This paper focuses on the Image Handler within the remote teaching scenario.

The remainder of this paper is organized as follows: First, an overview of the ICOA framework is given. Next, the concepts of the ICOA Image Handler are briefly described. Then, the Image Handler and its implementation as an integral part of the realization of a remote teaching scenario are outlined. Technological issues for this implementation of the Image Handler are addressed. Finally, a conclusion is given.

2 Image Communication Open Architecture

By looking at the whole field of images and image communication, a framework encompassing the various requirements on the handling of digital images can be derived. Within AMICS, such a framework, called the *Image Image Communication Open Architecture* (ICOA) [14] [15], was defined to illuminate the broad area of imaging and image communication. The conceptual building blocks of the ICOA can be elucidated as follows:

- The ICOA characterizes and addresses the widest range of digital images. To fulfill image communication requirements from a wide range of application areas including multimedia, the framework also addresses the relation of images to multi/hypermedia information and to international standardization efforts.

- For the establishment of the ICOA, several standards were examined with respect to images (information and data structures, communication and other aspects, such

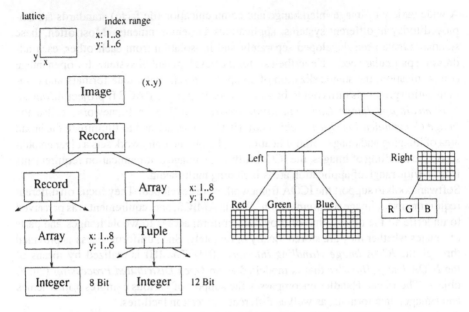

Fig. 1.: Example of a digital image defined by the IDM.

as storage, processing, management, and identification). Services were examined in order to identify specific requirements for image communication.

- A reference model for image formats which comprises a set of distinct criteria (image format characteristics) was derived, which can be used to characterize existing image interchange formats.

- A mathematical model of an image and the operations that might be carried out on that image was developed.

- Based on the mathematical model, a comprehensive, generic image data model (ICOA Image Data Model (IDM)) was defined. The IDM provides broad and flexible data structures which can be used to describe any kind of digital image. Furthermore, the IDM specifies the basic set of functionality necessary for the handling of images within open, distributed environments.

The IDM is in part based on the generic data model of the *Image Processing and Interchange Standard* (IPI) [7]. Extensions to the IPI data model have been developed and integrated into the IDM (e.g. the support of interlaced/time–variant data). The IDM uses a mechanism to constructively define image data types in order not to be limited to particular applications. A set of basic data types (constructors) together with a comprehensive set of attributes (including compression schemes) is provided. Comparable constructive descriptions can be found in well–known areas, e.g., the CSG method (*constructive solid geometry*) for computer graphics and CAD applications. Figure 1 gives an example how to apply the IDM constructors to define the structure of an image.

- To provide means for applications wishing to embed images into a multi/hypermedia environment, the relationship of the IDM to multi/hypermedia information was examined and addressed.

- Concepts of *Open Distributed Processing* (ODP) were examined and their impact on the ICOA was determined. The most significant ODP viewpoints from the ICOA perspective are the information viewpoint, the computational viewpoint, and the engineering viewpoint. The information viewpoint of the ICOA describes the information flow in the context of images. The computational viewpoint in form of the IDM defines the necessary data types and functions by applications for image handling. The engineering viewpoint of the ICOA discusses transparency and communication mechanisms that play a major role for digital images.

Since the ICOA covers these various aspects in regard to the modelling and handling of images, it enables the various standards and standardization activities in the imaging area, the image communication area, and in the multimedia area (in regard to images) to be related and the necessary support tools to be identified.

3 The ICOA Image Handler

The ICOA is a framework that relates the various standards for imaging and communication to satisfy the requirements for image communication. Tools to support that framework are referred to as *ICOA Software Tools*. The software tools focus on different requirements for image communication.
One requirement for image communication was perceived to underlie all the others, that of providing uniform access to whole images and parts of images whether they are stored locally or remotely. By providing such uniform access, communication services can be invoked as necessary, without the knowledge of the application, and other related services such as compression and conversion can be integrated. Such uniform access is provided through the *ICOA Image Handling Interface* (IHI) [16]. The IHI is realized by means of the *ICOA Image Handler* [13].
Within the ICOA the relationship between image, data format, compression, and conversion was carefully analysed and a clear model was developed. As a first step, the following representation layers in regard to the Image Handler were distinguished: (see Figure 2):

- *Application layer*: ICOA images are characterized by the property of 'open communicability'. They can be passed through the IHI between applications and between different parts of the same application.

- *ICOA Image Handling Interface*: The IHI provides the user — the application programmer — with a uniform interface for the handling of images, based on the IDM, that is independent on any distinct data format.

- *Format layer*: The data format layer focuses on the representation of images within a distinct (image) data file/interchange format.

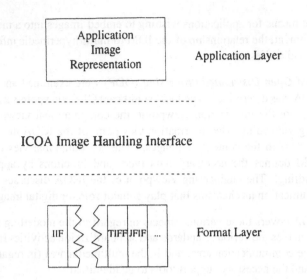

Fig. 2.: Representation layers in regard to the Image Handler.

Later, different types of compression and conversions between compression schemes were identified and incorporated into a more detailed model which also included encoding.

3.1 The ICOA Image Handling Interface

The IHI provides the basic set of operations necessary for the storage, retrieval, and control of images. It is based on the ICOA Image Data Model (IDM) thus providing broad and flexible data structures and attributes which can be used to describe any kind of digital image. The IHI establishes a common view on images stored in different data formats. Among other features it provides arbitrary access to parts of digital images and allows both compressed data (according to a variety of compression standards, e.g. JPEG, MPEG, etc.) and raw data (by which is meant an array of arbitrary pixel data type, according to the IDM) to be passed through the interface. An abstract specification of the IHI, that encompasses all data types and functions, was elaborated in the formal specification notation Z [12]. Concrete representations (or bindings) of the IHI in the C++ and C programming language were specified.

The IDM allows different data units of an image to be compressed differently. Conversion comes into play, if the compression scheme[2] used by the application does not match the compression schemes supported by the data format. In this case, either the application or the Image Handler has to provide the necessary conversion functionality. The Image Handler was designed with the goal of providing uniform access to a wide range of digital images and freeing applications from complex conversion tasks. As a consequence, the Image Handler incorporates the flexibility needed to access the variety of data formats supporting different compression schemes and conversion is provided by

---

[2]This includes the representation of raw data defined as a specific compression scheme.

the Image Handler underneath the IHI. The application may have control of conversion processes, by appropriately setting up the environment in which conversion takes place, but the details of conversion are hidden from the application. Equally, it is possible for applications to function totally unaware of any conversions taking place, if they so desire, as appropriate default settings for the conversion processes are provided.

3.2 Architecture

The Image Handler is modelled as an *Open Distributed Processing* (ODP) object, whose interface, the IHI, provides the basic set of operations for the handling of images whether they are stored locally or remotely. This encompasses the support of various compression schemes and (image) data formats, as well as different conversion facilities that can be performed either in software or with the support of specific hardware (e.g. the DIP chip).

In an ideal world, image communication would be simple. Every site a user wished to transfer images from would offer an ODP object Image Handler, that would process requests and return results through the IHI. Unfortunately, this is not the case. For this reason, an alternative approach to the problem of image communication was developed that provides the appearance of the ODP object Image Handler, although the underlying services may not all be in place. This allows applications to be written independently of services available at any one time. As a service becomes available, it is used transparently by the Image Handler, without any changes needing to be made to an application. Provision of such a concept makes use of ODP trading facilities for locating suitable service providers.

The following scenario may illustrate this concept: An application requests a subset of an image located at a particular site. The Image Handler requests the trader to find what services for transferring images from the given site are available. If an Image Handling Service is available, it can be used to transfer just the subset of the image, thus reducing the volume of data transferred. If an Image Handling Service is unavailable, the Image Handler may have to fall back to an available, common, lower level of service, such as FTP[4] or FTAM[9]. A bulk transfer of the entire image would be necessary and the subset would have to be extracted from the image using local services. From the application point of view, the request and the results would be the same, albeit the performance would probably be poorer in the latter case.

The outlined concept resulted in the decomposition of the Image Handler into functional building blocks as follows:

- For each data format a distinct ODP object is established, that provides the functionality of the Image Handling Service in regard to the respective data format. These objects are optimized in regard the handling of a specific data format since the requirements of the data formats in regard to image structure, image attributes, compression schemes, QoS parameters, etc. differ.

- An ODP object in regard to conversion is established. Conversion can be performed either in software or with the support of specific hardware (such as the AMICS DIP chip [3]).

The resulting architecture of the Image Handler is depicted in Figure 3.

369

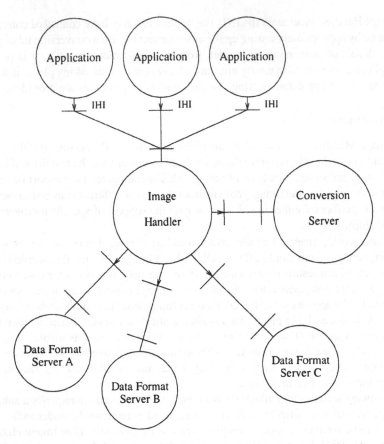

Fig. 3.: Decomposition of the Image Handler into specialized ODP objects.

3.3 Implementation Aspects

For the realization of the Image Handler as well as for the evaluation of its concepts data formats determined by the remote teaching scenario were chosen to provide a variety of digital images, i.e. moving images, multispectral images, binary images, etc. This encompasses the establishment of IIF[8], JFIF [5], and TIFF [1] Image Handling Services for still images, the realization of an MPEG–1 Video [10] Service and a Conversion Service capable to convert from raw data to JPEG/MPEG and vice versa.

ANSAware [2], a distributed computing environment for the development of distributed systems and services, was used for the establishment of the Image Handling and Conversion Services. ANSAware can be briefly characterized as an engineering platform of ODP providing — among other features — client/server concepts, trading and federation. The usage of ANSAware provided the following major benefits:

- The interfaces of the data format and conversion servers — as computational objects — could be specified in an object–oriented manner by using the *Interface Definition Language* (IDL) of ANSAware. These computational objects could be transformed with ANSAware compilers into an engineering specification where distribution is explicit. Thus, the communication objects used to mediate the interactions between clients and servers could be developed within the ANSAware environment.

- The Image Handler could make use of the facilities of the ANSAware trader. This was very useful for the establishment of the remote teaching scenario since data format and conversion servers could be dynamically started according to the actual distinct requirements within the realization and evaluation phase.

The services were integrated into the Image Handler that provides uniform access to local and remote images in form of the IHI.

4 The ICOA within a Remote Teaching Scenario

To demonstrate the concepts of the ICOA and the ICOA Software Tools, a remote teaching scenario was chosen. The scenario illustrates the accessibility of any image storage with any kind of digital image format on it within the multimedia communication environment. It is composed as follows:

- *Teaching Centre*: Classes are prepared and stored within the teaching centre. It is composed of multimedia workstations that provide editing facilities for the teachers, and a central database for information storage. Only teachers are authorized to edit the database, to store information and to design new lessons.

- *Network*: The network acts as a point to point link between the teaching centre and the students. A broadband network is desirable for efficient media transfer.

- *Students*: Students are able to consult the central database for their classes. They can use multimedia workstations and can access the database from everywhere on the network.

A complete specification of the remote teaching scenario is given in [11]. Figure 4 identifies the general building blocks of the teaching scenario. The access to media is provided by a multimedia communication service. To retrieve still and moving images and those parts of audio–visual information that are covered by the ICOA, the Image Handler is used.

4.1 The Teaching Demonstrator

The teaching demonstrator, focusing on those parts of the teaching scenario that relate to the student access to the system, was developed to demonstrate the benefits of the ICOA.

The reference structure of the lessons and all non–image parts are stored in a multimedia database at the teaching centre. The lessons are stored as multimedia documents. The modelling of the multimedia documents was strongly influenced by the forthcoming MHEG Standard [6]. However, in comparison to MHEG, the supported structure is much simpler (e.g. in regard to temporal and spatial relations).

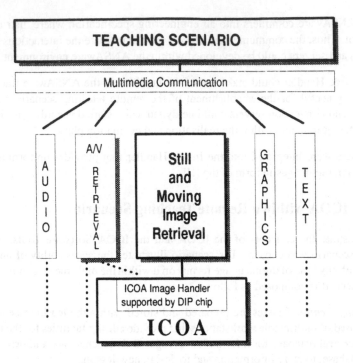

Fig. 4.: General building blocks of the remote teaching scenario.

A multimedia retrieval service, realized as a client–server approach, is used to retrieve the lessons. The service encompasses a generic image retrieval service, a retrieval service for audio–visual information, and a document retrieval service. The generic image retrieval service requests the IHI, as the interface of the Image Handler, to access all images (still and moving) that are stored on the image servers, applying the appropriate conversions through use of the conversion server.

Figure 5 illustrates the teaching demonstrator. The teaching centre and the Image Handler are located at one site (Frankfurt) while image format and conversion servers are located at two sites (Frankfurt and Darmstadt). The Image Handler could logically be placed elsewhere. The same holds for the student viewing station that is logically separated from the teaching centre. As there was no broadband network available, narrowband ISDN through its basic access (64kbit/s) was used to interconnect both sides. Over this network a TCP transport service was used. This approach offered a large degree of flexibility as, on the one hand, it is compatible with current computer communications, and, on the other hand, it allows an easy adaptation to broadband (e.g. ATM testbeds endowed with AAL5).

4.2 Evaluation of the Image Handler

All images within the wide area teaching demonstrator are accessed through the IHI that provides location and access transparency to the user. Thus, all images can be retrieved, stored and controlled from/at both sites without any changes to the application. Existing

Bosch Telecom Telenorma. Frankfurt

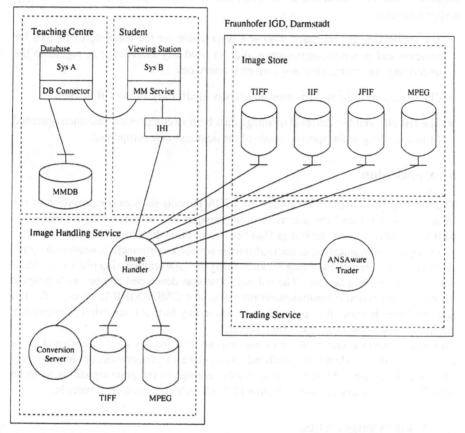

Fig. 5.: The AMICS wide area teaching demonstrator.

applications for the processing of images (e.g. presentation, editing, etc.) could be easily adapted to the IHI.

The Image Handling and Conversion were evaluated in the local infrastructures of Fraunhofer IGD (TCP/IP (ethernet, FDDI backbone)) and Bosch Telecom Telenorma (TCP/IP (ethernet)) as well as using the interconnection via narrowband ISDN. The evaluation showed that within the local infrastructures efficient compression techniques are needed for the transfer of moving images and huge multispectral still images (like satellite data) to provide adequate response times. For the interconnection via ISDN compression schemes are seen as a necessary prerequisite even for the transfer of small images due to the bandwidth constraints imposed. Since the Image Handler and its services support various representations of image data, the application may make use of its own specific requirements. The benefits of accessing image subsets locally and remotely are clear: only those parts that are requested by the application need to be transferred.

The supported set of IHI functionality well matched the requirements of the demonstrator scenario to retrieve, store, and control still and moving images that are stored locally and/or remotely:

- The application programmer, who was responsible for the handling of images for retrieval and presentation purposes, did not need any knowledge in regard to the underlying communication and conversion services.

- The IHI allowed the specification and access of arbitrary subsets of images.

- The IHI allowed the retrieval of image data both in compressed and uncompressed form according to the specific needs of the demonstrator components.

5 Conclusion

The teaching demonstrator illustrates the ICOA benefits such as the provision of the Image Handler for uniform access to digital images and subsets of digital images as well as the flexibility of the Image Handler in a multimedia communications environment. Open communication in the multimedia service environment is supported by the establishment of Image Handling Services using the ODP engineering platform ANSA-ware with its trading service. The software that was developed can be easily adapted to other object based communication protocols like OMG/CORBA. Through the IHI and the Image Handler the application is open to any further conversion functionality, compression scheme or image format server.

The general service environment of the remote teaching scenario is designed to be extended by further advanced broadband services like telecooperation, or video services like video telephony. Moreover, it is flexible enough to integrate signalling modules offering access to services which follow ITU–TS Q2931–signalling protocols.

6 Acknowledgements

The work reflected in this paper was performed within the RACE II project *Advanced Multimedia and Image Communication Services* (AMICS) supported by the Commission of the European Communities (CEC).

AMICS was performed by the following partners: Bosch Telecom Telenorma (TN), Frankfurt, Germany; Fraunhofer Institute for Computer Graphics (IGD), Darmstadt, Germany; Rutherford Appleton Laboratory (RAL), Didcot, United Kingdom; Tampere University of Technology (TUT), Finland; Universidad Politècnica de Madrid (UPM), Spain; and University of Sheffield (US), United Kingdom.

Many colleagues namely Christof Blum (IGD), David Duce (RAL), Irek Defée (TUT), Jürgen Dziumbla (TN), Narciso García (UPM), Harald Lindner (IGD), Luc Neumann (IGD), María J. Pérez–Luque (UPM), Hauke Peyn (TN), Peter Weis (TN), and Rob Yates (US) from the AMICS project have contributed to this paper in technical discussions.

References

[1] Aldus Corporation, Seattle. *Tag Image File Format Specification, Revision 6.0*, 1992.

[2] *ANSA: object pioneers*, volume 7. Technology Appraisals Ltd., February 1993.

[3] S. Evans, N.A. Thacker, R. Yates, and P.A. Ivey. A Massively Parallel Vector Processor for Image Communications. In *Proceedings of the 2nd International Conference on Image Communication (IMAGE'COM 93)*, pages 303–308, Bordeaux, March 1993. SEE, IREST, ADERA.

[4] *FTP (File Transfer Protocol)*, Request for Comments (RFC) 959. Internet Network Working Group, October 1985.

[5] E. Hamilton. *JPEG File Interchange Format, Version 1.01*. C–Cube Microsystems, December 1991.

[6] *ISO/IEC CD 13522–1: Coded Representation of Multimedia and Hypermedia Information Objects, Part 1: Base Notation (ASN.1)*. ISO/IEC, June 1993.

[7] *ISO/IEC DIS 12087–1: Information Technology—Computer Graphics and Image Processing—Image Processing and Interchange (IPI)—Functional Specification—Part 1: Common Architecture for Imaging (CAI)*. ISO/IEC, November 1992.

[8] *ISO/IEC DIS 12087–3: Information Technology—Computer Graphics and Image Processing—Image Processing and Interchange (IPI)—Functional Specification—Part 3: Image Interchange Facility (IIF)*. ISO/IEC, November 1992.

[9] *ISO/IEC IS 8571–1: Information Processing Systems — Open Systems Interconnection (OSI) — File Transfer, Access and Management (FTAM) — Part 1: General Introduction*. ISO/IEC, 1988.

[10] *ISO/IEC IS 11172–2: Coded Representation of Picture and Audio Information, Coding of Moving Pictures and Associated Audio for Digital Storage Media up to about 1.5 Mbit/s — Part 2: Video*. ISO/IEC, 1993.

[11] H. Peyn editor, C. Blum, R. Cordes, P.N. Courtney, I. Deféc, J. Dziumbla, S. Evans, N. García, G. Leicher, H. Lindner, L. Neumann, M.J. Pérez-Luque, R. Strack, S.B. Walker, P. Weis, and R. Yates. The AMICS Demonstrator. Deliverable R2056/TN/EWZ/DS/L/007/a1, RACE Project R2056 Advanced Multimedia Image Communication Services (AMICS), December 1993.

[12] J.M. Spivey. *The Z Notation: A Reference Manual*. Prentice–Hall, second edition, 1992.

[13] R. Strack, C. Blum, D. Duce, D. Sutcliffe, and N. García. Uniform Access to Images within Open Distributed Environments. In W. Herzner and F. Kappe, editors, *Multimedia/Hypermedia in Open Distributed Environments, Proceedings of the Eurographics Symposium, Graz, June 6–9, 1994*, Eurographics, pages 122–142, Wien, Austria, June 1994. Springer–Verlag.

[14] R. Strack, C. Blum, D. Duce, D. Sutcliffe, N. García, M.J. Pérez-Luque, E. Moeller, and H. Peyn. Image Communication Open Architecture. *Computer and Graphics*, 18(1):21–34, 1994.

[15] R. Strack editor, C. Blum, R. Cordes, I. Defée, D.A. Duce, N. García, G.R. Hofmann, R. Maybury, E. Moeller, M.J. Pérez-Luque, D.C. Sutcliffe, and R. Strack. Conceptual Building Blocks for an Image Communication Open Architecture (ICOA). Deliverable R2056/FhG/IGD/DS/P/002/b1, RACE Project R2056 Advanced Multimedia Image Communication Services (AMICS), April 1993.

[16] R. Strack editor, C. Blum, J. Cullan, D.A. Duce, N. García, R. Maybury, L. Neumann, M.J. Pérez-Luque, H. Peyn, R. Strack, and D.C. Sutcliffe. Final Report on the Image Communication Open Architecture (ICOA), Part B. Deliverable R2056/FhG/IGD/DS/P/005/b1, RACE Project R2056 Advanced Multimedia Image Communication Services (AMICS), January 1994.

Design and Implementation of a High Quality Video Distribution System using XTP Reliable Multicast

Bert J. Dempsey, Matthew T. Lucas, and Alfred C. Weaver

{bert,matt,weaver}@Virginia.EDU
Computer Science Department, University of Virginia, Charlottesville, VA 22903 USA

Abstract. In this paper we present the design and implementation of a novel protocol solution for distributing high quality compressed video streams to multiple receivers across a network. Our end-to-end protocol uses the connection-oriented multicast facility in the next-generation transport, the Xpress Transfer Protocol (XTP). XTP multicast provides in-order, multipoint delivery of packet streams with user-selectable options for enabling control algorithms and managing dynamic group membership in the multicast connection. XTP multicast gives the application flexibility in determining the robustness of the communication while insulating the application from managing multipoint delivery. In our design the multicast video application coding need only handle buffering for end-to-end synchronization. We have implemented our approach and provide performance measurements of this video distribution system in an FDDI network. We evaluate the performance of our system under different compression ratios when delivering a 640-by-480 8-bit color video stream at 30 frames/sec to four receivers.

1 Introduction

Recent technology trends have made the delivery of high quality digital video streams to the desktop increasingly feasible. Many emerging applications for digital video have a natural requirement for delivering a video stream from its source to multiple receivers across a network. We refer to point-to-multipoint distribution as *multicast video*.

Designing an end-to-end protocol for multicast video streams requires consideration of the following requirements and related networking issues:

- *Multipoint delivery* of packets in the video stream must be provided by the network. While point-to-point communication patterns have traditionally been dominant in data networking, support for multicasting has advanced dramatically in recent years. Multicast addressing and multipoint delivery mechanisms exist for a variety of protocol domains, including link layer multicast in IEEE 802 LANs, IP multicast, multicast in ATM networks, and transport layer multicast facilities. Due to the real-time nature of video streams, timely multipoint delivery of packets is an important consideration.

- *Group management* techniques are needed to create and control multicast receiver groups. Group management functionality in current multicast facilities is generally primitive, reflecting the immaturity of multicast applications. Especially useful for many video applications is dynamic group membership in which receivers can leave and join an in-progress video stream without disturbing the communication. Dynamic group membership should not preclude procedures for reliable transfer.
- *In-order delivery* of packets in the video stream is required to preserve the correct time sequencing of the video. ATM channels do enforce data sequencing. Connectionless multicast facilities, e.g., IP multicast, do not preserve sequencing, leaving this function to higher layer protocols or the application.
- *Error control* for the recovery of packets lost in the network is desirable for video, but error control procedures must be evaluated with the delay constraints of the stream in mind. Error control is not a strict requirement, but a quality issue—the fewer frames corrupted due to network packet loss, the better the quality of the network service. The sensitivity of the video stream to lost packets depends on the encoding used. Proposed mechanisms for the recovery of packet loss in delay-sensitive streams include forward error correction [1, 6], channel coding [5], or retransmission [3].
- *Synchronization* of the video source with its multiple receivers is needed in order to ensure continuous playback of the video frames at the receiving endsystems. Factors contributing to the loss of synchronization between the video producer and consumer include variations in network delay (*jitter*), the unpredictability of operating system scheduling, and the accuracy of timers. Buffering is required to compensate for these variations in delays.

In this paper we study a novel end-to-end protocol for high quality full-motion multicast video. Having implemented the protocol, we analyze its performance in an FDDI network. The protocol solution we study is based primarily on the connection-oriented transport layer multicast available with the next-generation transport the Xpress Transfer Protocol (XTP) [7]. XTP multicast provides in-order, multipoint delivery of packets with error control options such as retransmission-based recovery of lost packets, rate control for the prevention of packet loss, and dynamic group membership in the connection. XTP thus handles the functionality for timely, reliable multipoint distribution of the video stream in a manner transparent to the application. The application need only focus on buffering techniques to synchronize devices with the network transfer rate.

In the rest of this paper we first discuss the range of functionality available for multipoint delivery of packets using the XTP multicast algorithms. We then describe our multicast video system and present performance measurements that characterize the system under different data rates. The final section gives the conclusions of our study.

2 XTP Multicast

Our end-to-end multicast video protocol derives its simplicity (from the application's viewpoint) from the powerful XTP multicast algorithms underlying it. These algorithms are the outgrowth of the on-going experimentation with transport layer multicast since XTP was first established in 1987. In this section we highlight the important and unique aspects of XTP multicast functionality.

2.1 Connection Paradigm

The XTP designers chose a unique connection-oriented multicast paradigm in which the XTP sending context establishes a simplex, one-to-many connection with a set of receivers. Since endpoints have connection state, a connection-oriented multicast can incorporate significantly more functionality than connectionless multicast facilities. In XTP the functionality found in point-to-point (unicast) connections are extended, with few exceptions, to multicast connections. This approach has the advantage of allowing the protocol user to select from a consistent set of protocol options, whether the communication is unicast or multicast.

XTP multicast uses the IP multicast addressing scheme and expects multipoint packet delivery in the underlying network. When running over an IEEE 802 local area network, the IP multicast address is mapped onto a multicast IEEE 802 MAC address, as specified in RFC 1112 [2]. The IEEE 802 LAN standards support multicast addressing, and network interface hardware commonly supports filtering on several addresses simultaneously. On the FDDI network described in Section 3, for instance, the network interface boards can filter on ten multicast addresses at the same time.

As multicast routing matures, XTP multicast will use IP multicast or ATM multicasting for multicast connections with a receiver group spread across an internetwork. In the Internet, multicast routing protocols are not yet widely deployed, but an experimental multicast network, the MBONE, that uses tunneling across IP routers has proven the capability for large-scale multicasts. With ATM networks, high bandwidth switching will allow high quality video services, and XTP over ATM will provide flexibility and additional functionality to applications.

2.2 Control Algorithms

XTP provides its user with the ability to enable or disable the error, rate, and flow control procedures on a connection. Thus, the XTP multicast configuration with the lowest overhead is the one in which all control procedures are disabled, allowing the multicast transmitter to operate in a "fire-and-forget" manner Note that, even in this configuration, the service ensures correct sequencing of all data arriving at the XTP receivers.

If control algorithms are enabled, the crucial aspect affecting their robustness is the way in which connection state information from the receivers is gathered and processed at the multicast sender context. In a point-to-point connection, the sender and its single receiver use handshaking procedures to exchange connection state information. With multicast, the presence of multiple receivers complicates matters.

The XTP community has experimented with two methods for control algorithms: (1) a heuristic algorithm (the *bucket algorithm*) for timer-based processing of control information and (2) explicit processing of the state information from each receiver in the multicast group (*list-based multicast*). When using the bucket algorithm, the XTP multicast transmitter requests control information from the receiver group and then waits a user-selectable amount of time before processing the control packets that it has received. Control packets are processed without determining which receivers are issuing the packets, and there is no explicit knowledge of the receiver set at the transmitter. Consequently, the reliability of the connection is that, if the transmission completes, at least one receiver was delivered all the data.

The philosophy of the bucket algorithm is to allow the multicast facility to scale to large receiver groups by avoiding any need for explicit knowledge of the receiver set at the transmitter. However, our experience with the algorithm is that the timeout value is difficult to tune correctly, and the lack of a notion of group membership results in reliability semantics that are not applicable to many applications. Experience with the bucket algorithm prompted work on a list-based XTP multicast facility [4]. In this multicast configuration, the transmitter maintains a table of all active receivers and the control information from all receivers is processed for every connection state update. In this way reliable data transfer to a known set of receivers can be carried out.

2.3 Group Membership Control

With list-based multicast, the multicast group membership can be controlled at the transmitter by a user-specified policy. One semantic for list-based multicast is to determine the set of receivers in the multicast group at connection set-up and to ensure that all receivers in this original set report their control information for each update of the connection state at the transmitter. A variation is to permit members of the multicast group to notify the transmitter that are leaving the group. If a member gracefully leaves the group, the transmitter will continue the connection with the remaining members.

XTP defines a mechanism by which a receiver can join an existing XTP multicast connection. To perform an in-progress join, an XTP receiving context is opened and this context transmits a special packet on the multicast group address requesting an in-progress join. The transmitter responds to this request with a packet containing state information sufficient for the joining context to begin receiving data in the on-going multicast connection. The transmitter can control the point in the data stream at which the joining context begins receiving data.

3 Multicast Video Implementation

In this section we present an implementation of our end-to-end protocol for the distribution of high quality multicast video. After giving a description of the hardware and software components of the system, we present a set of experiments that provide performance data on our end-to-end protocol during the transmission of 30 frame/sec, 640-by-480, 8-bit color video at different compression levels.

3.1 Multicast Video System

Our testbed consists of a set of five Intel 486 nodes with EISA busses. Two machines run at 50 Mhz, two at 33 MHz, and one at 66 MHz. The machines are connected with a dual-ring Network Peripherals FDDI network. In addition, each machine is equipped with two video processing boards. A Bravado video board is used to encode NTSC video at the sender at 640x480 resolution with a color plane of 8 bits-per-pixel. At the receiver the Bravado card is used to decode the image and display it on a VGA monitor. The second board, a Rapid Tech Visionary card, is used for JPEG compression at the sender, and decompression at the receiver. The Bravado and Visionary cards are connected using a dedicated bus, thereby reducing contention and bandwidth on the EISA bus.

Testing with this hardware has established that 54 million bits per second (Mbs/sec) is an upper bound on the network throughput available through a raw FDDI interface. The XTP list-based multicast used here is part of the University of Virginia implementation of the Xpress Transfer Protocol. The upper bound on user throughput during reliable transfers using list-based multicast on this hardware has been measured at 37 Mbs/sec. This measurement represents throughput from application memory to application memory with a couple of receivers. This number is an upper bound on our video system's throughput, though the performance data below shows that XTP is not the limiting factor in the current testbed.

3.2 End-to-End Video Protocol

At the sending node an NTSC signal is fed to the Bravado board, which digitizes each frame before passing it to the Visionary board. Each frame consists of two *fields*, due to interlacing, and the video is digitized and compress on a field basis. The Visionary board periodically posts an interrupt that signals the availability of the next JPEG-compressed, digitized video field. The interrupt rate is determined by the frame rate of the video, which the application chooses. An application thread accepts a buffer from the Visionary board and immediately sends the buffer to the network with a call to an XTP multicast connection. If XTP has not completed transfer of the current buffer when the next buffer becomes available, then the application queues the buffer.

XTP multicast delivers each buffer to the set of receiver group using a list-based multicast with a group membership policy as follows:

- New receivers can join the multicast group using the XTP in-progress join procedure. The in-progress join procedure permits new receivers to tap into the in-progress video stream without disturbing the existing receivers. The transmitter controls the in-progress join such that new receivers join the stream on a video frame boundary.
- Existing receivers may gracefully leave the group. This procedure does not disturb the communication since the transmitter simply removes the terminated receiver from the receiver list.
- Existing receivers that leave the group in an ungraceful fashion do not terminate the multicast connection. Instead, the communication stalls for a user-selectable timeout period as the transmitter attempts to elicit control information from the terminated receiver. If the receiver does not respond by the end of this period of time, the transmitter removes the terminated receiver from its list of receivers and resumes the communication as before with the new receiver set.

Our system recovers even from a complete crash of a receiving node. At the MAC layer the FDDI protocol wraps the ring to reestablish communication between the remaining nodes on the ring while at the transport layer the XTP transmitter times out the lost receiver and then proceeds with reliable video distribution to the remaining receivers.

At each receiver a queue of buffers are filled at initialization and then the Visionary board is started. The board periodically interrupts for the next buffer to be decompressed and displayed. If the next buffer is available at the receiver, that buffer is handed to the board. Otherwise, the last received buffer is handed to the board. That is, if no buffers are available, the video display will freeze on the last received frame. Thus, for example, if a receiver leaves the group in an ungraceful manner, the effect at the remaining receivers is to see a single frame of video frozen on the screen, followed after the timeout period by a resumption of the video stream.

3.3 Parameters for Experiments

For our experiment we chose a ten-minute video sequence from a popular action movie, *The Terminator*. The ten-minute excerpt provides a variety of scenes, from static views of two people conversing to action-packed automobile chases.

The quality of the video is inherently determined by its spatial resolution, its frame rate, and the amount of compression. We want to study high quality video, which we define as 640x480 resolution at 30 frames/sec. For the JPEG compression [8] supported by our hardware, a parameter known as the *quantization factor*, or Q-factor, is a user-selectable input to the encoder that determines the lossiness of the compression algorithm. Specifically, the Q-factor determines the quantization step size for the Discrete Cosine Transform coefficients used in JPEG. Smaller values for the Q-factor yield higher quality video since the amount of loss due to the many-to-one quantization mapping is reduced.

In our experiments the ten-minute video is transmitted using a fixed Q-factor for each transmission. We consider Q-factors of 30, 45, and 60. These levels of compression provide a spectrum of quality. With a Q-factor of 30, the artifacts of the JPEG algorithm were not generally discernible, even with close inspection, in the images at the receivers. With a Q-factor of 60, the the receivers' displays show the "blocks of color" that are characteristic of the underlying JPEG compression algorithm. In each transfer there are four receivers in the multicast group.

3.4 Performance Measurements

Table 1 shows the data rates for our video under the different Q-factors chosen for the JPEG compression algorithm. Uncompressed, the video stream in our experiment would generate approximately 74 Mbs/sec. Thus the compression ratio over the entire transmission ranges from 7.1:1 to 12.6:1 in our experiments.

Q-Factor	Total Bytes	Average Data Rate	Max. Buffer	Min. Buffer
30	776 Mbytes	10.35 Mbs/sec	29,320 bytes	14,800 bytes
45	550 Mbytes	7.33 Mbs/sec	22,516 bytes	9652 bytes
60	439 Mbytes	5.85 Mbs/sec	18,692 bytes	7700 bytes

Table 1. Size of 10 Minutes of Compressed Video as a Function of Q-Factor.

Figure 1, Figure 2, and Figure 3 provide a more complete view of the video stream characteristics by showing the frequency count for the buffer sizes generated under each Q-factor. As mentioned earlier, each buffer represents the data for a field where a video frame consists of two fields. The horizontal axis is the field buffer size in bytes, and the vertical axis is the frequency count for that buffer size over the entire ten minutes, i.e. approximately 36,000 field buffers. The frequency count is discretized into histograms with a width of 100 bytes. The data shows the increase in the variance of the buffer sizes as the degree of compression decreases. This increase is explained by the wider range of frequencies admitted in images when less compression is used.

Table 2 shows the behavior of the synchronization buffer queue in the application code at the video source during each experiment. For each Q-factor the table shows the number of video buffers queued by the application for submission to the XTP layer. The queue size was measured each time the Visionary board delivered a new buffer to the application.

In our implementation we choose the maximum number of outstanding buffers to be four. If four buffers are outstanding when the application is notified that another video buffer is ready for transmission, the application overwrites the oldest buffer in the queue with the new buffer. Thus, Table 2 shows that no buffers were lost during the transfers with Q-factors of 45 and 60. When the Q-factor was lowered to 30, however, the video application loses 8% of all buffers generated due to the queue of outstanding buffers being full. This substantial loss rate indicates that the transmitting machine approaches its maximum sustainable throughput with the data rates generated under a Q-factor of 30, e.g., buffers of approximately 25 Kbytes (see Figure 1).

At each receiver, in our application-level buffer management scheme, four buffers are filled from the network before the Visionary video board is initialized. This buffer queue provides protection from a loss of synchronization between the delivery of new buffers from the network and the consumption of buffers by the video board. Table 3, Table 4, and Table 5 show the number of queued buffers at each of the receivers during our experiments. The column labelings represent, at the time the video board requests its next buffer, the number of buffers, including the one that XTP is in the process of filling, queued at the receiver.

In the tables, if the number of buffers is one, then the queue has drained and the last buffer played by the video board will be replayed. As seen in Table 3, for a Q-factor of 30, such pauses occurs at each of the receivers 8% of the time. Table 2 shows that the problems in maintaining the timing of the video stream at the receivers is due to performance degradation at the transmitter.

For the Q-factors of 45 and 60, Table 4 and Table 5 respectively show that the end-to-end jitter never drains the receivers' queues. In both of these cases, jitter variation is greater than a single buffer transfer only 1% (or less) of the time. The tables also show that the variation in the behavior of individual receivers is significant, but not surprising given the range of CPU processing speeds at the receivers, i.e., from 33-66 MHz.

4   Conclusion

In this paper we have presented an end-to-end protocol for the distribution of multicast video using a list-based multicast facility embedded in the Xpress Transfer Protocol. This protocol architecture for high quality video is an attractive solution. XTP list-based multicast allows the application to give the policies that determine the robustness of the communication, e.g., error control, rate control, and group membership management. Hence different communication profiles can be created, based on application requirements (e.g., video quality and preferred behavior when transient network conditions cause long delays) and knowledge of the system (e.g., number and location of the receiver group, endsystem hardware limitations, etc.). At the same time, the mechanisms to enforce the policies selected by the application are embedded in low-level (transport layer) communication algorithms where the overhead for packet exchanges is low.

We implemented our end-to-end multicast video solution on Intel-based machines equipped with JPEG compression hardware. We then measured this system with four receivers on an FDDI network. For our measurements we examined three compression ratios, from roughly 7:1 to 12:1, for the JPEG compression engine and transported a 10-minute movie sequence at 640-by-480 spatial resolution at 30 frames/sec with 8-bit color.

Our performance data shows the feasibility of high quality full-motion video distribution to a receiver set of four using our relatively modest hardware. In the testbed, endsystem hardware (i.e., bus bandwidth and CPU processing cycles) limit the bandwidth of our video system to a maximum of approximately 12 Mbps/sec. This throughput is about one-third of the maximum throughput for XTP list-based multicast. In future work we will move our end-to-end protocol to faster endsystems and consider supporting a zero-copy architecture, e.g., DMA buffers directly from the Visionary board to the FDDI card. We will also consider larger receiver sets, including receiver groups spread across multiple LANs connected by IP routers, in order to explore the behavior of XTP list-based multicast with large multicast groups.

Acknowledgments

We acknowledge the work of Fraser Street in producing the first version of this video system and of James McNabb who provided the XTP list-based multicast code.

References

1. E. Biersack. Performance Evaluation of Forward Error Correction in ATM Networks. *ACM SIGCOMM '92*, 22(4):248–258, August 1992.
2. S. Deering. Host Extensions for IP Multicasting, August 1989. RFC 1112.
3. B. Dempsey. *Retransmission-Based Error Control for Continuous Media Traffic in Packet-Switched Networks*. PhD thesis, CS-94-23, Computer Science, University of Virginia, May 1994.
4. B. Dempsey and R. Simoncic. Reliable Multicast and N-by-N Services in XTP. *XTP Forum Transfer Newsletter*, 6(4):12–17, July/August 1993.
5. M. Garrett and M. Vetterli. Joint Source/Channel Coding of Statistically Multiplexed Real-Time Services on Packet Networks. *IEEE/ACM Transactions on Networking*, 1(1):71–81, February 1993.
6. H. Ohta and T. Kitami. A Cell Loss Recovery Method using FEC in ATM Networks. *IEEE Journal on Selected Areas in Communications*, 9(9):1471–1483, December 1991.
7. W. Strayer, B. Dempsey, and A. Weaver. *XTP: The Xpress Transfer Protocol*. Addison-Wesley, July 1992.
8. G. Wallace. The JPEG Still Compression Standard. *Communications of the ACM*, 34(4):30–44, April 1991.

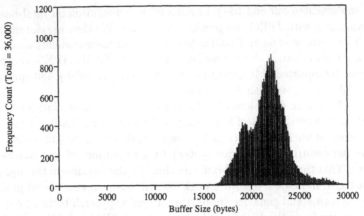

Fig. 1. Frequency Count of Video Buffer Sizes (Q-factor=30).

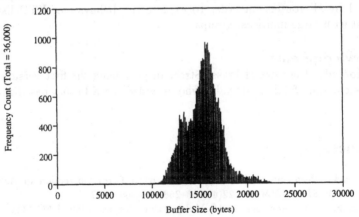

Fig. 2. Frequency Count of Video Buffer Sizes (Q-factor=45).

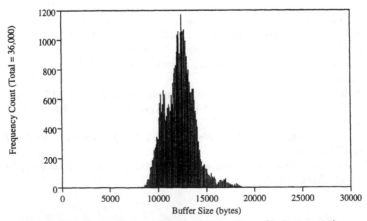

Fig. 3. Frequency Count of Video Buffer Sizes (Q-factor=60).

Q-Factor	0	1	2	3	4
30	15%	17%	5%	55%	8%
45	94%	5%	0.5%	0.5%	0%
60	98%	2%	0%	0%	0%

Table 2. Queue Depth at the Sender.

	1	2	3	4
Receiver 1	8%	52%	12%	28%
Receiver 2	8%	52%	11%	29%
Receiver 3	8%	52%	21%	19%
Receiver 4	8%	52%	11%	29%

Table 3. Queue Depth at the Receivers (Q-factor=30).

	1	2	3	4
Receiver 1	0%	1%	16%	83%
Receiver 2	0%	1%	6%	93%
Receiver 3	0%	0%	1%	99%
Receiver 4	0%	1%	5%	94%

Table 4. Queue Depth at the Receivers (Q-factor=45).

	1	2	3	4
Receiver 1	0%	0%	20%	80%
Receiver 2	0%	0%	14%	86%
Receiver 3	0%	0%	1%	99%
Receiver 4	0%	0%	9%	91%

Table 5. Queue Depth at the Receivers (Q-factor=60).

An Object-Oriented Implementation of the Xpress Transfer Protocol[1]

W. Timothy Strayer, Simon Gray, and Raymond E. Cline, Jr.

Sandia National Laboratories, California
{*strayer*| *sgray*| *rec*} *@ca.sandia.gov*

Abstract. Object-oriented design principles map well onto protocol implementations because protocols essentially manipulate two structures—packets and the states of the endpoints. In this paper we describe an implementation of the Xpress Transfer Protocol as a user-space daemon written in C++. The object-oriented model forces the programmer to properly place functionality and information ownership. The model facilitates porting to various platforms and greatly eases the task of building data delivery services.

1. Introduction

Transport protocols have at least two common components: a fundamental unit for information exchange, and a set of structures and procedures for managing the information as it is exchanged. The units of information exchange are called *packets*, and the state structures are called *protocol control blocks* or *contexts*. Protocols are distinguished by the amount of state information maintained, and by how the protocol behaves during its state transitions. For example, a simple unreliable datagram protocol keeps little state and does not react to lost packets, while a robust connection-oriented protocol maintains much more information about the data exchange, and notices and reacts to lost packets. Yet both types of protocols follow the same pattern of processing: a packet arrives, it is parsed, and the data, if present, is delivered to the client. Protocols can therefore be viewed in the abstract, where a specific protocol is an instantiation of this abstraction.

The object-oriented programming paradigm forces ownership of functionality. This is useful for any major software project. The paradigm also forces ownership of information. As a value is needed by an object, the programmer must decide how that value is to be conveyed. If the programmer cannot convey the information gracefully, there is a strong indication that the information is either ill-placed or unnecessary. Object orientation also provides well-defined modules for implementation hiding. This is useful for isolating services from the rest of an implementation, facilitating adaptability to a variety of services that are built with the same interface.

---

[1] This work is supported by Sandia Corporation under its Contract No. DE-AC04-94AL85000 with the United States Department of Energy, and CRADA No. 1136 between Sandia National Laboratories and AT&T Bell Labs.

This paper discusses the application of the object-oriented programming paradigm to the implementation of a transport layer protocol. Specifically, we have implemented the Xpress Transfer Protocol [1, 2] using C++. The compiled target of this protocol implementation is a user-space daemon process. Application processes load a user interface library (also object-oriented) that manages the interprocess communication between the application and the daemon. The daemon accepts user requests and incoming packets, and passes them on to a context container class that manages all of the contexts in the daemon. We describe the class structures and associated methods used to construct XTP, and we make observations that are relevant to object-oriented protocol implementation in general.

2. Protocol Implementation

Traditionally, protocols are implemented in the kernel of an operating system. TCP and UDP in Unix [3] are the most common examples. The x-kernel [4] and the Mach operating system [5] are examples of attempts to simplify the implementation of protocols in the kernel. Thekkath *et al.* [6] observe that monolithic implementations of protocol stacks in the kernel provide performance and security at the cost of making prototyping, debugging, maintenance, extensibility, and exploitation of application-specific information more difficult. They suggest implementation of protocols as user-level libraries, where an agent in the kernel is responsible only for demultiplexing the packets as they arrive.

Our implementation of XTP runs the protocol as an object-oriented user-level daemon process. This design provides rapid prototyping, portability, adaptability, configurability, and readability.

Rapid Prototyping. A user-level implementation of a protocol is typically faster to build than a kernel implementation for two reasons: it is easier to debug, and kernel programming requires an additional learning curve. Also, in theory, object-oriented programming forces the programmer to design the components of the software, and to assign functionality to these components, prior to writing code.

Portability. Kernel implementations of protocols are rarely portable. The Distributed Systems Research department at Sandia has a cluster of 50 workstations as a testbed for heterogeneous cluster computing research. This testbed has five vendors and six different operating systems; implementing a protocol for each operating system's kernel would be a difficult and time-consuming task. We coded the implementation in C++ since it is the most widely available object-oriented language.

Adaptability. We are interested in protocol characteristics in a variety of LAN and WAN environments. Consequently, we need to be able to rapidly switch between different data delivery services, such as IP, Ethernet, FDDI, ATM, and other solutions like the Scalable Coherent Interface (SCI). The modularity of the design of our implementation supports this.

Configurability. In addition to changing the underlying data delivery service, we want to be able to replace the various protocol control algorithms. Again, modular code construction supports this.

Readability. Kernel code is difficult to read since it is so deeply embedded in the operating system. Code written for user-level processes is (perhaps subjectively) easier to read and decipher. We expect our implementation source code to be used as a reference for XTP's protocol control algorithms.

In general, a protocol receives service data units from the service below it and demultiplexes their contexts, or protocol data units, to the various clients of the protocol. In transport protocols, these clients are the session layer (in the OSI view) or some user process (the Internet view). A protocol implemented in the kernel uses its knowledge of which processes have employed its services to do the demultiplexing. Protocols implemented outside of the kernel, in user space, must have some other means for managing multiple users. One approach is to implement the protocol in a library, and let each user process run the protocol. However, there is still a need for some agent, usually in the kernel, to demultiplex the incoming packets. Another drawback for library implementations is that the protocol is distributed, so control algorithms such as rate and congestion control must also be implemented in the common agent.

We use a user-level daemon as the single representation of the protocol. The daemon manages each of the client users, and demultiplexes incoming packets. In spite of the considerations listed above, there are drawbacks. This approach adds another process, requiring context switching for interprocess communication. Furthermore, a user-level process must invoke system calls in order to gain access to kernel-level services, such as timers and device drivers.

3. Meta-Transport Library

While guided by the above design goals, we added one more: hierarchical design. A useful tool in object-oriented programming is class hierarchies. While designing our implementation of XTP, we built abstract classes that would be applicable to most transport protocols. We call the collection of these abstract bases classes the *Meta-Transport Library* (MTL) because they can be used to build most transport protocols, not just XTP. The objective in designing MTL was to distill as many of the commonalities of transport protocols as possible into a set of classes that are made available as a library. Specific protocols are "derived" from MTL by implementing classes derived from the MTL base classes. A virtual function in the base class suggests that the MTL implementation for that method may not be sufficient, and additional protocol-specific processing may be necessary. A pure virtual function implies that an implementation must be provided by a derived class. These methods are mandated by MTL but require protocol-specific knowledge to implement.

Protocols derived from MTL have essentially the same performance advantages and disadvantages since they share a common foundation. In this way, comparison of derived protocols more accurately exposes the differences in the protocols rather than the differences in the skill of the implementors. Furthermore, since the "skeleton" of the protocol is already in place, an implementor simply needs to flesh out the implementation with the protocol specific-member variable and methods.

Fig. 1 shows the general model for protocols derived from MTL. A client process sends requests to the daemon via an IPC message queue (see the *ipcs*(1) Unix manual

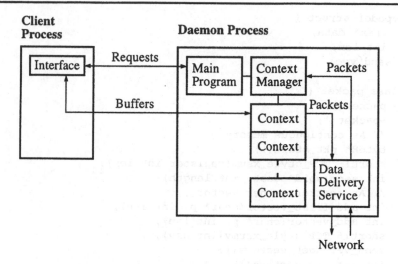

Fig. 1 MTL Client/Daemon Model

page) built into the interface object. The daemon returns the result of the request via the same IPC facility. User data, however, is written to and read from two buffers that are managed by a buffer manager in the interface. This separates the request/response activity from the maintenance of data buffers. Here we use shared memory between the daemon and client process; a buffer manager interface, however, hides these details. The main program in the daemon is a loop that accepts user requests and invokes the appropriate entry point into the actual protocol processing. These entry points are methods in the context manager. The context manager owns all of the contexts in the daemon, and steers incoming packets to the proper context. The contexts themselves implement the protocol-specific procedures, some of which generate packets. The daemon also owns a data delivery service object that manages the use of the network.

An MTL context is identified by a key value. This is actually a concept from XTP, but it is applicable to any protocol since the MTL model requires some manner of uniquely identifying a context data structure.

4. SandiaXTP

SandiaXTP is the XTP-specific protocol derived from MTL. We implement the latest version of XTP, XTP 3.7. This version changes a few packet formats and splits the control information into three distinct packet types, one for normal flow control information (CTNL), one to report error information (ECNTL), and one to negotiate the traffic shape information (TCNTL).

Below we discuss some of the division of responsibility between MTL and Sandia-XTP, and explain some of our design decisions.

```
typedef struct {
  char* data;
  int len;
} send_vec;

class packet {
  packet();
  ~packet();
  // As contiguous memory
  byte8* pkt_start();
  short16 TCP_style_xsum(register int len);
  int send(void* dest, int length);
  // As scatter-gather vector
  int add_first_vector(char* p, int len);
  int add_vector(char* p, int len);
  short16 TCP_style_xsumv(int nsv);
  send_vec* get_vectors();
  int get_num_vectors();
  int sendv(void* dest);
  // Virtuals
  virtual void host_to_net() = 0;
  virtual void net_to_host() = 0;
};
```

Fig. 2 MTL Packet Base Class

4.1. Packets and Packet Manipulators

Packets are the vehicle for data and information exchange between endpoints. Packets are sent and received by a data delivery service that treats the contents of the packet as uninterpreted payload. The protocol defines the structure of its packets, and information is placed or extracted only with knowledge of the structures.

The MTL packet base class, shown in Fig. 2, provides two ways to view a packet, as contiguous memory or as a vector of scatter-gather elements.[1] Since MTL's data delivery service assumes that packets are received as a contiguous piece of memory, the first form is generally used for processing a packet within the context. The size of a packet object is at least as big as the maximum protocol data unit length. The packet base class includes a member function that returns a pointer to the beginning of the packet object so that protocol-specific agents can read and write to offsets within the packet.

The scatter-gather vector is a set of pointer, length pairs. The packet base class provides member functions to set and retrieve the scatter-gather elements. This style is intended for constructing packets with minimal data copying.

---

[1] The types byte8, short16, word32, and word64 are unsigned integers with the specified number of bits.

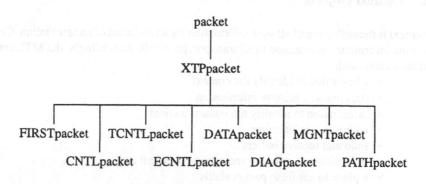

Fig. 3 SandiaXTP Packet Hierarchy

The packet base class provides an overloaded send method for both packet styles. In both cases, the packet::send() method calls the data delivery service send method. This ensures that the context, or any other agent constructing a packet, need know nothing of the data delivery service. There is no symmetric receive method, however, since receiving is not actually done *to* a packet in the same way that send is. Data simply arrives at the underlying data delivery service; that data is cast into a packet structure in order to retrieve the contents. For protocols with more than one distinct packet structure, this act of casting is probably done twice, once to retrieve the type, and again when the type is known.

In XTP all packets have a common header that holds the specific packet type. Methods for placing and extracting header information are defined in the class XTP-packet, which is derived from the MTL base class packet. Specific XTP packet types, such as FIRST, DATA, CNTL, etc., are derived from XTPpacket, and provide the type-specific manipulation methods. This hierarchy is shown in Fig. 3.

The packet base class has only two virtual functions, host_to_net() and net_to_host(). These are pure virtual functions since byte order conversion is only possible with knowledge of the packet structure. Other than these, there are no mandatory methods to be defined by the derived classes. Rather, functions common to all packets classes at one level are defined in the packet class one level above. For example, all packets must send, but sending is not a protocol-specific function since the data delivery service does not care what it is being given to send. So send() is defined at the MTL packet class level. All XTP packets must have access to the common header, so get_header() is defined in the XTPpacket class level.

MTL also includes two classes that manipulate abstract packet objects. For efficiency, a packet pool class is provided that pre-allocates and then manages packet shells. The packet FIFO class holds packets, then emits them in first in, first out order. If a key is given, the packets destined for the context with that key are emitted in FIFO order.

4.2. Context Objects

A context is the collection of all state information for an endpoint of an association. Certain state information is common to all transport protocols. Accordingly, the MTL context base class holds:
  • a key value to identify the context
  • the context's priority information
  • information to identify the context's client
  • addressing information
  • send and receive buffers
  • the maximum protocol and service data unit sizes
  • a place to get fresh packet shells
  • a place to store outgoing packets that are held by flow or rate control
  • a place where incoming packets are held until processed

The functions declared virtual in the MTL context base class are the functions whose inclusion is mandated by the class. These include methods to:
  • determine if the context is active or quiescent
  • cause the context to become quiescent
  • get and change the context's priority
  • initialize the context's state variables
  • bind an address to the context
  • process a packet
  • put data in the client's receive buffer
  • send data from the client's send buffer

The derived context class must redefine these virtual functions so that they represent the functionality specified by the protocol. The abstract class cannot mandate that error, flow, rate, or other control algorithms be present. In SandiaXTP, the send method checks flow and rate control parameters to decide whether a newly constructed data packet can be sent.

The derived protocol may need to implement some form of timer control to guard against lost packets or inactive connections. The XTPcontext class implements the XTP specification of these timers. Also, the protocol state machine is not included in the methods mandated by the context base class, although every protocol moves through a series of states over its lifetime. Since the states and their meanings are protocol-specific, the context base class mandates only that a test for quiescence be present.

4.3. The Context Manager Objects

The context manager is the container class for all of the contexts in a protocol implementation. The main purpose of the context manager is to match user requests and incoming packets to the appropriate context, so that the contexts can do the necessary protocol processing. To this end, the context manager base class provides functions that:
  • allocate a new context and key value
  • find an active context
  • initialize a specified context

394

```
class context_manager {
    context_manager(int nctxts, int npkts, int csize);
    virtual ~context_manager();

    // allocate a new context and key value
    virtual context* get_next_context(word32 prio);

    // find an active context
    context* get_context(word64 key);

    // initialize a specified context
    virtual int init_context(user_request* request) = 0;

    // bind an address to a context
    virtual int bind_context(user_request* request) = 0;

    // order active contexts by priority (head of priority list)
    context* get_head();

    // match incoming packet with context
    virtual void handle_new_packet(packet* pkt, void* from) = 0;

    // visit each context to do work pending
    virtual void satisfy() = 0;

    // release the context
    virtual int release(word64 key);

    // plus other utility functions...
};
```

Fig. 4 MTL Context Manager Base Class

- bind an address to a context
- order active contexts by priority
- match an incoming packet to the appropriate context
- visit each active context to satisfy pending work
- release the context

Since these functions are common to all protocols that would be implemented using the MTL model, they are mandated by the MTL context manager base class (shown in Fig. 4) through virtual functions. The derived class, XTPcontext_manager, redefines the virtual functions in terms of an XTP context rather than an abstract context. There are a few additional functions in XTPcontext_manager that reflect the way XTP handles incoming packets. Specifically, there is a function that finds the listening XTP context given the address segment of a FIRST packet. The fields and methods are not general enough to put this function in the base class, although most protocols would have some similar capability. Another function specific to XTP is the full context

```
class del_srv {
  del_srv();
  virtual ~del_srv();

  // Pure virtual functions to be filled in by derived classes

  virtual int install(address_segment* addr, void* dest) = 0;
  virtual void* alloc_addr_struct() = 0;
  virtual void free_addr_struct(void* asptr) = 0;
  virtual int sizeof_addr_struct() = 0;
  virtual int recv(char* data, int length, void* from) = 0;
  // send contiguous payload
  virtual int send(char* data, int dlen, void* dest) = 0;
  // send scatter-gather payload
  virtual int sendv(send_vec* sv, int nsv, void* dest) = 0;
  virtual int get_pdu_size() = 0;
};
```

Fig. 5 MTL Data Delivery Service Base Class

lookup. For packets carrying the key value defined at the destination host, the XTPcontext_manager can match the packet directly to the appropriate context. For others, a table lookup is required. The function of matching a packet to a context is common to all protocols, but how it matches is protocol specific, so this function is not included in the base class.

4.4. The Data Delivery Service

The data delivery service abstract class presents a common interface to the underlying packet transport, as shown in Fig. 5. Specific delivery services are derived from this abstract class. In particular, the derived classes must define receive and send methods, as well as a method for translating between the MTL address structures and the service-specific address structure. A method for determining the maximum protocol data unit size for the service are also included.

4.5. The Daemon

In SandiaXTP, the daemon, called xtpd, instantiates a context manager object and a data delivery service object, then blocks waiting for incoming user requests. The daemon unblocks when the client's request arrives, does a switch on the request type, awaits the result of the request, then sends the result back to the client. The user can specify that the request is a blocking request, where the daemon does not return the results until the request is satisfied. Either way, the daemon loops back and blocks waiting for another request.

Incoming packets interrupt this block. The packet is retrieved from the data delivery service, the destination context is determined by the key and other addressing information, and the packet is placed on a packet FIFO where it awaits processing by the context. When all of the newly arrived packets have been received from the data delivery service, the daemon invokes the context manager's satisfy routine, directing each context to do any pending work. Once all the active contexts have been satisfied, the daemon blocks again.

XTP_REG	register the client with the daemon, and allocate a context
XTP_BIND	bind the client's address structure with the context
XTP_LISTEN	listen for a FIRST packet to establish the association
XTP_REJECT	reject the association
XTP_GETADDR	get the client's address structure
XTP_GETSTATE	get the state of the context
XTP_SEND	send data in the send buffer
XTP_RECEIVE	get data from the receive buffer
XTP_UPDATE	synchronize the client's state with the daemon's
XTP_SENDCNTL	send a control packet
XTP_RELEASE	return any resources, release the context

Table 1 XTP Request Types

There are eleven types of XTP requests accepted by xtpd, as shown in Table 1. These requests correspond closely to the services offered by the XTP interface to the client. The intent is to offer request types that expose the flexibility of XTP without being overly complex. The rationale for the interface is given later.

Registering with the daemon causes a context, and the key that identifies the context, to be allocated for this client. The key will uniquely identify this context in all future requests (as well as future packet exchanges). The "bind" request sets the addressing structure. This must be complete before sending a FIRST packet, or, if listening, this address specifies the filter that discriminates which FIRST packets to accept. The "listen" request causes XTP to look for FIRST packets that match the address specified with the bind request. The "get address" request retrieves the fully specified address structure at any time after the association establishment. The "reject" request allows the user to reject an association if its parameters do not meet the user's needs.

Data placed in the send buffer is sent by issuing the "send" request, and data is retrieved from the receive buffer by issuing the "receive" request. Data is not actually moved by these commands; rather, the client, through its interface, and the context, through its buffer managers, read and write the data to the appropriate buffers. The "update" request synchronizes the client's and the context's buffer managers.

Since XTP is a transmitter-driven protocol, the control packets are not automatically generated by the protocol. The user has several mechanisms for causing control packets to be generated, such as using the SREQ and DREQ options field of a send request. The "send control" request is another such mechanism, giving the client direct access to generating a control packet.

The "get state" request returns the internal state of the client's context. The "release" request causes all of the resources allocated to this client to be returned, and the key associated with the context to become invalid.

4.6. Buffer Management

A client has two data buffers, one for sending and one for receiving. Each of the client's buffers is controlled by a buffer manager. When the client registers these buffers with the daemon, the context assigned to the client attaches its two buffer managers to the client's buffers. In this way, each buffer is "controlled" by two buffer managers—one at the user application side (and hidden in the user interface routines), and the other in the context associated with this user. In general, one of the managers writes to the buffer and the other manager reads from it, so there must be some way of coordinating the head and tail markers for the buffer. This is done via user request commands; when a send request is issued, the interface object places the new marker values into the request so that the context's buffer manager will know what data to send. A similar exchange happens for receiving data through the receive buffer.

The buffers themselves are implemented as segments of shared memory. The user application, through the user interface library, creates two shared memory segments. The identifiers for these segments are relayed to the daemon, where the newly initialized context for this user attaches the daemon process to the shared memory segment. This detail could, of course, have been accomplished with any number of IPC facilities, and replacement of the underlying mechanism for this object is easily done as long as its replacement uses the same class interface. Fig. 6 shows how the buffer managers, the shared memory segments, and the client and daemon are related.

4.7. The User Interface

The user interface is also implemented as a base class and a protocol-specific derived class, but it is not compiled into the daemon. Rather, it is targeted as a library to be compiled into the user's application code. The interface is instantiated as an object, although this is not entirely necessary. The important part of the interface is that it provides the user's application a means of issuing requests to the daemon and of managing the data in the user's send and receive buffers.

The user interface controls the user's end of the two buffer managers mentioned above. The user writes data into these buffers and issues the send command. The presence of the data is made known to the context, and the context's buffer manager pulls the data into the protocol. As packets with data arrive, the data is written into the receive buffer. As the user application asks for data through the receive request, the context

Client Process   **Daemon Process**

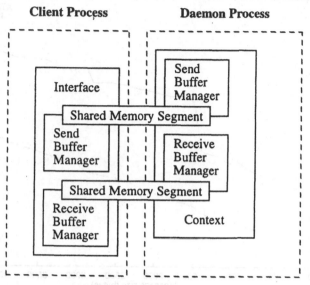

Fig. 6 MTL Buffer Management

informs the interface's receive buffer manager about the size and location of the received data.

The MTL interface base class, mtlif, has six member functions:

```
int install_buffers(int snd_buf_size, int rcv_buf_size,
                    char* sndaddr, char* rcvaddr);
int issue(reqbuf* req);
int inform(reqbuf* req);
buffer_manager* get_s_bm(); // send buffer
buffer_manager* get_r_bm(); // receive buffer
virtual void perror(int res, char* usr_msg);
```

The install_buffers() method creates the buffer managers. They are accessed via the get_r_bm() and get_s_bm() methods. User requests are sent to the daemon via the issue() method, which waits for the results, and the inform() method, which does not wait. The perror() method prints the error messages corresponding to return codes.

The SandiaXTP user interface derived from mtlif is called xtpif. This class uses the methods from mtlif to generate and issue the requests given in Table 1. These requests reflect the several basic things a user would need from XTP and, as such, they are fairly raw. Different interfaces can be constructed by deriving a class from the xtpif class. In this way, standard APIs, like sockets or TLI, can be emulated, and code written to use those APIs would not have to be modified to use XTP.

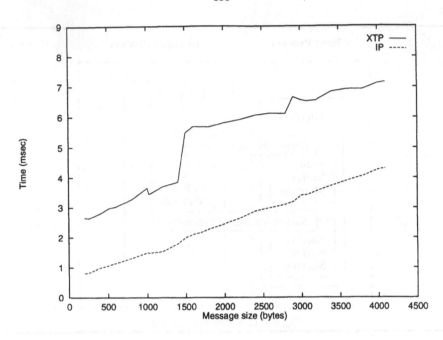

Fig. 7 SandiaXTP End-to-End Latency

5. Discussion

It is difficult for user-level protocol implementations to compete with kernel implementations, mostly because user-level processes have to use system calls to gain access to services which reside in the kernel. Crossing domain boundaries usually requires buffer copies. Without direct access to the network device, efficiencies such as flow-through checksumming are not possible. Another problem is that a context switch is required to change from the client process to the daemon process. On the other hand, user-level processes are much easier to build, debug, and configure. Further, the trend is to move functionality out of the kernel, provide better access to the functionality left in the kernel, and reduce the cost of context switching.

One of the data delivery services we have implemented is IP via raw sockets. In order to evaluate how expensive SandiaXTP is, we conducted a few latency tests for SandiaXTP and raw IP. Fig. 7 shows these two curves. Latency was measured from user process to user process on two SGI Indies over Ethernet. One process sent data to the other; that data was completely received, then the receiving process sent the same amount of data back to the original sender. The one-way latency is this roundtrip time divided by 2. To get statistically viable results, we took five samples of 50 iterations for each message size. Checksumming was enabled across the whole XTP packet.

Since SandiaXTP is running over the raw IP service, the difference between the curves is roughly the amount of time SandiaXTP is processing packets. From the graph we can see that SandiaXTP adds about 2 msec processing overhead for packets smaller than 1448 bytes. The size 1448 is the maximum PDU size used because we wanted an XTP packet to fit fully within one Ethernet frame. The characteristic jump in the graph is due to SandiaXTP segmenting and reassembling messages into two (at 1449 bytes) and three (at 2986 bytes) XTP packets. (Performance measurements of a kernel implementation of XTP versus IP, TCP, and UDP can be found in [7].)

There are several areas where improvements can be made to the SandiaXTP implementation. Checksumming is as costly as a data copy; inlining efficient assembly code could help reduce this cost. Similarly, the cost of conversion from host to network byte order can be reduced with more efficient methods. Currently, asynchronous packet arrivals and timer expirations are caught by trapping signals, yet we know that processing interrupts is fairly costly. Also, reducing the cost of crossing the user/kernel domain boundary will improve performance. Some of these improvements are simple, others, such as making the boundary less of a barrier, are more difficult. Nonetheless, our initial experience with implementing a protocol in user space has been satisfactory.

References

1 W.T. Strayer, B.J. Dempsey, A.C. Weaver: XTP: The Xpress Transfer Protocol. Reading, Mass.: Addison-Wesley 1993.
2 XTP Protocol Definition, Revision 3.6. PEI-92-10, Protocol Engines, Inc., January 1992.
3 S.J. Leffler, M.K. McKusick, M.J. Karels, J.S. Quarterman: The Design and Implementation of the 4.3BSD UNIX Operating System. Reading, Mass.: Addison-Wesley 1989.
4 N.C. Hutchinson, L.L. Peterson: The x-Kernel: An Architecture for Implementing Network Protocols. IEEE Transactions on Software Engineering 17(1), 64-76 (1991).
5 J. Boykin, D. Kirschen, A. Langerman, S. LoVerso: Programming Under Mach. Reading, Mass.: Addison-Wesley 1993.
6 C.A. Thekkath, T.D. Nguyen, E. Moy, E.D. Lazowska: Implementing Network Protocols at User Level. Proceedings of SIGCOMM '93, San Francisco, Ca., September 13-17, 1993, pp. 64-73.
7 W.T. Strayer, M.J. Lewis, R.E. Cline, Jr.: XTP as a Transport Protocol for Distributed Parallel Processing. To appear in Proceedings of the USENIX Symposium on High-Speed Networking, Oakland, Ca., August 1-3, 1994.

Acknowledgements

The authors wish to thank Paul S. Wang for his expert assistance in developing the object-oriented model for MTL, and the members of the XTP Forum for testing and suggesting refinements for MTL and SandiaXTP.

Development of a Multimedia Archiving Teleservice using the DFR Standard*

Thomas C. Rakow[†], Peter Dettling[‡], Frank Moser[†], and Bernhard Paul[‡]

[†] Gesellschaft für Mathematik und Datenverarbeitung mbH
Integrated Publication and Information Systems Institute (IPSI)
Dolivostraße 15, D–64293 Darmstadt

[‡] IBM Deutschland Informationssysteme GmbH
European Networking Center (ENC)
Vangerowstraße 18, D–69020 Heidelberg

Abstract. In this paper, we describe the development of an archiving system according to the BERKOM teleservice "Multimedia Archiving" defined recently. This teleservice is utilized for storage and retrieval of multimedia documents. It can be exploited in applications that require document pools, information services, and workflow management. As a basis of our system we use the ISO/IEC standard "Document Filing and Retrieval (DFR)". We have chosen the following approach for the integration of multimedia documents within the DFR environment. (1) A specific attribute for DFR-Documents is introduced to model multimedia documents. (2) A dual stack is used separating transmission of discrete data and continuous data like audio and video. An architecture based on this solution is outlined. Especially, the usage of an object-oriented database management system for the storage of multimedia documents is motivated. As an application scenario we describe a multimedia calendar of events. Finally, we evaluate our concepts and describe further developments.

1 Introduction

Multimedia teleservices in public networks can be utilized to make multimedia information available for a large number of users. The primary goal of the multimedia archiving teleservice (MMA) developed within the BERKOM II initiative is the management of multimedia documents in a public network. An archive is established which is constantly fed by information providers with information concerning specific subjects. Navigational and retrieval functions allow archive users to search for information

---

* This work is partially granted by DeTeBerkom GmbH, Berlin, as project "Globally Accessible Multimedia Archives (GAMMA)" within the BERKOM II initiative.

in the archive. Archive types can be classified according to their functionality. Archives for *document filing and retrieval* allow standardized storage of and access to documents by utilizing attribute oriented searching. Archives for *information services* additionally support partial access to information contained in the documents. More advanced archives with *work flow management* support collaborative document processing by "triggering" users to access document parts according to workflow specifications. All these types also provide functions for user administration and for the calculation of utilization costs.

The following functionality is defined within the BERKOM MMA teleservice [11]:

- storage and presentation of multimedia documents,
- navigational access within a hyperlinked structure,
- retrieval which allows for the selection of multimedia documents or document parts,
- concurrent access of several users and error recovery by transaction management, and
- administration of collaborative work process for a large number of users.

According to this definition, we have developed an MMA system based on the ISO/IEC standard "Document Filing and Retrieval (DFR)". The DFR standard defines *some* functionality needed for the MMA teleservice. Multimedia documents, however, are not supported by DFR. The main contribution of this paper is the description of how to design and construct an MMA system by using standard-compliant extensions to DFR and by integrating multimedia transport protocols into this environment. The DFR standard describes a distributed client and server architecture for storing documents in a *hierarchically structured pool*. The notion pool means that documents are handled independently of a specific exchange or document structuring format. The structure of the pool can be defined by the application.

Several problems occur if multimedia documents have to be stored in a DFR document pool. Complex structures of multimedia documents must be mapped to the information model of DFR. Real-time requirements for the presentation of continuous data like audio and video must be considered.

In our approach, a specific attribute for DFR-Documents is introduced to model multimedia documents. Even a content location model for text is supported. The transmission of discrete data and continuous data like audio and video is separated by using a dual stack approach. Based on these solutions we have developed an architecture for the MMA system. Especially, the usage of an object-oriented database management system (DBMS) for storage of multimedia documents is integrated. It supports object-oriented modelling and querying (which allows the usage of method calls within queries) and eases the development of the server functionality. Additionally, we support continuous multimedia transport protocols by the DBMS, but due to space limitations this is not described here. We demonstrate our developments by a kiosk system for a multimedia calendar of events.

Related work considers the storage of multimedia documents utilizing the DFR standard [18], but no specific multimedia transport protocol is supported. The BERKOM teleservice for multimedia mail (MMM) specifies access to a global store for multimedia parts of a mail document [10]. The MMA teleservice also includes access to multimedia parts of documents but additionally enables searching and storing of complete multimedia documents. We use the content location model of MMM and adopted it to the DFR standard. Thus, compatibility between these BERKOM teleservices is achieved. Another development of an MMA system uses the MMM teleservice for access to the archive [16]. Specific results on file systems for audio and video are reported in [9, 17]. These can serve as an enabling technology for storing and accessing multimedia parts in MMA. Also developments of high speed network protocols can be utilized (e.g. HEITS [4], BERKOM MMT [12]).

In *Section 2* we introduce the DFR information model. In *Section 3* the alternatives of handling multimedia documents within DFR are described. In *Section 4* the architecture for the multimedia archiving system is sketched. It incorporates an object-oriented database management system with multimedia extensions. In *Section 5* the application Calendar of Events is described. In *Section 6* we evaluate our approach and describe further developments.

2 The DFR Information Model

According to the information model of DFR all objects are arranged in a tree structured hierarchy [5]. The root of this tree is called *DFR-Root-Group*. This object exists exactly once in every *DFR-Archive* (i.e. the document pool). The inner nodes of the tree are build up by *DFR-Groups*. Every object belongs to exactly one DFR–Group (except the DFR-Root-Group). Information itself is stored in *DFR-Documents*. They contain uninterpreted bulks of data. Hence, storage of documents of different exchange or structuring formats is facilitated in one DFR-Archive. The content of DFR-Documents is described by a predefined but extensible set of *DFR-Attributes*. A DFR-Object can indirectly be contained in more than one group by using a *DFR-Reference*. Another purpose of a reference is the possibility to represent a *Distinguished Object Reference* (DOR) [6]. As an example a DOR could be used to store the description of a large multimedia data object in the archive, the data itself is handled in a remote store.

The DFR standard defines operations to *create* and *delete* DFR-Objects, to *list* members of DFR-Groups, to *search* objects via attributes, and to *modify* attributes and the content of a DFR-Object. The result of a search operation is stored in a specific DFR-Object type called *DFR-Search-Result-List*. DFR operations invoked after a search operation can refer to an explicitly stored result. The framework to define search conditions is the concept of filters in the X.500 standard. An example of a DFR-Archive is illustrated in Figure 1.

3 Multimedia Extensions of DFR

Due to the distinction of attributes describing the contents of DFR-Documents and the content data itself, the DFR-Standard can in principle be used for handling multimedia documents as well [18]. There are two problem domains that require a better support of multimedia data: transformation of complex structures in multimedia documents to the information model of DFR and requirements for the continuous presentation of multimedia data.

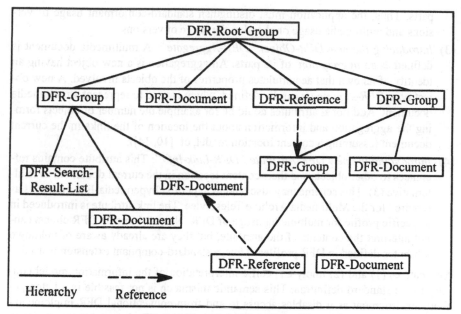

Fig. 1: A Sample DFR-Archive [5].

3.1 Introducing Multimedia Documents into DFR

A multimedia document is a collection of non-continuous and continuous multimedia document parts. In the DFR standard a document is treated as a single entity that is not typed. No content-specific structuring is available within the DFR access. To facilitate real-time dependent transmission of different multimedia data types both in a communication network as well as between the DFR-Server and secondary storage device the following requirements for a multimedia document can be formulated:

- it should be possible to read different parts of a document separately and
- every multimedia part should have a type information of its own.

The latter can be easily achieved by using the value of the DFR-Attribute "DFR-Document-Type" (contained in the DFR basic attribute set) in an appropriate way. To fulfill the first requirement a structuring mechanism is needed. The following alternatives can be distinguished:

(1) *Using DFR-Groups:* For every multimedia document a DFR-Group is introduced, grouping together all parts with a type identification of its own. This implies that a leaf group in the hierarchy would necessarily represent a multimedia document. The disadvantage of this approach is that the semantics of groups within the classification hierarchy of DFR is changed. Thus, navigation in the archive structure will become more difficult for an application.

(2) *Using the versioning mechanism of DFR:* This alternative is similar to the one mentioned above. The version graph of a DFR-Document contains all the multimedia

parts. Thus, the application must distinguish standard-conformant usage of versions and multimedia usage or abandon the usage of versions.

(3) *Introducing the new DFR-Object "DFR-Aggregate":* A multimedia document is defined as an *aggregation* of its parts. An aggregation is a new object having an identity of its own that accumulates properties of the objects involved. A new object type *DFR–Aggregate* can be defined, which is used to represent a multimedia document. Additional attributes could be for example the number of objects forming the aggregation and information about the location of the links in the current document (assuming a content location model, cf. [10, 14]).

(4) *Introducing the new DFR-Attribute "DFR-Link-Info":* This attribute contains references to other objects and their content location in the current document as in alternative (3). This scheme may also lead to hypertext/hypermedia-like messages as required for the Multimedia Archive Teleservice. The link attribute is introduced in a specific profile for multimedia usage of DFR. Non-multimedia DFR clients cannot interpret the contents of the attribute, but they are already aware of unknown attributes defined in DFR profiles. Thus, a standard-compliant extension is used.

Alternatives (1) and (2) imply a specific interpretation of the information model contrary to the standard definition. This semantic mismatch is not feasible in a heterogeneous environment as it disables access to and from conventional DFR implementations. From the modelling perspective, alternative (3) is most elegant, but it requires an enhancement of the DFR standard. Because this would be a long-term activity we choose alternative (4). It should be mentioned that one part of a single multimedia document can consist of multiple media data (e.g. audio and video combined). Intra-media synchronization is not covered by introducing the new DFR-Attribute DFR-Link-Info.

The syntax of the DFR-Link-Info attribute is given in Figure 2 (in ASN.1 [1]). The "dor" definition contains the reference to the linked object. This parameter has the type DOR (see Section 2). We do not use identifiers local to the DFR archive. Instead storing of multimedia parts outside the DFR-Archive is supported. The "link" definition contains the position where the referenced object are anchored in the document. This parameter is optional because it is currently specified for textual data only (cf. [10]). The "media-type" and the "format" parameter specify some attributes which describes the media. The "duration" parameter specifies the duration of the referenced object (given in seconds). This is only valid if the media has a duration in time. The other parameters are quite obvious.

3.2 The Dual Stack Approach

The multimedia parts of a document must be transferred from the server to the client and presented continuously. The DFR standard requires the usage of the OSI communication stack. Unfortunately, there is no concept like "STREAM" within the OSI reference model, i.e. the data has to be transferred in an all-at-once fashion. The "Remote-Operations-Service-Element (ROSE)" of the OSI stack handles an operation as one indivisible action, consisting of an operation identifier, arguments, and a result. Due to the possibly large size of multimedia data there will be a big delay between the request for a presentation and its beginning when using pure OSI. Sometimes, if after a few seconds the user decides not to view or to hear the whole multimedia data, most data have been transfered unnecessarily. In addition, the local capabilities of the client may not be able

```
                DFR–Link–Info ::= SEQUENCE OF SEQUENCE {
                dor          [0] DOR,
                link         [1] IA5STRING OPTIONAL,
                media–type   [2] MediaType,
                format       [3] Format,
                duration     [4] INTEGER OPTIONAL }

                MediaType ::= SEQUENCE {
                cm–type      [0] ContMedia–Type,
                audio–qos    [1] Audio–QoS OPTIONAL,
                                 — mandatory if cm–type = audio or av
                video–qos    [2] Video–QoS OPTIONAL }
                                 — mandatory if cm–type = video or av


                ContMedia–Type ::= ENUMERATED {
                audio (0), video (1), av (2), animation (3) }

                Audio–QoS ::= SEQUENCE {
                sampling–rate      INTEGER,
                bits–per–sample    INTEGER,
                nr–of–channels     INTEGER}

                Video–QoS ::= SEQUENCE {
                pic–per–sec        INTEGER,
                height             INTEGER,
                width              INTEGER,
                bits–per–pic       INTEGER,
                nr–of–colours      INTEGER,
                colored            BOOLEAN }

                Format ::= ENUMERATED {
                — other formats, also private ones, are possible
                mpeg (0), m–jpeg (1), pcm (2), adpcm (3), dvi (100) }
```

Figure 2: ASN1 Syntax of Profile Attribute DFR-Link-Info.

to completely store the multimedia data requested. The solution to these problems is to use communication stacks providing the transfer of multimedia data in a part-by-part fashion. Hence, applications are able to begin the presentation of multimedia data earlier. For example, after the first parts of the multimedia data are transferred, a forward presentation can be started.

One way of using such protocols is by implementing the upper OSI layers on top of them, but this will lead to a new communication stack. It will no longer be possible to use the standardized protocol of DFR to communicate with other DFR implementations which are using the OSI stack. So we have decided to follow another approach. The OSI stack is used in combination with a transport protocol for audio and video data called *AV-Protocol*. Additionally, to abstract from the used AV-protocol we have introduced a new protocol called Multimedia Control Protocol (MCP). The Multimedia Control Protocol is independent from DFR but uses the same OSI stack within another context. MCP provides functions like "Open", "Play", "Stop" etc. to initiate, start and terminate continuous data streams. There are some other operations to pause, replay, rewind etc. the object.

If the parts of the multimedia data arrive at a constant rate, i.e. the time between the single parts is constant, then this communication is called *isochronous*. Such protocols guarantee delivery of data in time. This functionality will be provided by advanced network and transport protocols (e.g. HEITS [4] or MMT [12]). The transfer of continuous data streams is done by isochronous multimedia transport protocols or in-use protocols like TCP. The latter is used in our development until a multimedia high-speed transport system is available.

3.3 How the Multimedia Control Protocol Works

Before MCP can be used the DFR-Client must have received a DOR in a "DFR-Link-Info" attribute (see subsection 3.1). The DOR contains the address of the MM-Server, to which the connection must be established, and the local identifier for the MM-Server to find the multimedia object (besides other information). When a multimedia data stream has to be presented a connection from the MM-Client to the MM-Server is established (bind) and an open request for the specified object is called by the MM-Client and sent to the MM-Server to initialize the data transfer. After successful initialisation one is in the position to start the presentation. To this end, the "Play" operation of MCP is called. An appropriate multimedia presenter is invoked, an AV-connection to the MM-Server is established, and the presentation starts. The presentation is stopped with a "Close" operation and the AV-connection is disconnected from the MM-Server. In our implementation MCP is an application on top of the OSI stack using the ACSE and ROSE service elements from layer 7. This is the same as DFR so both protocols can be administered by one server with different contexts.

Our AV-Protocol using TCP proceeds as follows. Depending on information contained in the DOR the MM-Client opens a connection to the MM-Server with suitable Quality of Service (QoS) parameters. After the connection is established, the MM-Client starts fetch-operations to get multimedia data packets of a specified length. The presentation tool presents the data fetched and initiates another fetch request by the MM-Client and so on. Afterwards the connection is closed. Note, that if an isochronous transport protocol is used the fetch-operation does not initiate sending a request over the communication network. It is the responsibility of the isochronous transport protocol to provide the data packets on the client's side just in time. If an asynchronous protocol is used, the client is responsible for requesting new packets in time and for caching data packets.

4 Architecture of the Archive System

In a DFR environment several *DFR-Clients* can be connected to several *DFR-Servers* via a *Communication Network*. An authorization mechanism identifies *DFR-Users* which are brought into the system by an *Administrator*. Figure 3 shows the architecture

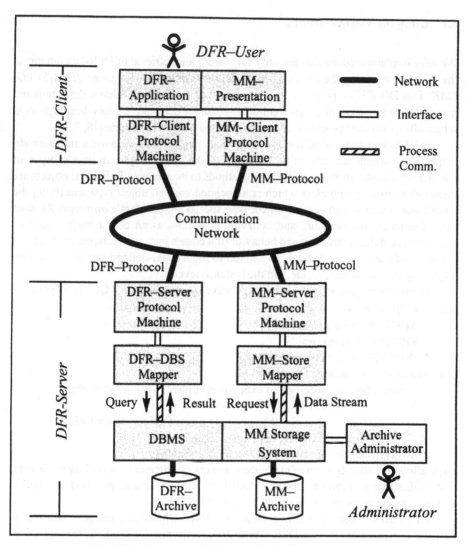

Fig. 3: Architecture of the Multimedia Archiving System.

using the dual stack approach in combination with the DFR environment. The *DFR-Application* is responsible for handling DFR-Documents and DFR-Groups. For multi-media documents the contents of DFR-Link-Info attributes are interpreted. Presentation is requested using a DOR by a media-type-dependent *MM-Presentation* tool. The *protocol machines* implement the DFR and MM protocol on the client's side and the server's side, respectively. *Mappers* separate communication and storing functionality to allow independent developments. The corresponding modules are connected via a process communication mechanism.

4.1 Using the VODAK DBMS

We have implemented the functionality for storing and retrieval of DFR-Documents at the DFR-Server using the object-oriented *database management system (DBMS) VO-DAK*. This DBMS is a proprietary development of GMD-IPSI. It has a definition and manipulation language of its own called *VML* and a descriptive query language *VQL* which allows for the invocation of object-oriented methods in queries [8, 7].

In VML structure and behavior of objects are described by *properties* and *methods*, respectively, which are defined in *object types*. An object type can be a subtype of another which *inherits* its properties and methods to its *subtypes*. Individual objects are instances of exactly one *class* which is associated with an object type specifying the class's *instance type* and another object type specifying the class's *own type*. An own type determines the structure and behavior of a class as an object itself, while an instance type defines structure and behavior of a class's instances. Classes itself can be instances of *metaclasses* that can be applied to model user-defined *semantic relationships* (e.g. part-of, role-of, ... etc.) on the instance level.

Query statements written in the declarative query language VQL. The syntax of VQL is SQL like:

ACCESS <access clause>
FROM < from clause>
WHERE <where clause>

The clauses are specidied as follows:

<access clause>   specifies an expression applied to the selected objects of the query,

<from clause>   specifies the classes or a set of objects against which the query is posed,

<where clause> specifies the selection conditions.

VQL allows the specification of method calls in query statements. It is a helpful feature, that VQL statements can be directly included in VML, e.g. in a method body defined in an owntype querying the corresponding class's instances.

Every DFR-Objecttype corresponds to a VML-Objecttype. Integrity constraints like the correspondence of a DFR-Object to exactly one DFR-Group (except DFR-Root-Group) is encapsulated in methods of these types. A Metaclass is used to encapsulate the semantics of the hierarchy relationship at one place in the database schema. Encapsulation allows precomputing of return values thus improving performance. For example, if an archive is mostly used for reading of documents (e.g. as the article pool of an electronic newspaper [15]) the value of the *DFR-Attribute* "Pathname" containing the identifiers of all DFR-Objects from the object up to the root can be computed at insertion time. In case an archive is modified rather frequently this value may be computed on request. The usage of VQL allows declarative querying on attributes of *DFR-Documents*. The usage of method calls inside queries eases the handling of complex operations. For example, searching recursively for all *DFR-Groups* starting at a specific *DFR-Group* down to a user-defined number of levels can be executed by a single method call.

4.2 Storage of Multimedia Data

Multimedia parts of the documents are stored in a specific *MM Storage System* using the AT&T Bell Labs *EOS* prototype of a storage manager [2]. Our implementation integrates the VODAK DBMS and the MM Storage system [13]. Hence, multimedia documents are stored at one node and can be manipulated as a whole (e.g. creation, deletion, moving to another DFR-Group). Further developments focus on the integration of the multimedia sorage system into the VODAK DBMS and on support of user interaction like interrupting a video presentation or changing its speed.

5 A Multimedia Calendar of Events

The Calendar of Events (*CoE*) is a sample application of the MMA teleservice. Event descriptions consist of texts, images, graphics, auditions, and video clips. Examples of these parts are a picture showing a theatre stage, a video clip about the announced top actor, digitized newspaper critics, and an audio sequence presenting the latest song of the performing interpret. The descriptions can be inserted into an archive by information providers. Customers can query descriptions according to their interests and can present them at their workstations. It is planned to enable customer feedback like ticket–orders or reservations, reports or critics per voice annotations.

Events of the CoE can be classified into different regions to simplify the user's orientation when retrieving a large pool of events. The hierarchical structure of the DFR information model allows a direct mapping of a specific region to a respective DFR-Group. For example, the DFR-Root contains the countries of *Germany*. These DFR-Groups contain cities as other DFR-Groups, e.g. *Hesse* contains *Frankfurt, Darmstadt* and *Kassel*. The meaning of a DFR-Reference in CoE is the consideration of regions which cannot be classified to only one subordinated region. For example, the DFR-Group *Main-Neckar-Area* under the DFR-Root-Group contains DFR-References to the DFR-Groups *Darmstadt* and *Heidelberg*. DFR-Documents contain the event descriptions. The content of a DFR-Link-Info attribute points to multimedia data in or outside the DFR-Archive related to the event. Multimedia data inside the DFR-Archive is stored in specific DFR-Groups related to the type of information (especially images and short audio annotations).

The standardized DFR-Attributes such as creation date and position in the hierarchy are related to documents independent of their type and purpose. For the CoE application we need some additional information. An event is decribed by the attributes event organizer (which may be different from the creator of the corresponding DFR-Document), an event category as concert, talk, exhibition, etc., a location of the event, start and end time, the entrance fee and some keywords describing the event. In addition to the multimedia profile we have defined a specific profile for the Calendar of Event application. These attributes are instantiated by the information providers and can be queried by the users of the CoE application. For example, a customer can query all concerts in *Darmstadt* taking place in the next week. Additionally, he or she can specify an upper limit for the entrance fee in the query.

6 Conclusions and Further Development

We have described the development of a multimedia archiving teleservice utilizing the DFR standard. The DFR standard is suitable for open access to documents independently of a specific document format. Additionally, the pool of documents can be structured in a hierarchy and become a hyperlinked network by the use of the DFR referencing mechanism. Profiles can be defined for application specific attributes like in the CoE application.

The need of storing multimedia data within documents require some additions both on the modelling level and wrt. to the architecture of the DFR environment. Our solution allows to store multimedia data in the DFR archive as well as in external stores which may use specific storage devices or support real-time access. The linking attribute allows reasonable modelling of multimedia documents. The dual stack approach enables the integration of specific AV transportation protocols. Thus, DFR can be used for new generation multimedia teleservices. Enhancements of the DFR standard for integrated support of multimedia application are introducing a DFR-Aggregate for referencing locally in the archive stored multimedia data and inclusion of MCP to control several types of AV-Protocols.

The archive system is implemented across different platforms (IBM RS/6000$^{®}$ and SUN SPARC$^{®}$ workstations) and the multimedia Calendar of Events is implemented as a sample application. A version of the prototype system has been presented at the CeBIT fair in March '94. In the second project phase, that is currently going on, the integration of a multimedia high speed transport system and support for efficient access to continuous objects in the VODAK DBMS are under development. Additionally, access to multimedia data in the presence of multiple clients will be facilitated.

Acknowledgement

The participation of *Johannes Rückert* from IBM-ENC in starting this project is gratefully acknowledged. The *EOS* team at AT&T Bell Labs, especially *Alex Biliris*, has been very helpful by supporting us.

References

1. CCITT Recommendation X.208 – Specification of Abstract Syntax Notation One (ASN.1), 1988.
2. A. Biliris: The Performance of Three Database Storage Structures for Managing Large Objects. Proc. ACM SIGMOD Conf., p. 276–285, 1992.
3. K. Böhm, K. Aberer: An Object-Oriented Database Application for HyTime Document Storage. Arbeitspapiere der GMD, Nr. 846, St. Augustin, June 1994.
4. L. Delgrossi, et. al.: Media Scaling for Audiovisual Communication for the Heidelberg Transport System. Proc. ACM Multimedia Conf., 1993.
5. ISO/IEC 10166: Information technology – Text and office systems – Document Filing and Retrieval (DFR) – Part 1 and Part 2, First edition 12/15/91, 1991.
6. ISO/IEC JTC1/SC18/WG4: Information technology – Text and office systems – Distributed Office Applications Model (DOAM), IS 10031, 1991.
7. G. Fischer: Updates in Object-Oriented Database Systems by Method Calls Queries. Proc. of 3rd ERCIM Database Research Group Workshop, Pisa, Sept. 1992.
8. W. Klas, K. Aberer, E.J. Neuhold: Object-Oriented Modeling for Hypermedia Systems using the VODAK Modeling Language (VML). Object–Oriented Database Management Systems, NATO ASI Series, Springer, Berlin, 1993.
9. T.D.C. Little, et al.: A Digital On-Demand Video Service Supporting Content-Based Queries. Proc. ACM Multimedia '93, Anaheim, Aug. 1993.

10. E. Möller, L. Neumann, G. Schürmann, S. Thomas, R. Weber, F. Wolf: The BER-KOM Multimedia-Mail Teleservice. Draft paper, January 1994, accepted for publication in Computer Communications, Butterworth.

11. Final report of the BERKOM Working Group "Multimedia Archives", forthcoming.

12. The BERKOM Multimedia Transport System, Rel. 2.2. DeTeBerkom, Berlin, July 1993.

13. F. Moser, T.C. Rakow: Database Support for the Access towards an Open and Multimedia Archive (in German). GI–FG Databases, Fall Workshop, Jena, Sept. 1993.

14. S.R. Newcomb, N.A. Kipp, V.T. Newcomb: The "HyTime" Hypermedia/Time–based Document Structuring Languag. Communications of the ACM, Nov. 1991, Vol. 34, No. 11.

15. W. Putz, E.J. Neuhold: is-News: a Multimedia Information System. Data Engineering Bulletin, Vol. 14, No. 3, Sept. 1991.

16. H. Thimm, T.C. Rakow: A DBMS-Based Multimedia Archiving Teleservice Incorparating Mail. Proc. Int. Conf. on Applications of Databases, Linköping, Sweden, June 1994.

17. P.V. Rangan, et al.: Techniques for Efficient Storage of Digital Video and Audio" Proc. of Multimedia Information Systems, Tempe, Arizona, p.68–85, 1992.

18. J. Rückert, B. Paul: Integrating Multimedia into the Distributed Office Application Environment. Datenbanksysteme in Büro, Technik und Wissenschaft, GI-Fachtagung Braunschweig, Springer-Verlag, 1993.

MOSS as a Multimedia-Object Server

Rolf Käckenhoff, Detlef Merten

Chair for Database Systems, University of Erlangen-Nürnberg,
Martensstraße 3, 91058 Erlangen, Germany
{kaeckenhoff, merten}@informatik.uni-erlangen.de

Klaus Meyer-Wegener

Chair for Database Systems, Institute for Software-Engineering II,
Faculty of Computer Science, Technical University of Dresden, 01062 Dresden, Germany
kmw@freia.inf.tu-dresden.de

Abstract. Multimedia requires new techniques for storage and retrieval of data objects. Based on the concept of abstract data types MOSS is presented as a multimedia-object storage system to experiment with new methods of managing so-called single-medium objects (SMOs; pertaining to one medium only). SMO-sets are introduced as a means to group SMOs and to connect them with the data in a DBS. They also form the basis for content-based retrieval. The open, layered architecture of MOSS allows the separation of all media-related functionality into modules for medium-specific abstract datatypes. This is especially aimed at the design of new access methods that represent some of the semantic richness of SMOs and the employment of novel storage techniques to meet the requirements of the new media. Another focus of our work is the design of general interfaces between the modules.

1   Storage and Retrieval of Multimedia Objects

Today, most multimedia (MM) applications store their data objects in a proprietary way, usually in files on some storage device in some format that is exactly tailored to the necessities of this one application. But MM data objects (MMOs) are expected to exist for a long time and to be accessed by many different applications. Additionally, the employment of new opportunities offered by the technical progress (new media, devices etc.) must be possible. The demands for *application neutrality* (application-independent data management) and *data independence* (abstraction from different storage formats, compression techniques and storage devices) are obvious. *Database management systems* (DBMSs) have been identified as the ideal means to guarantee these issues. The integration of MMOs into these systems leads to multimedia DBMSs (MMDBS) [Meye91].

The design of a MMDBS can be based mainly on two kinds of DBMSs: extensible relational systems of different flavors [Bato90, Care90, Stone87] and object-oriented DBMS [Kim90]. All existing systems, even if they are designed for complex objects, use only conventional datatypes or bit fields with no semantics defined on them in form of special operations [Käck93]. The storage and retrieval of MMOs demand for flexible mechanisms to incorporate new storage techniques and devices which are only partially available in these systems. Other important issues are the development of novel, medium-specific *access paths* for MMOs that reflect the semantically rich contents of MMOs. The design of generalized interfaces for the handling of MMOs is another challenge.

The multimedia-object storage system (MOSS) designed at the University of Erlangen is a standalone system that allows to focus on questions related to the storage and retrieval of media data. MOSS has a highly modular structure, which is ideal to compare different solutions for the issues mentioned above. It is integrated into a client-server architecture.

2 Access to Multimedia Objects in MOSS

2.1 Handling of Media Objects

MMOs consist of identifiable parts that show different characteristics and functionality depending on the kind of medium they belong to, the so-called *single-medium data objects* (SMOs). MOSS has been developed to store these SMOs.

Medium-related abstract datatypes (MADTs) like raster image, graphic, sound or video provide a solid foundation to express medium-specific behavior and to hide implementation details. MADTs offer operations to create, delete, present, evaluate, and search for SMOs. Manipulation on the contrary is seen as a matter of special editing tools, which usually are well-established and should not be reimplemented. In MOSS all MADT-operations provide format and storage device independent access to the media objects: in order to access SMOs one only has to specify a certain SMO and the format wanted.

Generally, SMOs consist of different kinds of data [Meye91]. *Raw data* like the pixels of a raster image represent their information content. They are the unformatted data part of an SMO. *Registration data* are necessary for the interpretation and identification of the raw data, so a major part of them are format information, otherwise usually concealed in file headers. *Description data* are a supplement to make the use of SMOs easier and especially faster. They include information about content and structure of the SMO that is generated in advance to decrease the effort of processing them. Otherwise, they would have to be obtained by complex and time consuming analyses which should be performed as seldom as possible.

As there are also formatted application data to be managed, these have to be stored in a separate system, e.g. a relational DBS. The connections between the data in the different systems are achieved by a unique identifier for each SMO (SMO-id). These are returned by the operations for creation and search of SMOs. Only the SMO-ids are stored in the records of the DBS and are used as a handle for the SMOs that logically belong to them. So, if you want to see for example the X-ray photo of a certain person in a medical archive, you have to take the identifier of the photo from the person's data record in the DBS and hand it over to MOSS with the request to display the picture belonging to this identifier. Figure 1 illustrates this procedure. Of course, this implies some kind of interoperation between MOSS and the DBMS. The transaction management of the DBMS has to be expanded to MOSS, using the two phase commit protocol (2PC) for instance.

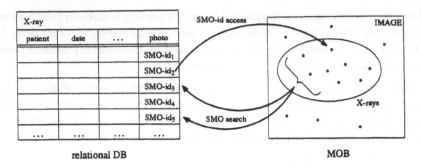

Figure 1: Cooperation of MOSS and a relational DBS

Media-object bases (MOBs), comparable to databases, form the top level unit of access. A MOB should contain all SMOs belonging to an application field, regardless of their MADT. An SMO can exist in a single MOB only, and an application can only work with one MOB at a time. So it is necessary to switch between MOBs, whenever access to another one is required.

2.2 Content-based Retrieval

Like database systems, MOSS also offers mechanisms for searching. Hence, some MADT operations refer to a very powerful concept in MOSS: the content-based retrieval.

Content-based search conditions that are really discriminating have to be expressed in a meta-medium. There must be means to formulate abstract queries like that for a picture with *any* old man with *any* green hat on, but not just with Mr. X with a particular green hat. The only medium to express concrete as well as abstract conditions is language. So the SMOs are extended by a description representing their content, which has to be specified by the user in (restricted) English language for convenience.

To improve the performance of the search both the content-based descriptions of the SMOs and the query provided in natural language are transformed into a predicate-like internal representation with the help of dictionaries. These dictionaries contain all words to be accepted in a description, their grammatical category and instructions for generating the predicate expressions representing them [Rowe91]. Thus, a dictionary covers all domain-dependent knowledge and has to be maintained quite carefully. Taking the example of figure 2, all three objects will meet the query "plants", due to the domain "landscapes" described in the corresponding dictionary.

2.3 Assembling Media Objects into Sets

Often it is necessary to group SMOs which have some characteristics in common or which are related in a way unknown to MOSS. Therefore, the concept of *sets* as collections of SMOs is introduced. One main issue of sets is to limit content-based search: only those SMOs of a specified set (i.e. reflecting the values of a relational attribute) will be examined with regard to a query.

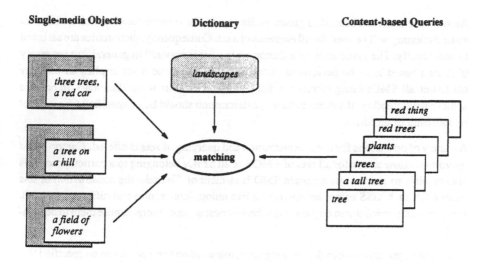

Figure 2: Content-based retrieval in MOSS

On its creation, each SMO automatically becomes member of the appropriate MADT-spe-cific *built-in set*, that is under system control. Additionally, it is possible to arbitrarily place SMOs into *application-defined sets*, that are fully under control of the user. The name of built-in sets is identical with that of their corresponding MADT, the name of application-de-fined sets is free of choice. The only restriction is, that set names have to be unique within their MOB. For performance reasons the set-id is used as an argument for most operations.

These sets can be created and destroyed whenever needed, and SMOs can be inserted as ele-ments and removed at will. It is possible to place an SMO into several sets. Thus, an applica-tion is free in organizing its SMOs in any manner. Figure 3 gives an example of some built-in sets (names in capital letters) and some application-defined ones (names in lower case).

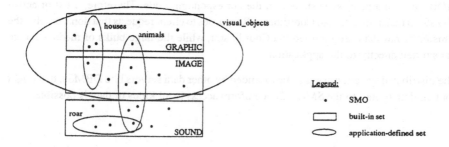

Figure 3: Application-defined and built-in sets

As sets have been introduced to group media objects for content-based search mainly, the same dictionary will be used for all elements of a set. Consequently, dictionaries are assigned to sets directly. The association of a dictionary to a set is optional in general, but necessary if content-based is to be performed. Attaching a dictionary to a set implies an integrity constraint: all SMOs being member of that set have to conform to the dictionary. In order to achieve predictability of system behavior, a description should be be specified for every set membership of an SMO.

A variety of operations for the *administration* and querying of sets is offered. *Cursor-based operations* allow to ask for all sets of a MOB or all SMOs belonging to a particular set. It is also possible to get the sets a certain SMO is element of. To make the construction of sets more efficient MOSS offers *set operations* like union, intersection and difference as well. Here again, attention has to be paid to the dictionaries: set operations require them to be identical.

In addition, applications can define *integrity constraints* on two sets: it can be specified that sets should always be disjoint or that one set should be a subset of the other. If an integrity constraint does not hold when it is defined, it will be rejected. At any time when an SMO is added to a set, MOSS will check all corresponding constraints. So integrity constraints are a simple, but flexible and efficient means to keep sets consistent. Obviously, the use of constraints on build-in sets is restricted.

2.4   Integration of MOSS into a Client-Server Architecture

Today's applications usually run in a networking environment, where functionality and data are distributed across several nodes. Therefore, MOSS is designed as a server for a MOB that treats requests originating from multiple clients. The distributed system is managed by a central component, the Coordinator, that may be located on a node separated from any MOSS. It serves as distributor for requests and responses, and maps names of MOBs to their nodes, so that location transparency is accomplished. Moreover, it handles access rights to MOBs. The overall architecture is sketched in figure 4.

The clients address themselves to the Coordinator specifying the name of the desired MOB. After checking the access rights, the request is directed to the MOSS already serving this MOB, or such a process is started on the corresponding node. Therefore, a list of active MOSSs is maintained by the Coordinator. Responses to client requests that don't involve the transfer of raw data are returned via Coordinator, while the large bulks of media data are transmitted directly to the application.

The distributed system can easily be extended by other data servers like DBMSs, that might contain data related to the SMOs. So a uniform access to all kinds of data is possible.

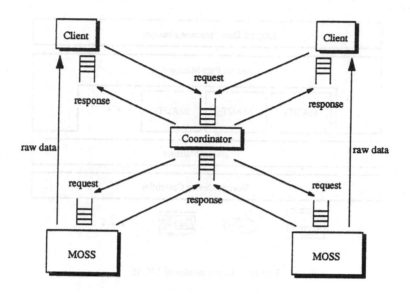

Figure 4: MOSS within a client-server architecture

3 MOSS Architecture and Implementation

3.1 General Overview

In MOSS, all MADT-specific operations are gathered in the respective MADT module, one for IMAGE, one for VIDEO etc. (see figure 5). All the other layers are MADT-independent. They form the MOSS Frame, which consists of the Logical Data Structure and the Access Path Manager at the top, and the Buffer Manager and the Storage Device Controller at the bottom.

The modularized structure makes MOSS an ideal platform for research: New MADTs can be plugged into the MOSS Frame with low effort. Existing implementations can be exchanged by new ones without any changes to the application programs or to MOSS, as long as their interfaces remain the same.

3.2 The MOSS Frame

The Logical Data Structure Manager at the top of MOSS offers MOBs, sets, integrity constraints and SMOs to the applications. The employment of these entities and their connections at application level are described in chapter 2 already. An overview of the logical objects representing them and their relations is given in figure 6.

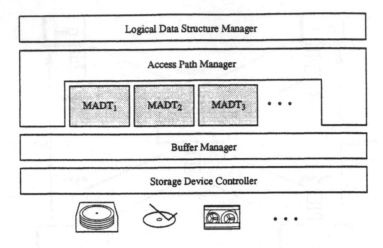

Figure 5: Layer model of MOSS

The Logical Data Structure Manager controls the necessary steps for storage and retrieval, and selects the corresponding physical access paths. For each new MOB this component initiates the creation of its physical data structures. It also checks, whether the SMOs and sets involved in an operation do exist. If not, the intended operation will be rejected here already. As some logical objects, like sets and SMOs, can be found in several physical data structures, their consistency has to be maintained. Another task is to keep track of the various cursors mentioned above.

The Access Path Manager stores a record for each MOB (entity "MOB" in figure 6) containing its name, the highest set-id and SMO-id delivered, and the directory for its secondary data. These internal data structures of a MOB concern all its sets, integrity constraints and SMOs. Records containing set-id, set name and dictionary (entity "Set" in figure 6) are organized in hash tables (relationship "contains_set"). Just in the same way the SMO records (entity "SMO") are stored (relationship "contains_SMO"). These are composed of SMO-id, MADT, and their physical location as an uninterpreted byte string, which has to be delivered to the MADT-specific module. The hash keys are set-ids and SMO-ids, respectively.

To maintain the connection between an SMO and all its sets (entity "Set Membership") a record for each SMO is provided in a hash table, and for the other direction, there exists a bit list for each set: setting bit i to 1 means that SMO with SMO-id i is element of the corresponding set. Thus set operations like union, difference and intersection can be reduced to cheap bit operations, and the storage amount is expected to be much less than it were, if the SMO-ids would be stored explicitly (assuming that a set contains more than 3.2% of all SMOs of its MOB, and that a SMO-id has a length of 32 bit; compressing the bit lists would yield an even lower threshold). The maintenance of data structures for both directions of the SMO set membership causes some storage overhead, but this is of little effect compared to the gain in performance. All these secondary data mentioned so far, beside others, are stored via the Buffer Manager.

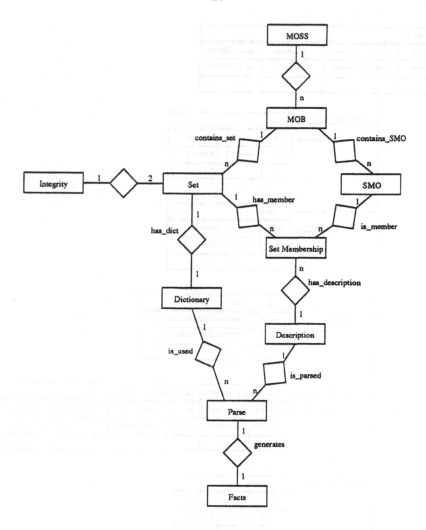

Figure 6: Entity Relationship Diagram of the logical data structures in MOSS

Other MADT-independent aspects concern set integrity constraints and the internal representations of the content descriptions, which therefore are also managed by the MOSS frame.
Assuming that there won't be many integrity constraints within a MOB, these are stored in
a sequential file (entity "Integrity"). The description module within the Access Path Manager
parses the natural language description of an SMO – and the content-based queries as well –
by applying the dictionary assigned to the set involved. It generates the internal content
representation in a predicate-like syntax (entity "Facts" in figure 6). These are collected into
a separate file for each set to accelerate content-based retrieval. Thus the matching process
needs not bother about the membership of SMOs to the relevant set, as it would be necessary
when collecting internal representations for each dictionary separately. A general survey of
the physical data structures is given in figure 7.

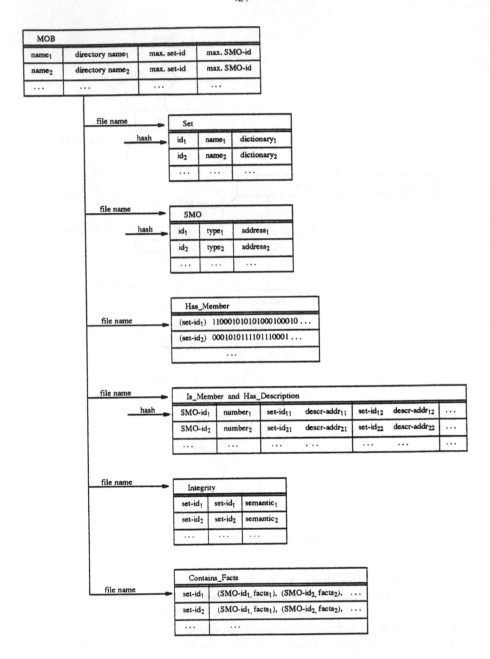

Figure 7: Physical data structures in MOSS

In order to answer performance needs, some aspects of the Buffer Manager have already been investigated. Because of the large bulks of SMO data and their time constraints the number of internal copy operations has to be kept at a minimum. In contrast to conventional DBSs, data locality cannot be expected. The same piece of SMO data will rather seldom be requested in the same format and quality within a time interval short enough to keep the data in main memory. Therefore, the main task of the Buffer Manager is to support time-critical I/O. A pair of buffers is used to pipeline reading and writing raw data from external storage devices, (de-)compression and conversion etc., and the access by a MADT module. These processes can be scheduled by deadlines depending on time constraints.

The Storage Device Controller has hitherto been covered by picking up existing operation system functionality and is therefore not discussed here any further.

3.3 The MADT Modules

The MADT module concept allows all medium-dependent requirements of SMOs to be treated in an appropriate way. In detail, such a module comprehends not only the MADT operations offered to the applications, but also some MADT-specific functions for internal use. These concern formats, compressions and the interpretation of the address information of its SMOs, provided by the Access Path Manager above. Raw, registration, and description data are written to persistent storage via the Buffer Manager in a manner decided within the MADT module. As the MADT operations cope with only one SMO at a time, the description data handled on this level cannot be involved in the content-based search within a set mentioned above. Though, they can be used inside the MADT modules for the evaluation of the SMO content. Typically, these data comprise information about the logical structure of the content.

3.4 Implementation of MOSS

MOSS is written in C and runs under VAX/VMS. The currently available MADTs are TEXT (line-oriented), IMAGE, AUDIO and VIDEO. Due to our system environment, video objects are stored by a RS232-controlled video recorder. Nevertheless, they are presented in a window on the workstation monitor, that is manageable in an X-Window-like fashion. MOSS is being ported to DEC-Alphas under OSF.

4 An Example Using MOSS

4.1 The Data Schema

To illustrate the functionality and features of MOSS a simplified example in the domain of journalism is presented. It deals with persons, sites and events, that are represented in a conceptual database schema comprising the following four tables:

```
PERSON    (No, Name, Birth_date, Nationality, Affiliation, ...,
           Passport_photo, Media_set)
SITE      (No, Name, Type, Located_near, ..., Media_set)
EVENT     (No, Name, Site, Begin, End, ..., Media_set)
PARTICIPATION    (Event, Person, Role)
```

Each person is described by a tuple containing the primary key No, the name etc., and more interestingly, the two attributes Passport_photo and Media_set of type integer. Passport_photo represents the SMO-id of a photo stored by MOSS, and Media_set the id of a MOSS set containing SMOs of potentially all supported media types that present – at least among many other things – an aspect of the person in question. This flexibility is comparable to an NF^2 extension of the conventional relational data model. Analogously, the attributes Media_set of SITE and of EVENT stand for sets of all the SMOs showing that site or event, respectively. PARTICIPATION reflects which person took part at which event. These four tables are managed by a usual relational DBS, while the sets and SMOs referenced are handled by MOSS.

Besides the obvious (and just presented) SMO sets comprising all media objects connected to a real world entity (relationally modelled as a tuple), there are additional sets thinkable that convey different aspects. Examples are sets containing all events concerning a special topic, like G-7 conferences or sets reflecting media subtypes like passport photos or X-rays.

4.2   Insertion of New Data

Suppose, a new site, the White House for instance, should be registered. The formatted data has to be appended to the table SITE by an appropriate SQL statement. On the multimedia side, the corresponding MOB "Journalism" has to be opened by a function sent to MOSS, if this hasn't been done already, so that the correct MOB session context is chosen. The new set (named "White House") is created by the MOSS operation

```
create_set ("White House").
```

The result value is the new set-id, i.e. –208, which has to be inserted into the SITE tupel for the White House by an SQL statement for later reference. If content-based retrieval is to be performed on this set, a dictionary must be assigned:

```
set_new_dictionary (-208, "Journalism-Dictionary")
```

SMOs showing the White House, for example photos, can be inserted by

```
create_image_from_file ("Tiff", <file name>)
```

where "Tiff" specifies the format of the file, which might have been created by scanning a photograph. The result value of the operation, say 4711, is a new SMO-id, which is inserted automatically by MOSS into the built-in set IMAGE. All memberships to application-defined sets must be stated explicitly by the user. In the example, the photo should at least be inserted into the White House set:

```
insert_set (-208, 4711)
```

Thereby MOSS checks all specified integrity constraints and aborts the operation, if any occurs.

The provision of a content description for the new SMO can be accomplished by

```
new_description(4711,
                "Front side of the White House in spring 1993")
```

4.3   Data Retrieval

An interesting scenario involves content-based retrieval: the search for all images and videos showing Bill Clinton and Helmut Kohl shaking hands in front of the White House for example. The set underlying this query is

White_House_Set ∩ (IMAGE ∪ VIDEO)

which can be obtained by the following steps (due to a simplified interface the scheme is a bit complicated). First, a search set has to be created, as a copy of the set IMAGE (whose id is -1):

```
copy_set ("search", -1)
```

The new set, whose id might be -251, is united with VIDEO (id -3):

```
set_union (-251, -3)
```

Then this set must be intersected with the White House set:

```
set_intersection (-251, -208)
```

Now, the content-based query can be started by

```
search_first (-251, "Bill Clinton and Helmut Kohl shaking hands")
```

which yields the id of the first SMO satisfying that condition, say 400. To present this SMO to the user, its type must be investigated in order to determine the adequate presentation operation

```
get_media_type (400)
```

In the case of an image (return value -1), the SMO can be displayed on screen by

```
display_image (400, <workstation id>).
```

The next SMO qualifying can be obtained with

```
search_next (-251)
```

At the end the search cursor has to be closed.

5   Conclusions and Outlook

MOSS is a prototype system for the storage and retrieval of single-medium objects that focuses on the novel aspects of the management of multimedia data. Of special interest are performance questions and issues related to semantic modelling and access. The modular architecture of MOSS is aimed at the separation of medium-dependent issues from those concerning the general management of data objects. Modules for medium-specific abstract datatypes, that allow an adequate treatment of SMOs, can be easily inserted into the MOSS Frame. Together with the Buffer Manager, these MADT modules guarantee data independence, especially concerning the employment of different storage devices, formats and compression algorithms.

In the future, work on performance issues, which are especially related to time-critical handling of large bulks of data, will be continued. The issue of transaction management, which has been put aside so far due to the assumption, that for MM data read operations are predominant, has to be tackled next. Furthermore, the design and integration of new MADTs is planned to be supported by a generic ADT mechanism. The means to describe relationships between SMOs have to be enhanced, so that the context of an SMO, which is to be represented by its relationships, can influence the way in which operations on it are executed.

References

Ande89   Anderson, D.P., Tzou, S.-Y., Wahbe, R., Govindan, R., and Andrews, M. *"Support for Continuous Media in the DASH System,"* Report No. UCB/CSD 89/537, University of California, Berkeley, Ca., October 1989.

Bato90   Batory, D.S., et al., "GENESIS: An Extensible Database Management System," in: S.B. Zdonik and D. Maier (eds.), *Readings in Object-Oriented Database Systems*, Morgan Kaufmann, San Mateo 1990, pp. 500–518.

Care90   Carey, M.J., et al., "The EXODUS Extensible DBMS Project: An Overview," in: S.B. Zdonik and D. Maier (eds.), *Readings in Object-Oriented Database Systems*, Morgan Kaufmann, San Mateo 1990, pp. 474–499.

Käck93   Käckenhoff, R., Merten, D., Meyer-Wegener, K. "Eine vergleichende Untersuchung der Speicherungsformen für multimediale Datenobjekte" (A Comparison of Storage Techniques for Multimedia Data Objects), in: Stucky, W., Oberweis, A. (eds.), *Datenbanksysteme in Büro, Technik und Wissenschaft*, Proc. of BTW93, Braunschweig, Springer, Berlin 1993, pp. 164–180 (in German).

Kim90   Kim, W., et al., "Architecture of the ORION Next-Generation Database System," *IEEE Transactions on Knowledge and Data Engineering*, Vol. 2, No. 1, March 1990, pp. 109–124.

Lisk74   Liskov, B., Zilles, S., "Programming With Abstract Data Types," in: *ACM SIGPLAN Notices*, April 1974.

Meye91   Meyer-Wegener, K., *Multimedia-Datenbanken (Multimedia Databases)*, Teubner, Stuttgart 1991 (in German).

Rang92   Rangan, P.V., Vin, H.M., and Ramanathan, S. "Designing an On-Demand Multimedia Service," in: *IEEE Communications Magazine*, Vol. 30, No. 7, July 1992, pp. 56–65.

Rowe91   Rowe, N.C., and Guglielmo, E.J. *"Exploiting Captions for Access to Multimedia Databases,"* Report No. NPSCS-91-012, Naval Postgraduate School, Monterey, Ca., April 1991.

Ston87   Stonebraker, M., and Rowe, L.A., "The Design of POSTGRES," in: *ACM SIGMOD Record*, Vol. 15, No. 2, June 1987, pp. 340–355.

Thim93   Thimm, H., and Rakow, Th.C. *"Upgrading Multimedia Data Handling Services of a Database Management System by an Interaction Manager,"* internal report, GMD-IPSI (Integrated Publication and Information Systems Institute), Darmstadt, July 1993.

The Universal Personal Telecommunication Service in a Multi-operator Environment

Marie-Pierre Gervais

Laboratoire PRiSM, Université René Descartes
45, Av. des Etats-Unis, F - 78035 Versailles Cedex

Abstract. The Intelligent Network architecture has been designed in order to facilitate the rapid introduction of services in telecommunications networks. The first implementations are currently realized by the telecommunications operators and new services are going to be developed, such as the Universal Personal Telecommunication service. Nevertheless these developments prove that there are some limits, especially due to the multi-operator context. For that, interworking of services is needed.
We present in this paper some mechanisms to achieve the Universal Personal Telecommunication (UPT) service in a muli-operator environment based on the analysis of the requirements for the UPT operation.

1   Introduction

The Intelligent Network architecture has been designed in order to facilitate the rapid introduction of services in telecommunications networks. The first implementations are currently realized by the telecommunications operators. Although they concern essentially the voice-oriented services, the architectural concept of the Intelligent Network can be applied to any kind of service. New services are going to be developed, such as the Universal Personal Telecommunication service. Nevertheless these developments prove that there are some limits, especially due to the multi-operator context; because the service developed by a telecommunications operator does not run beyond the geographical area covered by the network provided by this operator. For that, interworking of services is needed.
We present in this paper some mechanisms to achieve the Universal Personal Telecommunication (UPT) service in a muli-operator environment. First we recall the Intelligent Network functional architecture. Secondly we describe the UPT service, its architecture and its information model. Thirdly we present the requirements for the UPT operation in a multi-operator environment. Finally we give a scenario that illustrates this operation.

2   The Intelligent Network Functional Architecture

The introduction of the Intelligent Network will be realized in several phases, named Capability Sets. A functional architecture will be defined for each Capability Set. Each architecture will have to be in conformance with the conceptual model defined in [1]. The IN functional architecture for the first Capability Set (IN CS1) has been introduced by the ITU [2]. It is defined with Functional Entities (FEs). The FEs are pieces of software that are always in one single machine. The following Functional Entities are defined:

The *Call Control Agent Function (CCAF)* provides access to the network for users. It defines the interface between the user and the network.

The *Call Control Function (CCF)* provides basic bearer-connection capabilities. It takes care of the call/connection processing and control.

The *Service Switching Function (SSF)* is associated with the CCF and provides the set of capabilities required for the interaction between the Service Control Function and the Call Control Function.

The *Service Control Function (SCF)* contains call-related service logic. The SCF is a function that commands call control capabilities in the processing of IN-provided and/or custom service requests.

The *Service Data Function (SDF)* contains customer and network data for real time access by the SCF in the execution of IN-provided services.

The *Special Resource Function (SRF)* provides the specialized resources required for the execution of IN-provided service (e.g. digit receivers, announcements and conference bridge).

The *Service Creation Environment Function (SCEF)* allows to specify, to test and to introduce services on INs.

The *Service Management Access Function (SMAF)* provides an interface between the SMF and service managers.

The *Service Management Function (SMF)* allows deployment and provision of IN-provided service.

3   The Universal Personal Telecommunication Service

Universal Personal Telecommunication (UPT) enables access to telecommunications services while allowing personal mobility. It enables each UPT user to participate in a user-defined set of subscribed services and to initiate and receive calls on the basis of a unique, personal, network-independent UPT number across multiple networks at any terminal, fixed, (movable) or mobile, irrespective of geographic location, limited only by terminal and network capabilities and restrictions imposed by the network provider [3].

In order to offer users the capability of establishing and receiving calls on any terminal and at any location, the identification of UPT users is treated separately from the addressing of terminals and network access points. UPT user identification is achieved by means of a UPT number. The UPT user is therefore personally associated with his or her own UPT number, which is used as the basis for making and receiving calls. The UPT user may be assigned one or more UPT numbers.

In order to have available the UPT service, the UPT subscriber has to subscribe to it with the service provider. At this subscription time, the UPT service profile is defined, which is a record containing all the information related to the UPT user and the personal rules for handling his mobility. Access and operation on the service profile can be done by the UPT user/subscribers. This possibility of access to the service profile information has to be subscribed with the service provider, who decides which parameters the UPT user/subscribers are allowed to modify.

3.1 The UPT Functional Architecture

The functional architecture for UPT is defined by ETSI in [4]. It is based on the standard Intelligent Network (IN) architecture. ETSI has identified three phases in the definition of the UPT service, according to the Capability Sets of the Intelligent Network. Thus the UPT Phase 1 is based on the IN CS1. The functional architecture for the UPT Phase 1 defines three call-oriented network roles: the originating network is the network where the UPT service request is initiated (e.g. outgoing call, modification of service profile); the home network is the network that holds all the information related to the UPT user; the terminating network is the network where the UPT call terminates. The Figure 1 illustrates this UPT Phase 1 functional architecture.

The interconnection between the networks of the UPT functional model takes place between SCF and SDF functional entities, as indicated by the arrows on the figure. The interface between SCF and SDF is specified in IN CS1.

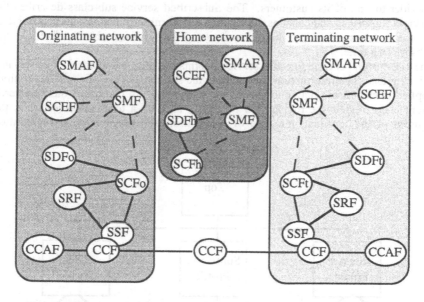

CCAF: Call Control Agent Function   SRF: Special Resource Function
CCF: Call Control Function   SCF: Service Control Function
SSF: Service Switching Function   SDF: Service Data Function
SMF: Service Management Function
SMAF: Service Management Agent Function
SCEF: Service Creation Environment Function

Fig. 1. UPT Phase 1 Functional Architecture

3.2   The UPT Information Model

The UPT information model is defined in [5] and a description can be found in [6].
The information model is not strongly dependant on the IN Capability Sets. It uses
an object oriented approach. According to the actors involved in the UPT service,
namely the service provider, the subscriber and the user, three main classes of objects
are defined: the *Service provider* class, the *Subscriber profile* class and the *User profile*
class. The other objects classes refine the information of these main classes (see
Fig. 2).

The *Service provider* class defines the identity of the service provider. The Provided
service sub-class describes the services offered by the service provider to its
subscribers. It is used by the home network. The Agreement sub-class describes the
service offered by the service provider to visiting users. It is required by the
originating network to determine if there is a roaming agreement with the home
network of the user requesting service.

The *Subscriber profile* class defines the nature of the contract binding the service
provider to one of its customers. The Subscribed service sub-class describes the
services subscribed and contains information on the subscription options applicable to
all the users covered by the subscription.   This information is fixed at the
subscription time.

The *User profile* class defines the service information attached to one of the users in a
subscription. The information contained in the sub-classes specifies the restrictions
applied to each user relative to the service subscribed. In particular, the Registered
agent class models the registration of a UPT user on a terminal access (e.g. the
address of the terminal, the type of registration and the maximum registration
duration).

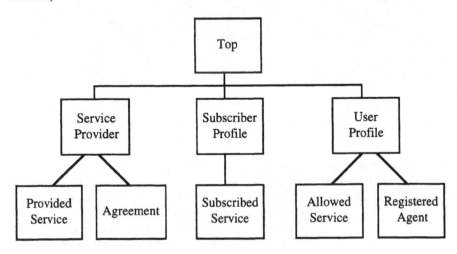

Fig. 2. The UPT Information Model

4 Requirements for the Operation in a Multi-operator Environment

As mentioned above, the functional architecture of UPT has introduced three network roles: the originating network, the home network and the terminating network. When a service provider, e.g. a telecommunications operator, offers the UPT service upon its own IN infrastructure, this network plays these three roles. In such a case, the limit is the scope of the service that is not provided to the users beyond the geographical area covered by the network.

In order to remove this limit, service providers have to be able to provide the user with an end-to-end service. The challenge is the ability of an IN to perform the UPT service that would be invoked by a user from another IN. For that, interworking of the UPT service is needed, i.e., the ability of services from different suppliers to work together upon several IN infrastructures. The UPT service should be able to operate in a multi-network and multi-operator environment. Therefore a global UPT service can be offered upon a global IN resulting of the different IN infrastructures [7, 8]. Each IN constitutes in fact a domain under the authority of a telecommunications operator. In a domain, the service information is accessed, modelled and controlled in a uniform way. Thus the interworking of service is achieved through a cooperation that takes place among the domains provided by the telecommunications operators. These ones have to decide through contract agreements the nature of their cooperation, namely to define protocols and interfaces used to enable inter-domain communications, to analyse the distribution of service functions and essentially to identify shared information. Because on the one hand, when separate database in multiple networks exist and an individual network's database contains only customer-specific service intelligence on its users, service intelligence of users in one network needs to be shared with other participating networks; on the other hand, there is a strong need to keep a control over the ownership of data which may be distributed over several networks, only one of which really owns it. In order to achieve cooperation, each authority domain has to allow the other ones to access its own information. Thus it has to define the set of information it is going to exchange. The set of this shared information is named the *Shared Service Knowledge* (SSK) [9]. The SSK must be negotiated between the partners of cooperation. It strongly depends on the UPT control architecture used. When several domains are involved into the provision of the UPT service, different kinds of control can be found [10]:

- The *local control* : the SCFo located in the originating network performs the service logic, uses information in its SDFo about the capabilities of the local SSFo and terminal, obtains the required information about subscribed services from the user's home SDFh and controls the SRF. The controlling SCFo uses the local SSFo to provide services to the UPT user.

- The *remote control* : the SCFh located in the home network performs the service logic, uses information in its associated SDFh to determine the user's subscribed services and uses information in the SDFo to determine the capabilities of the SSFo being used to deliver services. The controlling SCFh uses the local SSFo to provide services to the UPT user.

- The *home control* : the SCFh performs the same actions as in the local control. It uses the home SSFo to provide services to the UPT user.

A fourth kind of control can be introduced, which is the *mixed control*. In such a control, the SCFo and the SCFh share the execution of the service logic: some functions such as the identification and authentication of the UPT user can be performed by the SCFh while other functions such as the registration/deregistration processes are performed by the SCFo.

Depending on the type of control chosen by the partners of the cooperation, the nature of the shared information will be different. For example, if the local control is applied, the service profile is moved from the home network to the originating network while it is not necessary in the two other cases.

Negotiations on the Shared Service Knowledge are realised through the process of context negotiation. This process provides the creation of two schemas: the Export schema and the Import Schema. The *Export schema* defines the portion of information that a domain is willing to share with the other domains of the cooperation. This is a collection of service object classes of the information model denoting the service information to be exported to the other domains. An exported object class contains an attribute that is the domain name to which it is exported, with the access rights on this object. The Export schema contains in particular the "Agreement" object that describes the provided services by the domain to the visiting users. The *Import schema* specifies the service information that a domain desires to use from the other ones. This is the set of service object classes of the information model denoting service information that a domain imports from the other ones. An imported object class contains an attribute that is the domain name from which it is imported, with the access rights on this object. The Import schema contains in particular the "Agreement" object that describes the provided services by the domain to the visiting users.

So the process of context negotiation permits several domains to agree about the Export schema and the Import schema, which define the cooperation context.

Once this context of cooperation is established, the UPT service logic in a particular domain manipulates the local or imported information in the same way. If the information is encapsulated in an imported object, a mechanism similar to the client-server one is provided. The attribute "domain name" enables the identification of the target domain. The consultation of the imported "Agreement" object enables to verify the existence of the cooperation and the provision of the requested service. A request is therefore addressed to the remote domain by means of a communication that takes place between the domains through a service and protocol for the information transport. The remote domain uses its exported "Agreement" object to validate the request.

When the request concerns information modification, a mechanism of notifications is realised to ensure the consistence of the schemas.

5   A Scenario

Let us consider a user X whose the home network is H. The actions that X can perform when he is located in the network H concern the personal mobility (registration and deregistration), the UPT call handling (outgoing and incoming calls) and the UPT service profile management (interrogation and modification). When X is visiting another network O, the actions that he can perform in this network are depending on the cooperation between the networks H and O. Let us consider the following agreement of the cooperation policy between H and O:

   - the control is local, that means the SCFo performs the service logic,

- X can perform any actions when he is located in H, except the modification of security information (Personal Identification Number PIN).
According to this agreement, the process of context negotiation enables the creation of the import and export schemas, integrated in the information model of each domain (Fig. 3, Fig. 4).

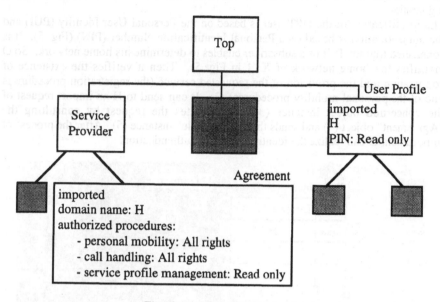

Fig. 3. O Information Model

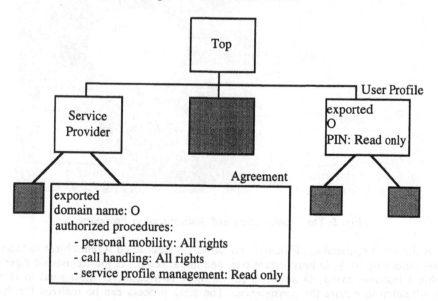

Fig. 4. H Information Model

The first action performed by X is the registration at the visiting network O. He dials a specific number, such as *XX, which is the UPT access code. The SSFo detects this access code and signals the SCFo. This one identifies an UPT demand and begins the UPT registration procedure. This procedure is composed of the identification and authentication of the UPT user, and the incall and outcall registration.

The identification of the UPT user is based on the Personal User Identity (PUI) and the authentication is based on a Personal Identification Number (PIN) (Fig. 5). It is considered that the PUI of a subscriber enables to determine his home network. So O identifies the home network of X (1 in Fig. 5). Then it verifies the existence of cooperation and the provision of the requested service (the registration procedure is one of the personal mobility procedures) (2). It can send to H an import request of the concerned object instance (3). H validates the request by consulting the "Agreement" object (4) and sends the "User Profile" instance (5). After the process of importation, O can realize the identification and authentication.

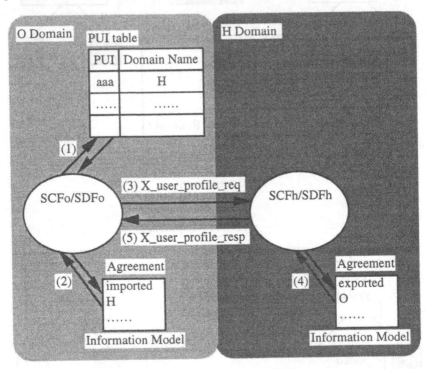

Fig. 5. The Identification and Authentication Processes

For the incall registration, if there is not already a Registered Agent object instance corresponding to X, O must create one, and initialises it. If the Registered Agent object instance exists, O modifies it. In these two cases, O must send to H a notification to ensure the consistency. The same process can be realized for the outcall registration.

6 Conclusion

Actually, in the real context of world-wide communication, the provision of international services is of great significance. It means that end-to-end network and service connectivity needs to be realised, especially in the context of the evolving Global IN services that cross national boundaries. We have proposed in this paper some mechanisms to realise a UPT service in a multi-operator environment. These mechanisms are based on the cooperation between the organisations, and the sharing of information. We have then illustrated these mechanisms with an exemple.

References

[1]   ITU-T Recommendation Q.1201
        Principles of Intelligent Network Architecture
[2]   ITU-T Recommendation Q.1214
        IN Distributed Functional Architecture for IN CS-1
[3]   ITU-T Draft Recommendation F.851
        UPT Service Description
[4]   Draft ETSI Technical Report NA-71303
        UPT Phase 1: Architecture and functionalities for interworking
[5]   Draft ETSI Technical Report NA-71205
        UPT Phase 1 subscriptions, service profile and information model
[6]   O. Boulot, G. Brégant, C. Vernhes
        Long-term IN architecture applied to UPT
        Proc. of the 4th TINA Workshop, L'Aquila, Italy, September 1993
[7]   S. Chen, M. Fujioka, G. O'Reilly
        Intelligent Networking for the Global Marketplace
        IEEE Communications Magazine, vol 31, n°3, March 1993
[8]   M. Fujioka et al.
        Globalizing IN for the New Age
        IEEE Communications Magazine, vol 31, n°4, April 1993
[9]   M.P. Gervais
        Interoperability in Intelligent Network
        Proc. of the 4th TINA Workshop, L'Aquila, Italy, September 1993
[10]   G. S. Lauer
        IN Architectures for Implementing Universal Personal Telecommunications
        IEEE Network Magazine, March/April 1994

On the Personal Communications Impacts on Multimedia Teleservices

[1] Tim Eckardt, [2] Thomas Magedanz

[1] Technical University of Berlin
Hardenbergplatz 2, D-10623 Berlin, Germany
e-mail: tim@cs.tu-berlin.de

[2] •De•Te•Berkom•
Voltastr. 5, D-13355 Berlin, Germany
e-mail: magedanz@fokus.berlin.gmd.d400.de

Abstract. The development of broadband networks and especially the design of emerging multimedia applications on top of these networks have to recognize a major trend in telecommunications: the evolution towards personal mobility and personalized communications. This paper gives an overview of the required capabilities for supporting user mobility and enhanced call management in future multimedia broadband environments. It introduces a generic Personal Communications Support System (PCSS) based on state-of-the-art management system technology, which integrates X.500 directory and X.700 management standards. The description of the PCSS includes a preliminary generic service user profile, which contains besides general user data user-specific location information and sophisticated communication control attributes for any communication services. Access to the profile data for both, user profile management and control of any multimedia communications applications will be realized via uniform management application programming interfaces.

1 Introduction

In the light of an emerging information society, the provision of intelligent communications services within future broadband environments, ranging from enhanced telephony services up to sophisticated multimedia conferencing services, is of pivotal importance, where in particular the permanent reachability of people and the customization of services in accord with specific user needs represent the key issues. This can be recognized when looking at the recent research activities and evolution trends of both, telecommunications and distributed computing.

Some recent work in computer science research focuses on distributed computer-augmented environments with titles such as *ubiquitous computing, augmented reality* and *virtual reality* [1]. Although these technologies differ, they are united in a common philosophy: the primacy of the physical world and the construction of appropriate tools that enhance our daily activities, particularly in distributed computing environments [2], [3]. Main areas of interest are multimedia applications, development of smart palm top devices such as personal digital assistants, personal mobility support and personalization of distributed computing environments.

In parallel the telecommunications environment is evolving towards an open serv-ices environment, where more intelligent communication services, offering enhanced customer control capabilities, and mobile communications represent the state-of-the-art. This process is driven by international standards in the field of Intelligent Networks (IN) [4] and Telecommunications Management Network (TMN) [5], which provide the frameworks for future service provision and management [6], but represent in addition the basis for mobile communications standards, such as Global System for Mobile Communications (GSM) and Universal Personal Telecommunications (UPT). An important aspect of the long-term activities is the concept of service integration, some-times referred to as "universal communications", which allows users to subscribe to a whole bunch of telecommunication services and to customize specific services in accord with their personal needs.

Comparing the targets of both, computing and telecommunications, a convergence of the concepts can be witnessed in respect to supporting user mobility and personali-zation of work environments in accord with specific user requirements. Whereas the computer sciences research focuses more on supporting mobile users within a single location area (e.g. within an organization or enterprise), the telecommunications world looks for global communication support of customers. Nevertheless both philosophies are based on the idea of using centralized knowledge, i.e. a *personalized service profile*, also referred to as "user profile", containing user-specific location information and optional customized service control parameters, which will be queried for call control. Hence, the management of user location information and service control data, compris-ing both user profile management (including user registration) and usage of that infor-mation for creation of enhanced communications services, adds a new dimension to future multimedia environments. Consequently the impacts of these trends have to be taken into account for the design of future multimedia applications, and in particular for multimedia teleservices, since these services represent only one part of the overall com-munications service environment.

Therefore this paper introduces a *Personal Communication Support System (PCSS)* designed to support personal mobility and service personalization for communication services, in the area of one or more Customer Premises Networks (CPNs). The PCSS will offer enhanced user profile management capabilities, including sophisticated reg-istration capabilities (based on electronic location technology), and enhanced call man-agement capabilities for various communications services, ranging from telephony up to multimedia services, in a uniform way. A variety of (tele-) communications applica-tions and in particular multimedia teleservices such as *Multimedia Mail (MMM)* and *Multimedia Collaboration (MMC)* may benefit from using such functionality.

The capabilities offered by the PCSS are strongly influenced by emerging TMN service management concepts and in particular by current IN service features [7], which provide the basis for advanced mobile communications services, such as Universal Per-sonal Telecommunications (UPT) [8] [9]. Nevertheless the outlined PCSS is based completely on current TMN management system technology, comprising X.500 direc-tory and X.700 management standards [10], [11]. This approach is based on the thesis, that emerging TMN service management concepts could be used for the realization of IN service capabilities in the near future [12], since IN services could be considered as specific short term implementations of management services for any (bearer) commu-nication services. The basic idea is to consider all service related control data as man-agement data, which can only be accessed via management protocols for both customer profile management and service execution control. Another important aspect in this context is to integrate the service control data for all communication services of a specific customer within a single generic profile, which unifies and hence simplifies the control of multiple services and allows for the integration and personalization of service environments.

Consequently the definition of a generic service user profile and the corresponding access capabilities form the heart of the PCSS. Since the information to be contained in the user profile comprises static data, such as user name, organizational information, personal number or ID, etc., as well as rather dynamic information, such as location information and service control parameters, the PCSS user profile implementation is based on current X.500 and X.700 standards, where the static information on users will be stored in the X.500 directory service and the information of a more dynamic nature, such as location information will be stored in an X.700 management information base. An integrated X.500 and X.700 information system [13] is used to provide transparent access to X.500 and X.700 information objects for both, user profile management (including user registration) and communication service execution control (e.g. translation of an incoming UPT call to the corresponding physical device in accord with the current user location and the available device capabilities).

In the following chapter we will addresses the potential impacts of mobility and service personalization on multimedia teleservices in more detail. Chapter 3 provides a general overview of the PCSS, whereas the definition of the PCSS Generic Service User Profile will be given in chapter 4. Finally a potential application of the PCSS for MM teleservices will be addressed in chapter 5, where in particular the MMC, a multimedia teleservice supporting audiovisual conferences and cooperative working on shared multimedia documents/applications [14], will be used as an example. It will be shown that the MMC could be supported by the PCSS in order to allow for personal mobility and personalization of the participant's service environment. That is, a required participant for a certain MM conference could be selected based solely on his/her personal number or a description of his/her business role. In case of the necessity of one participant to move during an ongoing conference, the ability of the PCSS to support the establishment of a personalized communication environment is useful.

2 Mobility and Personalization Impacts on Multimedia Teleservices

Future multimedia applications on top of broadband networks will be strongly influenced by the emerging mobility and personalization aspects of the evolving telecommunication services environment. Users of multimedia services have to be provided with the capability to organize their (multimedia) communication environment according to their specific needs. Another important aspect is personal mobility. It allows for mobility of the user in a fixed network environment. The notion of a user profile, containing all relevant information for service control/management and in particular location information on the users of such systems represent the starting point for the development of sophisticated, personalized communications services.The *knowledge of user location* will form the basis for a number of advanced information and communication services, resulting in more efficient communication between people. Current research on automatic user localization considers the use of active badges to realize an automatic tracing of users carrying that type of badge [15]. The central data structure to hold both, data on the current user's location as well as data on the personalization of the used telecommunications services will be the user profile.

Potential multimedia applications, that may be enhanced by the provision of user location information may be roughly grouped into two category, namely advanced location-aware user information services and intelligent/personal communication applications:

Before describing the development of adequate mechanisms for the collection, storage and access to location information in distributed broadband environments it seems worthwhile to spend a look on potential multimedia applications, profiting from the availability of user location information. Two categories of location-aware multimedia applications could be identified, namely advanced user information services and intelligent/personal communication applications:

- *Advanced user directory information services* involving the use of data provided by electronic location technology could be applied to present a variety of multimedia entries on various topics within a company. This may encompass the presentation of enterprise information to visitors: information on employees, the organizational structure, maps of the organization's site and location of rooms. This multimedia user information system could provide location information on the absence (e.g. holidays, business trips, etc.) or the current location (e.g. default office, library, remote office, etc.) of specific employees. This service may be used by the secretary or colleagues for locating a person, or by users external to an organization e.g. for remote queries or visitors at the entrance of a company. additionally, this service could also provide the querying user with the information on available communication options with the desired person. Hence this information service may be combined with one or more of the following communication services.

- *Multimedia Personal Communications Services:* The location information can be used to enhance a variety of communication services, ranging from simple telephony up to complex multimedia mail and conferencing services in order to support personal mobility. This means that for example PBXs could use the location information of a user for call setup or call forwarding purposes, where the dialled number will be translated into the appropriate number for call delivery to the current terminal. Also multimedia collaboration and conferencing services could use this location information for similar purposes. The integration of the MMC service with a Personal Communication Support System in order to extend the management and call control of that teleservice to include location-awareness of terminals and users is the logical consequence of the current trend in telecommunications towards ubiquitous personal mobility and individualization of services. Future generations of MM services may already be designed for an integrated management covering different telecommunications networks (fixed or mobile), different types of media, a multitude of advanced teleservices, and a wide scope of user-specific data (i.e. location information).

Before describing the development of adequate mechanisms for the collection, storage and access to location information in distributed broadband environments it seems worthwhile to spend a couple of words on the basic requirements on systems supporting personal mobility and service customization or personalization.

2.1 Automatic User Localization

The knowledge of the user's location is a central issue to both, mobility and personalization in a multimedia broadband environment. Basically, this information on the user's location may be provided manually by the user him-/herself (in advanced scenarios, cf. [16], this process will be supported by smart cards) or alternatively by a system of active badges to be carried by the users [17]. The active badges allow the user to be traced in a local environment with the help of room-based sensors. The automatic localization of a user involved in communications activities of a multimedia broadband environment enables a new, sophisticated style of personal communication. In the most extreme case a user wearing a personal active badge may enter a room and a multimedia workstation located in that room will immediately display the current user work environment on the screen, which has been a few seconds before on the screen of another workstation in the previously visited room.

An issue of considerable concern is the usage of user location information in enhanced information and communication applications potentially being harmful in respect to security and privacy. This may result in users being antipathetic to the storage of their location due to privacy reasons. They even may simply just don't want to be reachable all the time. Hence it should be possible to allow people to define different levels of access to their location information, ranging from being reachable all the time everywhere for everyone down to refusing their location monitoring generally.

2.2 User Registration

In the context of advanced communication services the aspects of user registration has to be briefly mentioned. *Registration* is the process allowing the user to register his/her current location with the personal communications support system in order to enable system-wide personal mobility. It may either be performed by manual registration or by the above mentioned methods of automatic user location involving the use of active badges. A third procedure may be based on user-defined schedules indicating the planned or anticipated location of the user. The registration-related data should be stored in the user's profile. Generally, registration supports advanced c all handling as being required in the context of service personalization.

2.3 Personalization of Call Control

With the concepts of service personalization and user mobility it seems straight forward to make use of enhanced call control capabilities. Such personalized call control capabilities allow users to define rules for forwarding incoming multimedia call establishing requests or simple phone calls to other users, e.g. a secretary or an answering machine, or to prohibit calls from specific callers by means of screening list (optionally on a per service basis). The user should be supported in defining his/her own customized call handling in a service-generic way to be applied to the communication establishment of all types of telecommunications services and multimedia services.

2.4 Integration of Services / Interworking

In a heterogeneous environment as seen from a user's perspective it is necessary to integrate present day communication services like telephone, fax and mail with advanced multimedia communication services like multimedia mail of multimedia conferencing applications. Only one basic user interface in respect to calling procedures should be sufficient for any type of tele/multimedia service. This implies that each user will have only one single logical personal number for any kind of telecommunication service as defined in UPT. A consequence of a strictly applied integrated handling of all types of services is the necessity to define interactions and mappings between different types of services, e.g. to forward incoming telephone calls to a multimedia collaboration workstation, etc. The area of service integration / interworking is still mostly for further study.

3 Overview of the Personal Communication Support System

The PCSS which has been specially designed for personal mobility and personalization of services in the area of Customer Premises Networks (CPNs) offers sophisticated registration capabilities (based on electronic location techniques), enhanced call management, and enhanced user profile management capabilities. Figure 1 displays the individual components within the PCSS and the main applications interacting with those components.

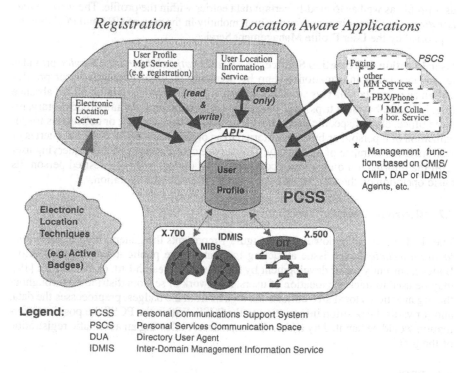

Fig. 1. Components of the PCSS and Related Areas

The three shaded regions in Figure 1 indicate the scope of the basic functional areas that potentially may be independently handled:

3.1 PCSS

The *PCSS* with its main components is in the focus of the current research. The central data structure within the PCSS is the *User Profile* containing all the information required to support personal mobility, personalization of communication services, and advanced user information services. The implementation of the User Profile applies the concepts of the *Inter-Domain Management Information Service (IDMIS)* integrating X.500 and X.700 [17]. It is the User Profile that coordinates and focuses the interworking of all applications in any of the three shaded areas indicated in Figure 1. Currently, the PCSS comprises the following applications that interact via an application programming interface (API) with the data contained in the user profile:

Electronic Location Server. The *Electronic Location Server* processes data provided by electronic location techniques and accordingly updates the user profiles with the current locations of the users. This functionality is a kind of *automatic registration* to be used for all types personal mobility.

User Profile Management Service. The *User Profile Management Service* provides an user interface to the User Profile. It is used to display any information contained in the user profile as well as to modify certain data entries within the profile. The *manual registration*, a basic requirement for personal mobility in the area of UPT and PCS, will be supported by the User Profile Management Service.

User Location Information Service. The *User Location Information Service* provides an advanced user information system by accessing the total number of user profiles within an organization. This service could provide location information on the absence (e.g. holidays, business trips, etc.) or the current location (e.g. default office, library, remote office, etc.) of specific users. It may be used by the secretary or colleagues for localization of a person, or by users external to an organization e.g. for remote queries or visitors at the entrance of a company. This service could also provide the querying user with the information on available communication options with the desired person. Its basic operation on the Generic Service User Profile is a read operation.

3.2 Electronic Location Techniques

The shaded area in the lower left of Figure 1 represents the general field of *electronic location techniques*, an issue attracting more and more public attention [18]. Active Badges, currently under development by the Olivetti Research Ltd. of Cambridge [19], may be used to track the location of users. A network of sensors distributed throughout the organization's local area collects the signals from the badges, preprocesses the data, and forwards the position information to the PCSS. Within the PCSS the position information would be handled by a *User Location Server* allowing an automatic registration of the user.

3.3 PSCS

The shaded area in the upper right corner of Figure 1 (labelled as *PSCS*) indicates the potential applications relying on the support provided by the PCSS. A multitude of telecommunications services ranging from simple telephone services to advanced multimedia collaboration services may profit by the user location information and advanced call handling prescriptions provided by the PCSS. The integration of teleservices in the general context of the PCSS points to a similar direction as the concept of *Personal Services Communication Space (PSCS)* currently under development in a number of RACE projects (e.g. MOBILISE [16]).

After this brief outline of the individual components comprising the basic features of the PCSS we will focus on the structure of the user service profile and the mechanisms to obtain and store that information. The User Profile is the main data structure in the focus of all the listed functionality. It contains all user-specific data necessary to personalize the communication of individual services as well as to personalize a large set of services in a service generic way. Note that the registration for outgoing communication, outgoing call screening and related security and accounting aspects will not be considered in the following, although these aspects are of high importance especially for global PCSSs.

4 Definition of a Generic Service User Profile

A prerequisite for enabling personal mobility and personal communication is the definition of a User Profile, containing all the essential data, such as personal number, current location, call control parameters etc., necessary for the flexible establishment of communication connections to and from a specific end-user.

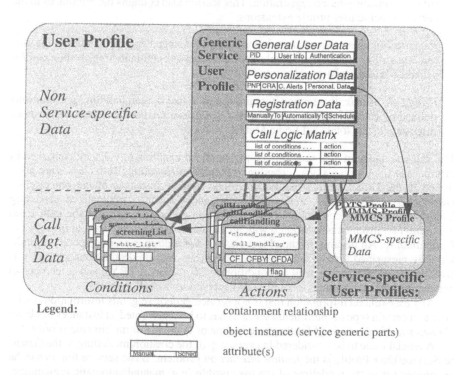

Fig. 2. Generic Service User Profile

The central part of the User Profile is the *Generic Service User Profile* containing all user-related information that may be applied to any teleservice the user may want to use. The basic idea to be applied to the modelling of the Generic Service User Profile has been to integrate as much non-service specific information as possible into one general type of user profile. Thus, instead of having repeatedly to define the same information in a number of service-specific user profiles, all the service-generic information is focused into the generic service user profile where it only once has to be defined. This data may be used for advanced organization-wide information services as well as all types of telecommunication applications/services intended to be used in a personalized way.

Figure 2 shows the user profile as being defined for the PCSS. It is composed of a special object class *GenericServiceUserProfile*, containing a number of non-call specific attributes grouped into the following sections:

- *General User Data*: this section contains attributes for user identification, user authentication, and user description. In detail, this may be the personal number, general user data (e.g. postal address, etc.), and a password or PIN to be used of user authentication.

- *Personalization Data*: this section contains all data specifying the personalization of the user's communication space: Private Numbering Plan (PNP), Customized Recorded Announcements (CRA), and for example a personal schedule specifying rules for schedule-based registration. This section also contains the references to the service-specific user profile extensions.

- *Registration*: this section comprises frequently changed entries containing the current registration (done by manual registration, and/or automatic registration, and/or schedule-based registration).

- *Call Logic Matrix:* this section contains the user-specific call handling prescriptions defined by rules of the structure *chained conditions / action*. A set of zero or more rules defines the total Call Logic Matrix.

The call management data specifying a personalized handling of calls are contained in instances of the object classes *ScreeningLists* and *CallHandling*. Both classes are positioned in the Directory Information Tree (DIT) below instances of the object class *GenericServiceUserProfile*. *ScreeningList* instances specify personalized screening lists that allow for a customized call handling based on the origin of the call. Each *ScreeningList* instance is associated with an instance of the second class of objects, i.e. *CallHandling*. An instance of the object class *CallHandling* is used to specify an individually prescribed type of call handling. Its entries include destinations for call forwarding (unconditional as well as conditional: CF, CFDA, CFBY), and references to prerecorded announcements to be played/displayed to the calling party.

Any user-related information that is not service generic (e.g. this information is specific to a certain type of multimedia service) has to be represented as instances of object classes specifically defined for that certain type of telematic or multimedia service.

A crucial issue to be considered in every step of the creation/modelling of the Generic Service User Profile is the *feature interaction* problem. Basic service features to be accounted for in the modelling of the user profile (e.g. manual/automatic registration, call forwarding, call screening, etc.) may potentially interfere with one another. Inconsistent specification of the general mapping from a personal number or personal ID to a physical terminal by registration, call forwarding and screening lists directives have to be resolved or generally avoided. The modelling of the Generic Service User Profile intended to minimize the number of possible conflicts and to specify strategies for resolving any remaining conflicts. The ultimate goal is to define a set of orthogonal operations specifying how the reference to the person to be addressed has to be mapped to a specific network address/terminal.

Our approach to handle the feature interaction problem builds on a simple rule-based mechanism which specifies the individual personalized call handling. Common IN Service Features or Services such as *originating call screening, call forwarding, call forwarding conditional, in-call registration* may be realized and even combined within the PCSS in terms of specifically defined objects of the super classes *Actions* or *Conditions*. The provisioning of any of the above mentioned Service Features requires the prior definition of corresponding object classes. A set of sequentially prioritized rules which are constructed by one or more condition instances and exactly one action instance prescribe the complete call handling specific to one user.

We developed an information schema extension for the X.500 directory service defining the Generic Service User Profile as a number of object classes, attributes, matching rules, and a corresponding DIT structure. Since at least part of the information to be registered in the user profile is subject to frequent changes, the modelling of the user profile will be ported to an integration of X.500 / X.700 called *Inter-Domain Management Information Service* (IDMIS) [17].

5 Application of the PCSS for the MMC Teleservice

This section indicates how the PCSS would be applied for providing personal mobility and service personalization to the BERKOM Multimedia Collaboration (MMC) teleservice. The BERKOM MMC service is a workstation-based conferencing system [20]. It allows users to engage in multimedia conferences. Participating in a conference, the users can share applications and engage in audiovisual communication. Generally, the application of PCSS functionality to the Multimedia Collaboration teleservice does not involve the direct control of any local switches. Rather, it applies some enhancements to the initial addressing of a user at the beginning or joining of a conference.

In the chosen example application of the PCSS (enhancing the MMC service) we will concentrate on the conferencing aspects, particularly on invitations to a conference. The capability to share applications between members of a conference will not be influenced in any way by the interaction with the PCSS. It is out of the scope of the descriptions contained in this chapter.

Figure 3 displays the main components of the MMC teleservice. Components related to application sharing are not displayed.

- *Conference Interface Agents* (CIAs) serve as user interfaces to the MMC. Each user requires exactly one collocated CIA.

- *Invitation Broker* (IB) passively receives invitations to conferences for one or more CIAs.

- *Conference Directory* (CD) is the central database for all MMC-related information (data on users, user groups, and conferences).

- *Conference Manager* (CM) is the core module to manage and control conferences.

- *Audiovisual Manager* (AVM) coordinates the establishment of communication channels between participants of a conference according to the directives from the Conference Manager.

- *Audiovisual Components* (AVC) which exist on all sites participating in a conference are used to establish audio and video connections between participants. Their operation is centrally coordinated by the AVM.

A CIA sends a **CMAP_OpenConference** request (1.) to the Conference Manager (CM) to initiate a multimedia conference. This results in the CM sending an **IBAP_DeliverInvitation** request (4.) to the Invitation Broker (IB) corresponding to the invited user's CIA. If the invited user accepts the invitation the corresponding CIA will reply with a **CMAP_JoinConference** request (5.) to the CM. This sequence (1, 4, 5) is a rather simplified description of the activities necessary for the invitation of users to a conference without any interference of the PCSS. The invitation phase corresponds to the call control phase in the PCSS.

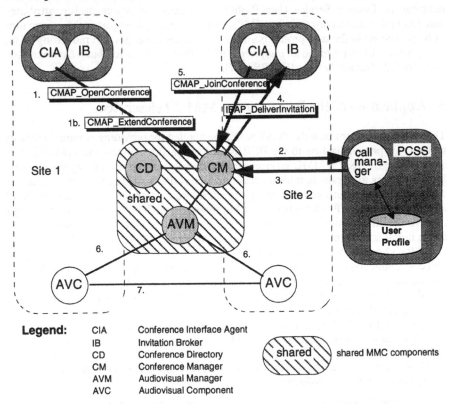

Legend:

CIA	Conference Interface Agent	
IB	Invitation Broker	
CD	Conference Directory	
CM	Conference Manager	
AVM	Audiovisual Manager	
AVC	Audiovisual Component	

shared   shared MMC components

Fig. 3. Components of the Multimedia Collaboration Service

A PCSS user's subscription to a certain teleservice (e.g. MMC service) will result in the creation of instances of service-specific user profile extension containing service-specific data related to the particular user. In the context of a PCSS user subscribing to the Multimedia Collaboration service, the MMC-specific User Profile within the PCSS would contain references to the Conference Directory (an X.500 distinguished name) and to one ore more Conference Manager(s) of the MMC service. Later releases of this service may allow some of the data currently contained in the Conference Directory to be placed in MMC-specific User Profile extensions of the PCSS. Potential candidates for such a placement in User Profile extensions are all those CD entries which are user-specific and which the user is allowd to modify. The MMC-specific User Profile may also specify that an automatic registration of the user (e.g. using Active Badges) or a manual registration may result in the user being accessible for conference participation at that location. This requires two simple activities:

- at first the user must be automatically logged on to an appropriate workstation at that location (the design of a heuristic algorithm which selects an appropriate worksta-tion from among a set of workstation placed at a given location is not trivial; it may involve the use of a Trading Function [21]);

- next, a CIA instance and an IB instance must be created at the workstation where the user has been logged into. (In an environment allowing for process/object mi-gration, a migration of both instances to the new location would be the preferred al-ternative).

After the instantiation of CIA and IB in the course of an automatic or manual registra-tion, the CIA would bind to an appropriate Conference Manager using addresses and passwords contained in the user's MMC-specific User Profile extension. Similarly, the IB registers with the CD. Since this is a rather dynamic procedure, i.e. the changing of the user's location requires a modification of the IB's registration within the CD, the CD may be better implemented using an IDMIS approach (cf. [13], [17]).

In the current stage of development of the PCSS we limit the scope of research to the enhancement of *incoming* calls. That is, a user's User Profile specifies how calls directed to this user should be handled. Currently, the User Profile does not specify how calls originating from the owner of the User Profile (i.e. *outgoing* calls) should be han-dled. The only exception may be the definition of a private numbering plan. In the con-text of the MMC service an incoming call corresponds to a *Conference Invitation*. Therefore, the PCSS will not interfere with a CIA actively joining a conference.

There are only two requests issued by the CIA to the Conference Manager (i.e.,CMAP_OpenConference and CMAP_ExtendConference) that may result in a conference invitation issued to a user's IB (i.e., IBAP_DeliverInvitation) and therefore have to be processed by an interaction with the PCSS. Figure 3 shows the sequence of activities including an interaction of the MMC service with the PCSS (i.e. 2. and 3.). Both requests, CMAP_OpenConference and CMAP_ExtendConference (1. or 1b.) contain a list of invited users as an attribute. The Conference Manger passes the list of invited users and additional information on the source of the invitation and/or the conference to the *Call Manger* of the PCSS (2.) for an enhanced call management. The PCSS Call Manager accesses the User Profiles corresponding to the users contained in the list of invited users and retrieves their personalized Call Logic Matrix. Based on the rules contained in the Call Logic Matrix of the PCSS the PCSS Call Manager decides to which Invitation Broker (IB) the conference invitation (i.e. the IBAP_DeliverInvitation request) actually will be routed (4.). The first rule of the PCSS's Call Logic Matrix that 'fires' determines the call handling. An exemplary call handling may be *call blocking* resulting in a CMAP_InvitationRejected event being sent from the Conference Manager of the MMC to all CIA's participating in the conference. The conditions for this particular rule to become 'true' may include Screening Lists specifying one or more invitation initiators ('calling parties' in the terminology of the PCSS). An alternative call handling specified in the rule may define the addressing of an appropriate IB based on the PCSS registration data or a call forwarding to a different user's (e.g. a secretary) IB.

6 Summary

Enhanced communication support in the light of emerging personal communications requires the knowledge of the location of mobile users and additional service control parameters. In order to offer end-users in a broadband environment the capability to organize their (multimedia) communication services according to their own preferences in respect to time, location, medium, quality, cost, security and privacy a corresponding user profile has to be defined. This paper has described the design principles for a preliminary generic user profile, which represents the heart of a Personal Communication Support System based on current X.500 and X.700 standards. This profile concept and the related registration and profile management procedures can be used by a variety of location aware applications. These applications include besides traditional communication applications, like PBXs, also future multimedia directory information systems and multimedia communication services, such as MMC. By indicating the application of the PCSS for providing personal mobility in the MMC teleservice as a specific example it has been stressed that future multimedia teleservices should enable personal mobility and service personalization. This could be achieved by the proposed PCSS.

7 Acknowledgements

The ideas presented in this paper have been developed within the BERKOM II project *"IN/TMN Integration"* carried out at the Department for Open Communications Systems at the Technical University of Berlin for Deutsche Telekom BERKOM.

8 Acronyms

CD	Conference Directory
CF	Call Forwarding (unconditional)
CFBY	Call Forwarding on Busy
CFDA	Call Forwarding on Don't Answer
CIA	Conference Interface Agent
CM	Conference Manager
CRA	Customer Recorded Announcements
DIT	Directory Information Tree
IDMIS	Inter-Domain Management Information Service
IB	Invitation Broker
IN	Intelligent Network
MM	Multimedia
MMC	Multimedia Collaboration Service

MMM	Multimedia Mail Service
PNP	Private Numbering Plan
PCSS	Personal Communication Support System
PSCS	Personal Services Communication Space
PTN	Private Telecommunication Network
UPT	Universal Personal Telecommunications
TMN	Telecommunications Management Network
X.500	OSI Directory System
X.700	OSI Management System

9 Literature

1 Communications of the ACM Journal, "Computer Augmented Environments: Back to the Real World", Vol.36, No.7, July 1993

2 S. Elrod et.al.: "Responsive Office Environments", Communications of the ACM, Vol. 6, No. 7, July 1993

3 Mark Weiser: "Some Computer Science Issues in Ubiquitous Computing", Communications of the ACM, Vol. 6, No. 7, July 1993

4 ITU-T Recommendations Q.1200 series: "Intelligent Network ", Geneva, March 1992

5 ITU-T Recommendation M.3010: "Principles of a Telecommunications Management Network", Geneva, 1992

6 T. Magedanz: "IN and TMN Providing the Basis for future Information Networking Architectures", in Computer Communications, Vol. 16, No. 5, April 1993

7 ITU-T Recommendation Q.1211: "Introduction to Intelligent Network Capability Set 1", Geneva, March 1992

8 ITU-T Draft Recommendation F.851: "Universal Personal Telecommunications - Service Principles and Operational Provision", November 1991

9 ETSI ETR NA-70102:"UPT, Principles and Objectives", 1992

10 ITU-T Recommendation X.500/ISO/IEC/IS 9594: Information Processing - Open Systems Interconnection - The Directory, Geneva, 1988

11 ITU-T Recommendation X.700/ISO/IEC/IS 7498-4: Information Processing - Open Systems Interconnection - Basic Reference Model - Part 4: Management Framework; Management Framework for CCITT Applications"

12 T. Magedanz: "Towards a Common Platform for Future Telecommunication and Management Services - Some Thoughts on the Relation between IN and TMN", Invited Paper at Korea Telecom International Symposium (KTIS´93), Seoul, Korea, November 1993

13 M. Tschichholz et.al.: "A Service for Administering Management Information - VPN Inter-Domain Management support using X.500 and X.700", ISINM, San Francisco, April 1993

14 BERKOM Technical Report, "The BERKOM Multimedia Collaboration Teleservice", Release 2.0, October 1992

15 A. Harter, A. Hopper: "A Distributed Location System for the Active Office", IEEE Network, Special Issue on Distributed Applications for Telecommunications, January 1994

16 M. Guntermann et.al.: "Integration of Advanced Communication Services in the Personal Services Communication Space - A Realisation Study", Proceedings of the RACE International Conference on Intelligence in Broadband Service and Networks (IS&N), Paris, November 1993

17 M. Tschichholz, W. Donnelly: " The PREPARE Inter-Domain Management Information Service", RACE IS&N Conference, Paris, November 1993

18   D. Pountain: "Track People with Active Badges", in the 'Features' section of Byte December 1993, Volume 18, Number 13

19   R. Want, A. Hopper, V. Falcao, J. Gibbons: "The Active Badge Location System", ACM Transactions on Information Systems, Vol. 10, No. 1, Jan. 1992

20   *The BERKOM Multimedia Collaboration Teleservice,* Release 3.1, BERKOM Technical Report,December 1993

21   ISO/IEC JTC1/SC21 N6084: "Working Document on Topic 9.1 - ODP Trader", WG7 Working Document, 31 May 1991

Author Index

Lecture Notes in Computer Science

For information about Vols. 1–792
please contact your bookseller or Springer-Verlag

Vol. 830: C. Castelfranchi, E. Werner (Eds.), Artificial Social Systems. Proceedings, 1992. XVIII, 337 pages. 1994. (Subseries LNAI).

Vol. 831: V. Bouchitté, M. Morvan (Eds.), Orders, Algorithms, and Applications. Proceedings, 1994. IX, 204 pages. 1994.

Vol. 832: E. Börger, Y. Gurevich, K. Meinke (Eds.), Computer Science Logic. Proceedings, 1993. VIII, 336 pages. 1994.

Vol. 833: D. Driankov, P. W. Eklund, A. Ralescu (Eds.), Fuzzy Logic and Fuzzy Control. Proceedings, 1991. XII, 157 pages. 1994. (Subseries LNAI).

Vol. 834: D.-Z. Du, X.-S. Zhang (Eds.), Algorithms and Computation. Proceedings, 1994. XIII, 687 pages. 1994.

Vol. 835: W. M. Tepfenhart, J. P. Dick, J. F. Sowa (Eds.), Conceptual Structures: Current Practices. Proceedings, 1994. VIII, 331 pages. 1994. (Subseries LNAI).

Vol. 836: B. Jonsson, J. Parrow (Eds.), CONCUR '94: Concurrency Theory. Proceedings, 1994. IX, 529 pages. 1994.

Vol. 837: S. Wess, K.-D. Althoff, M. M. Richter (Eds.), Topics in Case-Based Reasoning. Proceedings, 1993. IX, 471 pages. 1994. (Subseries LNAI).

Vol. 838: C. MacNish, D. Pearce, L. Moniz Pereira (Eds.), Logics in Artificial Intelligence. Proceedings, 1994. IX, 413 pages. 1994. (Subseries LNAI).

Vol. 839: Y. G. Desmedt (Ed.), Advances in Cryptology - CRYPTO '94. Proceedings, 1994. XII, 439 pages. 1994.

Vol. 840: G. Reinelt, The Traveling Salesman. VIII, 223 pages. 1994.

Vol. 841: I. Prívara, B. Rovan, P. Ružička (Eds.), Mathematical Foundations of Computer Science 1994. Proceedings, 1994. X, 628 pages. 1994.

Vol. 842: T. Kloks, Treewidth. IX, 209 pages. 1994.

Vol. 843: A. Szepietowski, Turing Machines with Sublogarithmic Space. VIII, 115 pages. 1994.

Vol. 844: M. Hermenegildo, J. Penjam (Eds.), Programming Language Implementation and Logic Programming. Proceedings, 1994. XII, 469 pages. 1994.

Vol. 845: J.-P. Jouannaud (Ed.), Constraints in Computational Logics. Proceedings, 1994. VIII, 367 pages. 1994.

Vol. 846: D. Shepherd, G. Blair, G. Coulson, N. Davies, F. Garcia (Eds.), Network and Operating System Support for Digital Audio and Video. Proceedings, 1993. VIII, 269 pages. 1994.

Vol. 847: A. L. Ralescu (Ed.) Fuzzy Logic in Artificial Intelligence. Proceedings, 1993. VII, 128 pages. 1994. (Subseries LNAI).

Vol. 848: A. R. Krommer, C. W. Ueberhuber, Numerical Integration on Advanced Computer Systems. XIII, 341 pages. 1994.

Vol. 849: R. W. Hartenstein, M. Z. Servít (Eds.), Field-Programmable Logic. Proceedings, 1994. XI, 434 pages. 1994.

Vol. 850: G. Levi, M. Rodríguez-Artalejo (Eds.), Algebraic and Logic Programming. Proceedings, 1994. VIII, 304 pages. 1994.

Vol. 851: H.-J. Kugler, A. Mullery, N. Niebert (Eds.), Towards a Pan-European Telecommunication Service Infrastructure. Proceedings, 1994. XIII, 582 pages. 1994.

Vol. 852: K. Echtle, D. Hammer, D. Powell (Eds.), Dependable Computing – EDCC-1. Proceedings, 1994. XVII, 618 pages. 1994.

Vol. 853: K. Bolding, L. Snyder (Eds.), Parallel Computer Routing and Communication. Proceedings, 1994. IX, 317 pages. 1994.

Vol. 854: B. Buchberger, J. Volkert (Eds.), Parallel Processing: CONPAR 94 – VAPP VI. Proceedings, 1994. XVI, 893 pages. 1994.

Vol. 855: J. van Leeuwen (Ed.), Algorithms – ESA '94. Proceedings, 1994. X, 510 pages. 1994.

Vol. 856: D. Karagiannis (Ed.), Database and Expert Systems Applications. Proceedings, 1994. XVII, 807 pages. 1994.

Vol. 857: G. Tel, P. Vitányi (Eds.), Distributed Algorithms. Proceedings, 1994. X, 370 pages. 1994.

Vol. 858: E. Bertino, S. Urban (Eds.), Object-Oriented Methodologies and Systems. Proceedings, 1994. X, 386 pages. 1994.

Vol. 859: T. F. Melham, J. Camilleri (Eds.), Higher Order Logic Theorem Proving and Its Applications. Proceedings, 1994. IX, 470 pages. 1994.

Vol. 860: W. L. Zagler, G. Busby, R. R. Wagner (Eds.), Computers for Handicapped Persons. Proceedings, 1994. XX, 625 pages. 1994.

Vol: 861: B. Nebel, L. Dreschler-Fischer (Eds.), KI-94: Advances in Artificial Intelligence. Proceedings, 1994. IX, 401 pages. 1994. (Subseries LNAI).

Vol. 862: R. C. Carrasco, J. Oncina (Eds.), Grammatical Inference and Applications. Proceedings, 1994. VIII, 290 pages. 1994. (Subseries LNAI).

Vol. 863: H. Langmaack, W.-P. de Roever, J. Vytopil (Eds.), Formal Techniques in Real-Time and Fault-Tolerant Systems. Proceedings, 1994. XIV, 787 pages. 1994.

Vol. 864: B. Le Charlier (Ed.), Static Analysis. Proceedings, 1994. XII, 465 pages. 1994.

Vol. 865: T. C. Fogarty (Ed.), Evolutionary Computing. Proceedings, 1994. XII, 332 pages. 1994.

Vol. 866: Y. Davidor, H.-P. Schwefel, R. Männer (Eds.), Parallel Problem Solving from Nature-Evolutionary Computation. Proceedings, 1994. XV, 642 pages. 1994.

Vol 867: L. Steels, G. Schreiber, W. Van de Velde (Eds.), A Future for Knowledge Acquisition. Proceedings, 1994. XII, 414 pages. 1994. (Subseries LNAI).

Vol. 868: R. Steinmetz (Ed.), Multimedia: Advanced Teleservices and High-Speed Communication Architectures. Proceedings, 1994. IX, 451 pages. 1994.

Vol. 869: Z. W. Raś, Zemankova (Eds.), Methodologies for Intelligent Systems. Proceedings, 1994. X, 613 pages. 1994. (Subseries LNAI).

Vol. 870: J. S. Greenfield, Distributed Programming Paradigms with Cryptography Applications. XI, 182 pages. 1994.

Vol. 871: J. P. Lee, G. G. Grinstein (Eds.), Database Issues for Data Visualization. Proceedings, 1993. XIV, 229 pages. 1994.